WHAT ABOUT DARWIN?

What about Darwin?

All Species of OPINION from Scientists,
Sages, FRIENDS, and ENEMIES Who Met, Read,
and Discussed the NATURALIST
Who CHANGED the World

Thomas F. Glick

THE JOHNS HOPKINS UNIVERSITY PRESS
Baltimore

The Johns Hopkins University Press
2715 North Charles Street
Baltimore, Maryland 21218-4363
www.press.jhu.edu

Library of Congress Cataloging-in-Publication Data

Glick, Thomas F.
    What about Darwin? : all species of opinion from scientists, sages, friends, and enemies who met, read, and discussed the naturalist who changed the world / Thomas F. Glick.
        p. cm.
    Includes bibliographical references and index.
    ISBN-13: 978-0-8018-9462-6 (hardcover : alk. paper)
    ISBN-10: 0-8018-9462-X (hardcover : alk. paper)
    1. Darwin, Charles, 1809–1882—Influence—History. 2. Celebrities—Attitudes—History. 3. Evolution (Biology)—Philosophy—History. 4. Evolution (Biology)—History. 5. Evolution (Biology)—Religious aspects—History. I. Title.
    QH360.5.G58 2010
    576.8′2092—dc22   2009028858

A catalog record for this book is available from the British Library.

Frontispiece: Illustration by Alice Du, courtesy of Boston University Charles Darwin Bicentennial Committee

CONTENTS

Preface   xv

Introduction   xix

A   1

Henry Adams, American historian • Louis Agassiz, Swiss-American naturalist • Amos Bronson Alcott, American author and educator • Sholem Aleichem, Russian Yiddish author and satirist • Grant Allen, English science writer and novelist • William Allingham, Irish poet • Duke of Argyll (*see* George Douglas Campbell) • Matthew Arnold, English poet and critic • Herbert Henry Asquith, English prime minister • Edward Aveling, English journalist

B   9

Walter Bagehot, English political theorist • Alexander Bain, Scottish philosopher • James Mark Baldwin, American psychologist • Arthur James Balfour, English prime minister • George Bancroft, American historian • John Kendrick Bangs, American editor and satirist • P. T. Barnum, American circus impresario • Henry Bates, English naturalist • William Bateson, English geneticist • Lydia Ernestine Becker, English suffragist and botanist • Henry Ward Beecher, American clergyman • Clive Bell, English art critic • Saul Bellow, American novelist • Ruth Benedict, American anthropologist • Henri Bergson, French philosopher • Isaiah Berlin, English philosopher and historian • Annie Besant, English social reformer • Elizabeth Bishop, American poet • Otto von Bismarck, German statesman • Franz Boas, American anthropologist • Alice Bodington, English naturalist and science writer • Wilhelm Bölsche, German nature writer • James Bonar, English economist • Francis Bowen, American philosopher • Charles Loring Brace, American clergyman and reformer • Franz Brentano, Austrian philosopher • Robert Bridges, English poet • Paul Broca, French anatomist and anthropologist • Robert Browning, English poet • Orestes Augustus Brownson, American essayist and editor • Ferdinand

Brunetière, French author and critic • William Jennings Bryan, American politician • William Cullen Bryant, American poet • James Bryce, English jurist and statesman • Ludwig Büchner, German materialist • Henry Thomas Buckle, English historian • Arabella Buckley, English author and science popularizer • Edward Bulwer-Lytton, English novelist and politician • Luther Burbank, American botanist and horticulturalist • John Burroughs, American naturalist • J. B. Bury, English historian • Samuel Butler, English novelist • Thomas Butler, English clergyman

C 50

Elizabeth Dwight Cabot, American social reformer • George Douglas Campbell, Duke of Argyll, Scottish politician and author • Alphonse de Candolle, Swiss botanist • Jane Welsh Carlyle, English woman of letters • Thomas Carlyle, English author • William Benjamin Carpenter, English naturalist and physiologist • Lewis Carroll, English author • Emilio Castelar, Spanish politician and president • Willa Cather, American novelist • John W. Chadwick, American Unitarian minister • Anton Chekhov, Russian author • G. K. Chesterton, English author • Winston Churchill, English prime minister • Carl Claus, Austrian zoologist • Georges Clemenceau, French statesman • William Kingdon Clifford, English mathematician and philosopher • Edward Clodd, English banker and author • Frances Power Cobbe, Irish social reformer and animal-rights activist • Ferdinand Cohn, German biologist • Edwin Grant Conklin, American biologist and eugenicist • Moncure D. Conway, American clergyman • Calvin Coolidge, American president • Edward Drinker Cope, American paleontologist • Mandell Creighton, English historian and prelate • João Cruz e Sousa, Brazilian poet

D 84

James Dwight Dana, American geologist • Richard Henry Dana, American author and lawyer • Nikolai Danilevsky, Russian naturalist • John William Dawson, Canadian geologist and paleobotanist • Hugo DeVries, Dutch botanist • John Dewey, American philosopher • Charles Dickens, English novelist • Emily Dickinson, American poet • Benjamin Disraeli, English prime minister • Anton Dohrn, German zoologist • Ignatius Donnelly, American novelist and politician • Fyodor Dostoevsky, Russian novelist • Arthur Conan Doyle, English author • Simon Dubnow, Russian

Jewish historian • Eugène Dubois, Dutch anatomist and paleontologist • Emil Du Bois-Reymond, German physiologist, and Albert Wigand, German botanist • Pierre Duhem, French philosopher of science • George Du Maurier, English novelist • Émile Durkheim, French sociologist

E 104

Theodore Eimer, German zoologist • Albert Einstein, German-American physicist • George Eliot, English novelist • T. S. Eliot, English poet • Havelock Ellis, English sociologist • Ralph Waldo Emerson, American philosopher • William Empson, English critic and poet • Friedrich Engels, German political theorist • Reginald Baliol Brett Esher, English politician and historian

F 114

Jean Henri Fabre, French entomologist • Michael Faraday, English physicist • Frederic William Farrar, English prelate • Sandor Ferenczi, Hungarian psychoanalyst • John Fiske, American social philosopher • Edward Fitzgerald, English author and translator • F. Scott Fitzgerald, American novelist • Gustave Flaubert, French novelist • Benjamin Orange Flower, American journalist • Antonio Fogazzaro, Italian novelist and senator • Edward Forbes, English naturalist • Ford Maddox Ford, English novelist, critic, and editor • E. M. Forster, English novelist • Harry Emerson Fosdick, American clergyman • Anatole France, French novelist • Edward A. Freeman, English historian • Sigmund Freud, Austrian psychoanalyst • Robert Frost, American poet

G 133

Francis Galton, English psychometrician and eugenicist • Mahatma Gandhi, Indian statesman • James A. Garfield, American president • Hamlin Garland, American author • Wendell Phillips Garrison, American editor and author • Marcus Garvey, African American leader • Elizabeth Gaskell, English novelist • Patrick Geddes, English biologist and social scientist • Carl Gegenbaur, German anatomist • Archibald Geikie, Scottish geologist • James Geikie, Scottish geologist • Isidore Geoffroy Saint-Hilaire, French zoologist • Alfred Giard, French marine biologist • James Cardinal Gibbons, American prelate • André Gide, French author • W. S. Gilbert, English poet and lyricist • Charlotte Perkins Gilman, American

novelist and feminist • George Gissing, English novelist • William Glad-
stone, English prime minister • Ellen Glasgow, American novelist • Ed-
mund Gosse, English poet • Stephen Jay Gould, American paleontologist
• Mountstuart E. Grant Duff, Scottish politician and author • Asa Gray,
American botanist • John Richard Green, English historian • John Thomas
Gulick, American missionary and naturalist • Ludwig Gumplowicz, Polish-
Austrian sociologist

H 164

Ahad Ha-ʿam, Russian Jewish essayist • Ernst Haeckel, German zoolo-
gist • James Duncan Hague, American mining engineer • J. B. S. Haldane,
English biologist • Granville Stanley Hall, American psychologist •
Thomas Hardy, English novelist and poet • William Henry Harvey, Irish
botanist • Willet M. Hays, American plant breeder • Lafcadio Hearn,
American journalist and author • Werner Heisenberg, German physicist •
Hermann von Helmholtz, German physicist • John Stevens Henslow,
English botanist and clergyman • John Herschel, English astronomer •
Theodor Herzl, Hungarian Jewish journalist and Zionist • Thomas Went-
worth Higginson, American clergyman and abolitionist • Charles Hodge,
American theologian • Thomas Hodgkin, English historian and banker •
Harald Hoffding, Danish philosopher • Oliver Wendell Holmes Sr.,
American physician and author • Oliver Wendell Holmes Jr., American
jurisprudent • Joseph Dalton Hooker, English botanist • Gerard Manley
Hopkins, English poet • Fenton John Anthony Hort, Irish theologian •
William Dean Howells, American author and critic • William Henry
Hudson, Argentine-English author • Aldous Huxley, English author •
Julian Huxley, English biologist • Thomas H. Huxley, English zoologist

I 201

Index of Prohibited Books (The Vatican) • Robert G. Ingersoll, American
freethinker and orator

J 203

Joseph Jacobs, English Jewish folklorist and anthropologist • Henry James,
American novelist • William James, American psychologist • Richard Jef-
feries, English nature writer • William Stanley Jevons, English economist
• Sarah Orne Jewett, American novelist • Ernest Jones, Welsh neurologist

and psychoanalyst • John Jones, English geologist • David Starr Jordan, American zoologist • James Prescott Joule, English physicist • Benjamin Jowett, English classicist and theologian • James Joyce, Irish author • John Wesley Judd, English geologist • Carl Jung, Swiss psychologist

K 217

Paul Kammerer, Austrian biologist • Karl Kautsky, German Marxist theorist • May Kendall, English poet and social reformer • John Maynard Keynes, English economist • Benjamin Kidd, English sociologist • Martin Luther King Jr., African American clergyman • Charles Kingsley, English clergyman and novelist • Henry Kingsley, English novelist • Rudyard Kipling, English author • Alexander O. Kovalevski, Russian biologist • Sergei Mikhailovich Kravchinsky, Russian author and revolutionary • Peter Kropotkin, Russian revolutionary and geographer

L 232

Jacques Lacan, French psychoanalyst • Paul Lafargue, French Marxist author and revolutionary activist • Friedrich Albert Lange, German philosopher • Sidney Lanier, American poet and professor • E. Ray Lankester, English biologist • Harold Laski, English political theorist • D. H. Lawrence, English novelist • Joseph Le Conte, American geologist • Vladimir Lenin, Russian revolutionary • George Henry Lewes, English philosopher • C. S. Lewis, English author • George Cornewall Lewis, English statesman and man of letters • Henry Parry Liddon, English theologian • Francis Lieber, German-American jurist • Abraham Lincoln, American president • Cesare Lombroso, Italian criminologist • Henry Wadsworth Longfellow, American poet • James Russell Lowell, American poet and diplomat • John Lubbock, English prehistorian and naturalist • Charles Lyell, Scottish geologist

M 259

Thomas Babington Macaulay, English historian and politician • Ramsay MacDonald, English prime minister • Anne Lumb Macdonell, English aristocrat • Ernst Mach, German physicist and philosopher • William Sharp Macleay, English-Australian entomologist • Alexander Macmillan, English publisher • Henry Sumner Maine, English legal historian • Osip Mandelstam, Russian poet • Mao Tse-tung, Chinese Communist leader •

Othniel C. Marsh, American paleontologist • Alfred Marshall, English economist • Harriet Martineau, English philosopher and feminist • Karl Marx, German political theorist • David Masson, Scottish author and literary historian • Maxwell T. Masters, English botanist and horticulturist • Henry Maudsley, English psychiatrist • Frederick Denison Maurice, English theologian • Justin Huntly McCarthy, Irish politician • James McCosh, Scottish-American philosopher and theologian • William McDougall, English psychologist • Margaret Mead, American anthropologist • H. L. Mencken, American journalist • Mikhail Aleksandrovich Menzbir, Russian zoologist • John Edward Mercer, Anglican clergyman • Herman Merivale, English civil servant and historian • Elie Metchnikoff, Russian microbiologist • Léon Metchnikoff, Russian Orientalist • John Stuart Mill, English philosopher • Henry Hart Milman, English historian and cleric • Peter Chalmers Mitchell, English zoologist • St. George Jackson Mivart, English biologist • Cecil James Monro, English mathematician • Maria Montessori, Italian educator • Aubrey Moore, English cleric and Christian Darwinian • G. E. Moore, English philosopher • Alexander Goodman More, English botanist • Lewis Henry Morgan, American anthropologist • Thomas Hunt Morgan, American geneticist • Edward S. Morse, American zoologist • Angelo Mosso, Italian physiologist • James Bowling Mozley, English theologian • John Muir, American naturalist • Friedrich Max Müller, German philologist and Orientalist • Fritz Müller, German zoologist • Hermann Müller, German botanist • Lewis Mumford, American historian • Roderick Impey Murchison, Scottish geologist • Robert Musil, Austrian novelist

N   312
Jawaharlal Nehru, Indian statesman • Dorothy Nevill, English horticulturalist and hostess • Simon Newcomb, Canadian-American astronomer and mathematician • John Henry Newman, English prelate • Alfred Newton, English zoologist • Friedrich Nietzsche, German philosopher • Max Nordau, Austrian social theorist • Marianne North, English naturalist • Charles Eliot Norton, American author and critic

O   327
Margaret Oliphant, English novelist and essayist • Eugene O'Neill, American playwright • Eleanor Ormerod, English entomologist • Henry Fair-

field Osborn, American paleontologist • William Osler, Canadian physician • Nikolaus Osterroth, German working-class leader • Wilhelm Ostwald, German physical chemist • Richard Owen, English anatomist

P 335

James Paget, English surgeon and pathologist • Francis Turner Palgrave, English poet and critic • Theodore Parker, American clergyman • Theophilus Parsons, American lawyer and jurisprudent • Walter Pater, English critic and essayist • Charles Henry Pearson, English-Australian historian • Karl Pearson, English statistician and eugenicist • Dom Pedro II, Brazilian emperor • Benjamin Peirce, American mathematician • Charles Sanders Peirce, American philosopher • Benito Pérez Galdós, Spanish novelist • Dmitry Pisarev, Russian social critic • Augustus Henry Lane Fox Pitt Rivers, English archaeologist • Pius IX, Roman Catholic pope • Frederick Pollock, English jurist • Edward Bagnall Poulton, English biologist • Ezra Pound, American poet • Baden Powell, English mathematician and theologian • John Wesley Powell, American geologist and explorer • John Price, Welsh naturalist • Michael Pupin, Serbo-American electrical engineer • Edward Bouverie Pusey, English theologian • James Jackson Putnam, American neurologist

Q 355
Armand de Quatrefages, French naturalist

R 356

Santiago Ramón y Cajal, Spanish neurohistologist • Andrew Crombie Ramsay, Scottish geologist • Winwood Reade, English explorer and novelist • Ernest Renan, French philosopher and historian • Hans Richter, Austrian conductor • Charles Valentine Riley, American entomologist • Anne Thackeray Ritchie, English author • David George Ritchie, Scottish philosopher • William Barton Rogers, American geologist and educator • George Rolleston, English anatomist and physiologist • George John Romanes, English biologist and comparative psychologist • Theodore Roosevelt, American president • Josiah Royce, American philosopher • Clémence Royer, French philosopher and translator • John Ruskin, English essayist and critic • Arthur Russell, English politician • Bertrand Russell, English philosopher

S 377

George Santayana, American philosopher • Francisque Sarcey, French journalist and drama critic • Domingo Faustino Sarmiento, president of Argentina • Minot Judson Savage, American Unitarian minister • August Schleicher, German linguist • Arthur Schopenhauer, German philosopher • Joseph A. Schumpeter, Austrian-American economist • Albert Schweitzer, Alsatian physician, musician, and philosopher • Adam Sedgwick, English geologist • Karl Gottfried Semper, German zoologist • Giuseppe Sergi, Italian anthropologist • Nathaniel Southgate Shaler, American geologist and paleontologist, and William Stimpson, American zoologist • George Bernard Shaw, English dramatist • Henryk Sienkiewicz, Polish novelist • George Gaylord Simpson, American paleontologist • Richard Simpson, English Catholic author and editor • Isaac Bashevis Singer, Polish-American Yiddish novelist • Osbert Sitwell, English essayist • George Washburn Smalley, English journalist • Mary Somerville, Scottish popular astronomer and physical geographer • Herbert Spencer, English political and social theorist • Franklin Monroe Sprague, American clergyman • Joseph Stalin, Russian dictator • Thomas Stebbing, English clergyman and naturalist • John Steinbeck, American author • Leslie Stephen, English author and critic • Robert Louis Stevenson, Scottish author • William Stimpson, American zoologist (*see* Shaler and Stimpson) • James Hutchison Stirling, English philosopher • Lytton Strachey, English author • David Friedrich Strauss, German theologian • George Templeton Strong, American lawyer and diarist • Arthur Sullivan, English composer • Billy Sunday, American evangelist • Algernon Charles Swinburne, English poet • J. M. Synge, Irish playwright

T 420

Hippolyte Taine, French critic and historian • James Monroe Taylor, American educator • William Bernhardt Tegetmeier, English naturalist and pigeon breeder • Pierre Teilhard de Chardin, French paleontologist • Alfred Lord Tennyson, English poet • William Roscoe Thayer, American historian • D'Arcy Wentworth Thompson, English mathematical biologist • Henry David Thoreau, American naturalist and philosopher • George Ticknor, American literary historian • Kliment Timiriazev, Russian biologist • Edward Bradford Titchener, American psychologist • Leo Tolstoy, Russian novelist • Paul Topinard, French anthropologist • Mary

Treat, American botanist • Henry Baker Tristram, English ornithologist and clergyman • Anthony Trollope, English novelist • Leon Trotsky, Russian politician • Mark Twain, American author • Edward Burnett Tylor, English anthropologist • John Tyndall, Irish physicist and philosopher

U 444
Miguel de Unamuno, Spanish philosopher

V 446
Georges Vacher de Lapouge, French anthropologist • Hans Vaihinger, German philosopher • Paul Valéry, French author • Nikolai Vavilov, Russian geneticist • Thorstein Veblen, American economist • Jules Verne, French novelist • Carl Vogt, German-Swiss zoologist

W 454
Richard Wagner, German composer • Alfred Russel Wallace, English naturalist • Harry Marshall Ward, English botanist • Lester Ward, American sociologist and biologist • Mary Augusta Ward, English novelist • Howard C. Warren, American psychologist • Robert Penn Warren, American poet and novelist • Booker T. Washington, African American educator • Erich Wasmann, SJ, German Jesuit biologist • Hewett Cottrell Watson, English botanist • John B. Watson, American psychologist • Alfred Weber, German economist • Max Weber, German sociologist • August Weismann, German biologist • H. G. Wells, English novelist • Edith Wharton, American novelist • William Whewell, English philosopher and polymath • Andrew Dickson White, American educator • Alfred North Whitehead, English mathematician and philosopher • Walt Whitman, American poet • Josiah Dwight Whitney, American geologist • William Dwight Whitney, American linguist • Julius von Wiesner, German botanist • Albert Wigand, German botanist (*see* Du Bois-Reymond and Wigand) • Samuel Wilberforce, English bishop • Oscar Wilde, English author • Woodrow Wilson, American president • Alexander Winchell, American geologist and paleontologist • Nicholas Patrick Wiseman, English prelate • Ludwig Wittgenstein, Austrian-English philosopher • Thomas Vernon Wollaston, English entomologist • Virginia Woolf, English author • Chauncey Wright, American philosopher • George Frederick Wright, American geologist and evangelical minister • Wilhelm Wundt,

German physiologist and psychologist • Jeffries Wyman, American naturalist

Y   497
Edmund Hodgson Yates, English journalist and novelist • William Butler Yeats, Irish poet • Edward Livingston Youmans, American science writer and publisher

Z   500
John Zahm, American priest and science professor • Chaim Zhitlovsky, Jewish socialist, philosopher, and critic • Emile Zola, French novelist

Index   505

PREFACE

This book has multiple functions: to display the breadth and depth of Darwin's interests; to serve as a guide to how Darwin may have touched each figure's career and thought; to serve as a source book; to demonstrate the "seep" of Darwin's ideas into diverse areas of thought; to offer some insight into what engaged the minds and interests of people at various places and times and what they talked about. Some of this material cannot be easily obtained elsewhere (for example, Ferdinand Cohn's account of his visit to Darwin at Down House, which appears here in English for the first time). I have tried to keep the mood light, looking for passages that project a sense of who Darwin was, how he affected people, and, later on, how various authors marshaled the icon.[1]

When I began this project I expected to register the obvious quotations and then proceed to make, I thought, a "systematic search." It never happened. Instead, I became keenly aware that, particularly in the earliest phases of reception, Darwin's ideas were diffused among networks of individuals. For example, a reference to a soirée in the rooms of the publisher Alexander Macmillan in January 1860 records "much fine talk" on Darwin's *Origin of Species* from T. H. Huxley, Charles and Henry Kingsley, Frederick Denison Maurice, Thomas Hughes, and David Masson. Henry Kingsley was a novelist, the lesser-known brother of Charles, clergyman and novelist. Both, along with the theologian Maurice, appear in this volume. Hughes and Masson, the latter a famous Milton scholar and the first editor of *Macmillan's Magazine,* were the presumed authors of a spoof with a witty reference to *Origin of Species* titled "Colloquy of the Round Table," which appeared in the first volume of *Macmillan's Magazine* (1859–

1. Two recent, admirable books present commentary on Darwin selected according to somewhat different criteria than those informing the passages reproduced here: *Charles Darwin: Interviews and Recollections,* ed. Harold Orel (New York: St. Martin's, 2000), a collection of English sources; and *Charles Darwin: Evolutionary Writings,* ed. James A. Secord (New York: Oxford University Press, 2008), wherein each of Darwin's works is followed by a section entitled "Reviews and Responses," covering a wide range of cultures and personalities.

60), right after Huxley's review of *Origin of Species* (see entry on Masson). I wanted to unearth connections among the actors in this story. I developed a strong feeling that the Victorians and the Bostonians were tightly knit circles and that the two circles were closely linked to each other. It was exciting to see the networks come to life and the links between them connected. To be sure, following reference chains induced a kind of distortion in that everyone turned up was a member of the literary or scientific elite. But that was the primary locus of the debate.

The book is organized as a biographical dictionary, encouraging the impulse of a prospective reader to look for a familiar name and then, where possible, to trace by means of asterisks (*) a particular circle of interaction, an affinity group among whose members information about Darwin and his theory circulated (for more on this, see the introduction). An unanticipated consequence of the format is that there is, nonetheless, a consistent narrative; moreover, the same narrative appears no matter in what order the entries are read. This is because of the Darwinian habit of mind established among these figures. I imposed on myself the requirement that each figure must have mentioned Darwin at least once. That criterion satisfied, I sometimes use passages obviously inspired by Darwin even though Darwin or Darwinism may not be mentioned explicitly. Figures who completely lacked the habit of mind or who did not engage it do not appear here, because searches of their works did not turn up references to Darwin. Opponents of Darwin are included in the field of reception because they contribute to the overall Darwinian discourse (whereby they unwittingly contribute to the further diffusion of Darwinism in its many forms). Three hundred eighteen individuals (72% of the total) are English-speakers, including 166 English, 128 Americans, 10 Scots, 9 Irishmen, 3 Canadians, and 4 Welshmen.

I have supplied first names for most persons mentioned by surname only in various entries, except in the case of classical figures in the history of science whose names are widely known, such as Galileo Galilei, William Harvey, Johannes Kepler, and Sir Isaac Newton. The surnames of three classical figures who influenced Darwin—Georges Cuvier, Jean-Baptiste Lamarck, and William Paley—also appear frequently. Cuvier (1769–1832), the most influential biologist of the early nineteenth century, promoted a theory of biological archetypes, four of them in the animal kingdom, which became the cornerstone of the scientific argument for the fixity of species. He viewed organisms as pliant and flexible, capable of developing new varieties, but always within the constraints of their "type." In 1830, in a celebrated debate, Cuvier bested French evolutionists in the tradition of Lamarck (1744–1829), who had proposed an evolutionary theory based on the inheritance of acquired characteristics. Because Lamarck was the only "big name" in evolutionary theory in the early nineteenth century, his name was constantly invoked in the Darwinian debate by Darwin's friends and foes alike.

In English natural history, design was given material form in Cuvier's types. The English theologian William Paley (1743–1805) influenced every English naturalist in the first half of the nineteenth century. In his famous book *Natural Theology,* the only book Darwin claimed to have enjoyed reading while a student at Cambridge, Paley presented an argument for the existence of God by demonstrating how nature might be read as a book revealing God's design. Paley's natural theology was focused on adaptation: each creature was created perfectly adapted to its environment. Paley and his followers turned up countless examples of ingenious "contrivances" displayed by animals and plants in their modes of adaptation, evidence that Darwin was able to put to use in his own theory, in exactly the same form in which naturalists in the Paleyian mode had presented it.

Several figures in the Darwin debate for whom there are entries in this volume are mentioned so frequently that supplying their first names in every instance would only annoy the reader, so I do not supply them. These include the geologist and biologist Louis Agassiz, the botanist Joseph Hooker, the geologist Charles Lyell, the sociologist Herbert Spencer, the co-discoverer of natural selection Alfred Russel Wallace, and the zoologists Ernst Haeckel and Thomas H. Huxley. While there are entries not only for T. H. Huxley but also for two of his grandsons, Aldous and Julian, all cross references are to the senior Huxley.

I have in most instances changed British standard orthography to American in order to introduce some coherence into the presentation of the texts. Thus I have changed *ou* to *o (favor,* not *favour)* and *-ise* to *-ize (baptized,* not *baptised).* In addition, I have deleted the spaces often found in Victorian books before colons, semicolons, exclamation points, and question marks. Ellipses added by me are in brackets, to distinguish them from ellipses present in the original.

In order to make the early chronology of reception easier to trace, I include full dates before selections covering the period from November 1859 through the spring of 1860. These dates serve to highlight the speed and intensity of the early debate over the *Origin of Species*. I occasionally include contemporary descriptions of how individuals thought of or interacted with Darwin or of the broader trend of reception, that is, indirect as well as direct evidence. Source notes and additional bibliography follow individual entries.

Two ongoing Darwin projects at Cambridge University have been indispensable. The first is the Darwin Correspondence Project (www.darwin project.ac.uk), founded by an old friend, Frederick Burkhardt (1913–2007). Cambridge University Press has to date published seventeen volumes of letters to and from Darwin for the years 1821–69, cited throughout the book as *The Correspondence of Charles Darwin*. The project now has several editors; I am particularly indebted to Alison Pearn. The second project is Darwin Online (Darwin-online.org.uk), directed by John van Wyhe. It is a treasure trove of documents, manuscripts, articles, and books by Darwin or about him.

This has been a long labor of love during which I logged many hours in solitude in the stacks of Mugar Library at Boston University and Widener Library at Harvard, as well as in front of my computer, endlessly surfing for "John Smith and Darwin," "Jones visits Down," "Life and Letters," and the like. Among colleagues who provided leads along the way are Robert Cunningham, James Johnson, Christina Matta, Allison Palm, Simon Rabinovitch, Jon Roberts, and Elinor Shaffer. Alice Du's drawing of Darwin on page ii was co-winner of a Darwin logo contest conducted by the Boston University Charles Darwin Bicentennial Committee and is reproduced by permission of the committee.

Readers are invited to visit the Facebook page for this book, www. facebook.com/pages/What-about-Darwin/316747122973?v=app_23730727 38&ref=ts#!/pages/What-about-Darwin/316747122973?v=wall&ref=ts, to comment on the book in its current form, to contribute new quotations by individuals of note around the world, and to participate in an ongoing conversation on Darwin and his influence.

INTRODUCTION

What *about* Darwin? He was hero and villain, idol and devil. Those who expressed a negative opinion of Darwin's mind, character, or personality were usually persons who did not know him, such as most critics abroad. There was remarkable uniformity in the judgment of those who were personally acquainted with him, who referred to "the proverbial modesty and singular simplicity and sweetness of his character" (Charles Valentine Riley), on the one hand, and his honesty and seriousness of purpose, on the other. This book is about perception, mood, intangibles, context, the drone against which the melody or melodies of reception play out. Thus there is no real distinction between the scientific channel and the literary channel: they are cut from the same cloth of common culture and common experience.

The reception of Darwin's *Origin of Species* produced a radical break in patterns of thought everywhere. Evolutionism, usually with Darwin's name attached to it, spread quickly and relentlessly, so that in England, the United States, Germany, and Russia[1]—where the early debate was liveliest—it was impossible for intellectuals to dodge it or to fail, sooner rather than later, to come to grips with its implications. Darwinism hit them in the gut in such a way that they could not avoid reacting. Indeed, in a study of Darwin's influence in the works of Thomas Hardy, E. M. Forster, and D. H. Lawrence, Roger Ebattson concludes that for them Darwin's theory became a kind of "mental habit." My book explores that habit of mind, which was not limited to creative writers but affected the intelligentsia, worldwide.[2]

Thus the idea informing this book is that Darwin influenced a vast range of people representing many walks of life. We see his influence on naturalists, of course, but also on clerics (8% of the sample, from Henry

1. Of the 442 individuals for whom there are entries in this book, 48 are German or Austrian, 24 are French, and 20 are Russian, these constituting 20.8% of the total.

2. Roger Ebbatson, *The Evolutionary Self: Hardy, Foster, Lawrence* (Totowa, NJ: Barnes & Noble, 1982), x.

Ward Beecher to Pope Pius IX), politicians (Lincoln, Gladstone, Disraeli, Clemenceau, and, in the twentieth century, Theodore Roosevelt, Woodrow Wilson, Lenin, Stalin, Mao, Gandhi, and Nehru), novelists (10%, including Charles Dickens, George Eliot, John Steinbeck, and Émile Zola), poets (6%, including Emily Dickinson, William Empson, and Thomas Hardy), musicians (Richard Wagner, Hans Richter), economists (John Maynard Keynes, Alfred Marshall, James Bonar), social theorists (Max Weber, Émile Durkheim, Ludwig Gumplowicz, Max Nordau), and philosophers (Bertrand Russell, George Santayana, Alfred North Whitehead). Also influenced were women naturalists of the Victorian era (Alice Bodington, Arabella Buckley, Mary Somerville, Dorothy Nevill).

I have been particularly drawn to two kinds of documents. First are reminiscences or accounts of first readings of Darwin, which are a leitmotif in the autobiographies of Victorian writers. Early first readings, in November and December 1859 and the spring of 1860, record the initial shock waves and document the quickening of debate. Second are reports of visits to Darwin. These are fascinating for their detail (see especially those of Moncure Conway and Kliment Timiriazev), reflecting not only the personalities of the actors but also life in a Victorian household. These accounts are also striking because of the unanimity of opinion regarding Darwin's personality. He is uniformly described as kind, gentle, good, and, of course, very intelligent.

No matter how Victorian readers reacted to his theory, they almost always took Darwin seriously and treated him with respect. This was owing, first, to his considerable reputation, built up popularly from generations of readers familiar with *The Voyage of the Beagle,* a book that was recognized early on as a classic of nature writing. The collected volumes of geological findings of the *Beagle* voyage, for which Darwin was made a fellow of the Royal Society, and the later volumes of zoological findings of the voyage added to his reputation among naturalists. Darwin was also the beneficiary of the willingness of prestigious individuals—arbiters of culture such as Alfred Lord Tennyson and George Henry Lewes—to support him publicly. Finally, the careful and artful writing—the book is a model of Victorian prose—and the clear, simple, and logical presentation of his ideas made his argument both eminently accessible and very compelling. Darwin's strategy of starting the book by appealing to the familiarity of Englishmen of his class with animal and plant breeding paid off.

Moreover, the logic underlying natural selection was difficult to refute head-on. Natural selection rests on two facts well appreciated by breeders: variation and culling; unfavorable variations cause their carriers to be culled from populations, leaving a higher proportion of individuals with favorable variations. It is interesting to note that naturalists who rejected Darwin's conclusions nevertheless deferred to his powers of observation, putting them into the uncomfortable position of having to rebut him with tortured reasoning and wishful thinking. This is because the logic underlying natural selection is so simple that it is impossible to argue against it without resorting to contrived arguments. Or to no argument at all, but simply an affective rejection on the basis of ideology, for example.

The Earliest Reception

The *Origin of Species* was published by John Murray on November 22, 1859, the edition of 1,250 copies selling out the first day. However, there were already several naturalists who, because they had heard Darwin's paper read at the Linnean Society on July 1, 1858, or read it when it was published in late August, had already become Darwinians—persons, that is, who did not belong to the circle of Darwin's intimates who knew what he was doing and were favorably disposed (e.g., Joseph Dalton Hooker, Asa Gray, etc.). One such naturalist was the zoologist Alfred Newton: "I know that I sat up late that night to read it; and never shall I forget the impression it made upon me. Herein was contained a perfectly simple solution of all the difficulties which had been troubling me for months past."[3] Another early convert was the ornithologist Henry Baker Tristram, an Anglican canon, who was the first zoologist (encouraged by Newton) to publicly accept Darwin's views. Later on, perhaps under ecclesiastical peer pressure, he recanted. Darwin had sent advance, prepublication copies to select friends and important personages. We know that the clergyman and novelist Charles Kingsley already had his on November 18, for he wrote to Darwin on that day thanking him and commenting favorably. On November 26 the publisher Alexander Macmillan wrote to Alfred Lord Tennyson bidding him look at Thomas Henry Huxley's article in the *London Times* of that day. Tennyson, meanwhile, was already reading

---

3. Sources for quotations not otherwise identified can be found in the entries for the individuals cited.

the *Origin,* a copy of which had been sent to him by Francis Turner Pal-grave as early as the nineteenth. Another day in November, the philosopher Alexander Bain was dining at Trinity Lodge with William Whewell when Darwin's old geology professor, the theological conservative Adam Sedgwick, appeared, announced that he had been reading the *Origin,* and delivered a diatribe, perhaps his first, against it.

On December 1 the veteran geologist Roderick Murchison wrote George Gordon that he had been "much startled by the apparition of Darwin's book on the *Origin of Species.*" On December 2 Michael Faraday received his copy. On December 6 the novelist George Eliot wrote to Madame Bodichon:

> We have been reading Darwin's book on the *Origin of Species* just now: it makes an epoch, as the expression of his thorough adhesion, after long years of study, to the Doctrine of Development—and not the adhesion of anonym like the author of the *Vestiges,* but of a long-celebrated naturalist. The book is sadly wanting in illustrative facts—of which he has collected a vast number, but reserves them for a future book, of which this smaller one is the *avant-coureur.* This will prevent the work from becoming pop-lar as the *Vestiges* did, but it will have a great effect on the scientific world, causing a thorough and open discussion of a question about which people have hitherto felt timid.[4]

From December 6 to December 8 Charles Bunbury visited Charles Kingsley, who "talked much of Darwin's new book on species, expressing great admiration for it, but saying that it was so startling that he had not yet been able to make up his mind as to its soundness." On December 21 Darwin complained to Charles Lyell that he had heard that John Herschel had denounced his theory as "the law of higgledy-piggledy."[5] On the twenty-third Henry Darwin Rogers wrote to his brother William Barton Rogers, a geologist, that "the only matter of any interest" to report from Glasgow was the appearance of Darwin's book. On the twenty-fourth Sedgwick finally wrote to Darwin his negative reactions to the *Origin,* which he had, of course, read the previous month. On the twenty-sixth

---

4. *George Eliot's Life as related in her Letters and Journals,* ed. J. W. Cross, 3 vols. (New York: Harper, 1885), 2:164–65.

5. On Herschel's denunciation, see the entry for Edward Clodd.

Harriet Martineau wrote George Holyoake enthusiastically about the *Origin*, a copy of which had been given to her by Darwin's brother, Erasmus.

The quickening so easily detected in the contacts made in November and December among key members of the elite had by January 1860 reached a feverish pitch. In the preface I mentioned the gathering at Alexander Macmillan's rooms on Henrietta Street, where Huxley, the Kingsley brothers, and others engaged in "much fine talk" on the *Origin*. On January 9 Charles Lyell wrote to the literary historian and Hispanist George Ticknor, in Boston, that the only ideas he had had to exchange recently pertained to "that question which my friend Charles Darwin's book has brought before the British reading public, both scientific, literary, and theological." No sooner had Lyell's letter arrived (in February) than the American geologist William Barton Rogers mentioned to his brother that he had heard that "Ticknor has just heard from Lyell, who speaks approvingly of Darwin's views." Word got around; it was the custom to lend letters to friends or to read them aloud at literary salons.

On January 17 the historian Henry Thomas Buckle wrote to a friend, Mrs. Woodhead, advising her to tell her husband to read the *Origin* and "master it." The same day, Huxley reported to William Barton Rogers that "Darwin is the great subject just at present, and everybody is talking about it." On the twenty-eighth Jane Welsh Carlyle wrote Mary Russell thanking her for the *Origin*, "which will be welcome to several of my acquaintances. There is quite a mania for geology at present, in the female mind."

These examples illustrate the rapid circulation among the English elite of the book itself and of the ideas it contained. All of these excerpts, as well as many more that pick up the story in England and the United States from February 1860, can be found in the selections herein.

The reception of ideas is about "affinity groups" and how ideas circulate. The great speed with which news of Darwin's theory got around is, in the first place, a function of the small size and high density of interlocking affinity groups. The educated elite in England was small enough that everyone knew everyone else, resulting in a hyperconductivity of ideas and the extreme rapidity with which they circulated. Entire affinity groups, such as the American transcendentalists (Emerson, Thoreau, Theodore Parker, Bronson Alcott) and the Bloomsbury Group (Virginia Woolf, Lytton Strachey, Clive Bell, John Maynard Keynes), appear here *in toto*, or almost so. Affinity groups can be perceived materially here by following the asterisks (*) used to mark individuals for whom there are entries in this volume. Moreover, the density of asterisks in any one entry provides a visual illustration of the connectedness of the person in question with others of the same "habit of mind."

From the examples presented we learn that ideas travel in various ways: by reading, by personal communication, and even by the writing of letters whose contents are read aloud to others. In the English reception of Darwin, moreover, a number of intellectuals played key roles. Tennyson and Thomas Carlyle, for example, had huge circles of acquaintances; they seem to have known everybody, and they appear in this book as conduits for the transmission of ideas. William Kingdon Clifford, a mathematician and philosopher who rarely appears in the vast literature of Darwin and Darwinism, was another such axial person: the people who gathered regularly at his house included Huxley and George John Romanes, two key Darwinian biologists, plus the author and critic Leslie Stephen (Virginia Woolf's father) and the jurist Frederick Pollock; less frequent attendees were the novelist and science writer Grant Allen and Thomas Hardy. As Pollock remarked,

> Clifford was not content with merely giving his assent to the doctrine of evolution: he seized on it as a living spring of action, a principle to be worked out, practiced upon, used to win victories over nature, and to put new vigor into speculation. For two or three years the knot of Cambridge friends of whom Clifford was the leading spirit were carried away by a wave of Darwinian enthusiasm: we seemed to ride triumphant on an ocean of new life and boundless possibilities. Natural Selection was to be the master-key of the universe; we expected it to solve all riddles and reconcile all contradictions.[6]

The earliest phase of reception in England, roughly 1859–70, was that of Darwin's initial impact among persons born no later than 1830, all of whom had to come to terms with Darwin's theory having been raised, for the most part, with a traditional world-view. The realization of, or assertion that, the episteme, or world of thought, had changed because of Darwin (e.g., George Eliot's initial reaction, in 1859) is a leitmotif. It is a com-

6. Frederick Pollock, "Biographical Introduction," in William Kingdon Clifford, *Lectures and Essays,* ed. Leslie Stephen and Frederick Pollock, 2nd ed. (London: Macmillan, 1886), 1–32, on 24.

monplace to think of the Wilberforce-Huxley exchange at Oxford in July 1860 as a significant turning point in the reception or acceptance of Darwinism in England. But in England the battle had long since finished: most of the elite had been won over in the first months. By March or April most of the old-style naturalists left were vicars of the Church of England, and many of them as well had become evolutionists.

The "eclipse of Darwinism" that historians of science talk about and the "Age of Evolutionism" of the literary historian Tom Gibbons both ran from 1880 to 1920.[7] However, a subperiod, roughly 1870–90, embraced the experience of the first generation of persons who grew up in a world where Darwinian assumptions were taken for granted, those born between 1830 and 1850. For example, Edward S. Morse declared in 1887 that "to a naturalist it may seem well-nigh profitless to discuss the question of evolution since the battle has been won."

The "eclipse" had to do with the profusion of hypothetical mechanisms of evolution in the absence of any way of confirming natural selection in a laboratory or in nature. So alongside a core group of Darwinians, Lamarckism and the inheritance of acquired characteristics were revived; orthogenesis (straight-line evolution) got a new lease on life; and so forth. But here is how Gibbons defines the Age of Evolutionism:

> Looked at in the very broadest possible way, the period 1880–1920 was one
> of full-scale ideological reaction from the scientific materialism, atheism,
> determinism and pessimism of the mid-nineteenth century. During the
> last quarter of the nineteenth century these gave way on many sides to
> their opposites: philosophical idealism, religious transcendentalism, vitalism and optimism.[8]

My view is different: vitalism never went away; high culture, including science, which participates in broader trends, oscillates between the poles of vitalism and mechanicism, of naturalism and romanticism, and indeed

7. Peter J. Bowler, *The Eclipse of Darwinism: Anti-Darwinian Evolution Theories in the Decades around 1900* (Baltimore: Johns Hopkins University Press, 1983); Tom H. Gibbons, *Rooms in the Darwin Hotel: Studies in English Criticism and Ideas, 1880–1920* (Nedlands: University of Western Australia Press, 1973). The term *eclipse of Darwinism* was coined by Julian Huxley.

8. Gibbons, *Rooms in the Darwin Hotel,* xi. The period is too inclusive, in any case; in science the "full-scale" reaction begins about 1900, with Planck's quantum of action, Freud's *Interpretation of Dreams,* and the "rediscovery" of Mendelian laws of genetics.

whatever antimonies inform high culture. I agree with Roger Ebbatson that the acceptance of evolution required "the idea of instability of the present order and of the paradox that a state of change is the invariable characteristic of natural systems and human institutions. Intellectuals digesting the implications of the theory were constrained to accept it as a theory of knowledge, in the sense that it meant understanding change from one order to another."[9]

That is, evolutionary thinking in the Darwinian mold was both an ontology—it told you how nature as we experience it came into being—and an epistemology—the ecological intuition that existed only after, and because of, Darwin—that laid down the paths by which we understand nature.

What a historian sees when surveying the literary scene in the Age of Evolutionism—and what the passages here from novelists and poets of that period reveal—is that Darwin, with an avalanche of appearances in novels, has now become iconic. This is when Ebbatson's "habits of mind" become entrenched, when the image of "Darwin" ceases to refer to the real person and becomes an icon.

It cannot be expected that historians of science and of literature will periodize the reception of Darwin in exactly the same way. But if the reception is considered to be one unified phenomenon, with scientific and literary registers side by side and intertwined, then unfamiliar landscapes may appear. For example, there may have been a systematic misrepresentation of the reception of Darwin in France, where among historians of science it is typically held that Darwinism was *not* received there in the nineteenth century, owing to a number of factors, the principle one being the prior standing of Lamarck. But it can certainly be argued that the *literary,* as opposed to the scientific, reception of Darwin in France was significant and perhaps more normative, more like Darwin's literary reception elsewhere. The first Darwinian novel published anywhere was, in fact, written by a Frenchman, Jules Verne's *Voyage to the Center of the Earth,* of 1864; then came Anatole France, Flaubert, Gide, Proust, Zola, and so forth. For novelists, Darwin was a signpost, a reality check, the unshakable symbol of the material world that would not go away, indeed could not. Darwin, like Freud, invites us to take a hard look at ourselves and, some-

---

9. Ebbatson, *Evolutionary Self,* xi.

what paradoxically, an awe-filled look at the organic world, nature. Meanwhile, the social sciences in the Age of Evolutionism became so thoroughly Darwinized that no theorization in any one of them was taken seriously unless it had a Darwinian component.

The Earliest American Reception

The chronology of the early American reception, with its hub in Boston, parallels that of the British reception. Similar events unfolded on both sides of the ocean. On November 27, 1859, Hooker wrote to Asa Gray that "Darwin's book is out & created a tremendous furore on all hands."[10] On December 9 another Londoner, Francis Boott, reported: "I have not seen Hooker. I suppose he has full faith. Lyell has not, nor Herschel nor Sedgwick. I hear that the last told Darwin he had to hold his sides with laughing."[11] Charles Eliot Norton describes visiting Jeffrey Wyman's laboratory in the Harvard Physiological Museum one chilly night sometime before December 26, 1859. He had gone there with the botanist John Torrey and James Russell Lowell. They sat around the fireplace and talked of Darwin and Louis Agassiz's reaction to his book.[12] Then, on January 1, 1860, Charles Loring Brace, a minister and social reformer from New York City who had come to Cambridge, Massachusetts, to visit Asa Gray, his brother-in-law, borrowed Gray's copy and took it to Concord, where he dined with Franklin B. Sanborn, along with Henry David Thoreau and Bronson Alcott, and their conversation focused on Darwin and his book. Thoreau was so excited that he obtained a copy and took pages of notes.[13]

On January 2, 1860, William Barton Rogers told his brother that he was looking into Darwin's argument. Gray wrote Hooker on January 5 praising Darwin's book. Addison Emery Verrill, a student (and later a

10. A. Hunter Dupree, *Asa Gray* (New York: Atheneum, 1968), 267.

11. Ibid.

12. Norton to Elizabeth Gaskell, 26 December 1859, in *Letters of Charles Eliot Norton*, ed. Sara Norton and M. A. DeWolfe Howe, 2 vols. (Boston: Houghton Mifflin, 1913), 2:201–2. That Norton wrote this to Gaskell, a relative of Darwin's and another axial individual in English literary circles, is significant.

13. See Alfred I. Tauber, *Henry David Thoreau and the Moral Agency of Knowing* (Berkeley and Los Angeles: University of California Press, 2001), 263–64; and Robert D. Richardson Jr., "Thoreau and Science," in *American Literature and Science*, ed. Robert Scholnick (Lexington: University of Kentucky Press, 1992), 110–27, on 123–24. Richardson notes that John Torrey, by contrast, waited three years to read Darwin.

distinguished biologist), recorded in his diary a walk with Agassiz, who commented on Darwin's "pleasing but false" theory.[14] On the twenty-third Gray wrote to Darwin describing a review of *Origin* that he had just written for the *American Journal of Science*. On January 25 Agassiz attacked Darwin at a meeting of the American Academy of Arts and Sciences, and he denounced him again on February 15 at a meeting of the Boston Natural History Society. On February 18 the diarist George Templeton Strong recorded slitting the pages of his copy of *Origin*. The same day, Simon Newcomb, then a student at Harvard, got permission from the mathematician Benjamin Peirce to attend another American Academy session on Darwin. Theodore Parker, in Rome, reported on February 24 that he had heard of the importance of *Origin;* on the same day, William Barton Rogers predicted Darwin's triumph in Europe. A sermon of March 11 was the first in which Henry Ward Beecher mentioned evolution. On March 27 the Harvard philosopher Francis Bowen attacked Darwin at American Academy of Arts and Sciences, and sometime in March, Elizabeth Cabot reported that her husband, James Elliot Cabot, had just finished reading *Origin*.

The vast majority of participants in the Darwin debate in Boston and in England were personally acquainted with one another. Two nonscientists who played a significant role in the Boston reception were George Ticknor, a literary historian who knew nothing about science but conducted a lifelong correspondence with Charles Lyell, and Charles Eliot Norton, a professor of art history at Harvard, who knew the principal players in England who were actively involved in the discussion of *Origin of Species* through his sister-in-law, Sara Sedgwick, who was the wife of William Darwin, a banker, one of Darwin's sons. William Darwin did not contribute any ideas to the debate, but his home and his circle of acquaintances became, via Norton, agencies that accounted for the rapid dissemination of the entire English debate to Boston. Because the circle was literary as well as scientific, it drew in unexpected persons, such as Henry James (via Norton) and Henry Wadsworth Longfellow (via Tennyson), both of whom had met Darwin personally. Many of the Bostonian protagonists had wide circles of acquaintance in England and circulated freely in intellectual circles there. Asa Gray's role was, of course, crucial because of his

14. Dupree, *Asa Gray*, 271.

closeness to Darwin, his prestige among American proponents of evolution, and his leadership of resistance to Louis Agassiz's creationism. Yet it could be argued that the humanist Charles Eliot Norton was the key link between the English players and the Americans.

A host of Americans turned up at Darwin's house. One could argue that the intelligentsia of both nations were knit together as a broadly interacting group. One measure of this is the incidence of Americans who corresponded with Darwin. Gray's interests and the uses Darwin made of him as a correspondent naturally turned Darwin's eyes toward New England, while the excitement about Darwin in America found easy channels of communication with like-minded persons across the ocean. One might argue that science generally and Darwinism in particular gave a distinctive tone to this generation of American intellectuals. Travel to England ceased to be a trademark of New England intellectuals in the next generation, who had other concerns and who had reached a certain level of cultural self-assuredness.

Darwin's Persona

Persons who reported having met Darwin or having known him personally were surprisingly consistent in their characterization of his personality and temperament: "a pleasant jolly-minded man" (William Allingham); "The kindness of the great man, his sympathy and charm . . . left a deep impression on my youthful mind" (Arthur Balfour); "I never met a more simple, happy man—as merry and keen as Dr. [Asa] Gray, whom he loves much" (Charles Loring Brace); "His varied, frank, gracious conversation, entirely that of a gentleman, reminded me of that of Oxford and Cambridge *savants*. The general tone was like his books, as is the case with sincere men, devoid of every trace of charlatanism" (Alphonse de Candolle); "a more charming man I have never met in my life" (Thomas Carlyle); "a tall, strong, grey bearded stature, full of life and cheerfulness and heart-winning amiability" (Anton Dohrn); "that perfect frankness and guileless simplicity of manner" (John Fiske); "Darwin's own hearty manner, hollow laugh, and thorough enjoyment of home life with friends" (Joseph Hooker); "beautifully benignant, sublimely simple" (Henry James); "I saw him as a very kindly and gentle man and felt that I had known him for a long time" (Konstantin Timiriazev); "of all eminent men that I have ever seen he is beyond comparison the most attractive to me. There is

xxx

something almost pathetic in his simplicity and friendliness" (Leslie Stephen). A magazine satire singled out Darwin as the most pacific member of a squabbling set of naturalists: "The whole neighborhood was unsettled by their disputes; Huxley quarreled with Owen, Owen with Darwin, Lyell with Owen, Falconer and Prestwich with Lyell, and Gray the menagerie man with everybody. He had pleasure, however, in stating that Darwin was the quietest of the set."[15]

Darwin famously remarked in his autobiography that from the time he opened his first notebook on the species question in July 1837 he had worked "worked on true Baconian principles, and without any theory collected facts on a wholesale scale." By the time Ramsay MacDonald was pondering Darwin's work about 1910, the question of Darwin's method, whether inductive or deductive, had become a commonplace of Darwinian debate: "We hear occasional disputes as to whether the Darwinian method was inductive or deductive, whereas, as a matter of fact, it was a scientific blending of both."

When the German zoologist Anton Dohrn visited Darwin in 1870, he raised the issue of methodology:

> During this conversation I asked him, "Excuse me, if I ask a perhaps very stupid question: how do you begin your studies at all?"
>
> "I'll tell you: I begin always with *a priori* solutions. If anything happens to impress me, I have hundreds of hypotheses before I know the facts. I apply one after the other, till I find the one that covers the whole ground. But I am exceedingly careful and slow in printing."

It would appear from this exchange that Darwin's approach was nearly the opposite of Baconian inductivism. Darwin may have been thinking of the voyage of the *Beagle,* when he simply gathered facts and observations (in zoology and botany; in geology, by contrast, he carried with him Lyell's *Principles of Geology,* which gave him a theoretical basis for ordering his observations). From the time he became an evolutionist, in July 1837 (when John Gould told him that Galapagos finches were not varieties but sepa-

15. *Public Opinion,* 23 April 1863. In *The Descent of Man* (1871) Darwin references a plate depicting an antelope reproduced in Dr. John Edward Gray, *Gleanings from the Menagerie and Aviary at Knowsley Hall* (Knowsley, 1846). The plate was from a series of drawings made by Edward Lear from living animals in the menageries of the Earl of Derby.

rate species), until October 1838 he maniacally collected everything he could bearing on the "species question." In October 1838, when he read Malthus's *On Population* and intuited natural selection, that became his working hypothesis, and his method thereafter, at least with respect to the species question, became largely deductive.

From the 442 assessments of Darwin that follow it is possible to build a composite picture of various aspects of his life, his work, his personality, and the way his ideas were received by a wide variety of individuals. In the aggregate they describe the emergence of a distinctive "habit of mind" regarding the natural world.

HENRY ADAMS (1838–1918)
AMERICAN HISTORIAN

At that moment Darwin was convulsing society. The geological champion of Darwin was Sir Charles Lyell*. [...] Sir Charles constantly said of Darwin what [Francis Turner] Palgrave* said of [Alfred Lord] Tennyson*, that the first time he came to town, Adams should be asked to meet him, but neither of them ever came to town, or ever cared to meet a young American, and one could not go to them because they were known to dislike intrusion. [...] Adams was content to read Darwin, especially his *Origin of Species* and his *Voyage of the Beagle*. He was a Darwinist before the letter; a predestined follower of the tide; but he was hardly trained to follow Darwin's evidences. Fragmentary the British mind might be, but in those days it was doing a great deal of work in a very un-English way, building up so many and such vast theories on such narrow foundations as to shock the conservative, delight the frivolous. The atomic theory; the correlation and conservation of energy; the kinetic theory of gases, and Darwin's Law of Natural Selection, were examples of what a young man had to take on trust. The ideas were new and seemed to lead somewhere—to some great generalization which would finish one's clamor to be educated. That a beginner should understand them all, or believe them all, no one could expect, still less exact. Henry Adams was a Darwinist because it was easier than not, for his ignorance exceeded belief, and one must know something in order to contradict even such triflers as [John] Tyndall* and Huxley*. [...]

He never tried to understand Darwin; but he still fancied he might get the best part of Darwinism, from the easier study of geology; a science which suited idle minds as well as though it were history. Every curate in England dabbled in geology and hunted for vestiges of Creation. Darwin hunted only for vestiges of Natural Selection, and Adams followed him, although he cared nothing about Selection, unless perhaps for the indirect amusement of upsetting curates. He felt, like nine men in ten, an instinctive belief in Evolution, but he felt no more concern in Natural than in unnatural Selection, though he seized with greediness the new volume on the *Antiquity of Man* which Sir Charles Lyell published in 1863 in order to support Darwin by wrecking the Garden of Eden. Sir Charles next brought out, in 1866, a new edition of his *Principles,* then the highest textbook of geology; but here the Darwinian doctrine grew in stature. Natural Selection led back to Natural Evolution, and at last to Natural Uniformity. This was a vast stride. Unbroken Evolution under uniform conditions pleased everyone—except curates and bishops; it was the very best substitute for religion; a safe, conservative, practical, thoroughly Common-Law deity.

Henry Adams, *The Education of Henry Adams: An Autobiography* (Boston: Houghton Mifflin, 1918), 224–25.

Secondary Literature: Ernest Samuels, *The Young Henry Adams* (Cambridge, MA: Harvard University Press, 1965), 128–67; idem, *Henry Adams: The Major Phase* (Cambridge, MA: Harvard University Press, 1964), 36–40, 383, 478–79; Christopher Berkeley, "Henry Adams," in *Encyclopedia of Historians and Historical Writing,* ed. Kelly Boyd, 2 vols. (London: Fitzroy Dearborn, 1999), 1:2–4.

LOUIS AGASSIZ (1807–1873)
Swiss-American Naturalist

I have for Darwin all the esteem that one has to have; I know the remarkable work that he has accomplished, as much in Paleontology as in Geology, and the earnest investigations for which our science is indebted. But I consider it a duty to persist in opposition to the doctrine that today carries his name. I indeed regard this doctrine as contrary to the true methods that Natural History must inspire, as pernicious, and as fatal to progress in this science. It is not that I hold Darwin himself responsible for these troublesome consequences. In the different works of his pen, he never made allusion to the importance that his ideas could have for the point of view of classification. It is his henchmen who took hold of his theories in order to transform zoological taxonomy. The different incarnations of that influence are felt on the general conceptions of Paleontology and more directly on those of Zoology.

Louis Agassiz, *De l'espèce et de la classification en zoologie* (Paris: G. Baillière, 1869), pt. 3, chap. 7, "Le Darwinisme—Classification de Haeckel," 375–91. This passage and others have been translated into English by P. J. Morris in "Louis Agassiz's Additions to the French Translation of his Essay on Classification," *Journal of the History of Biology* 30 (1997): 121–34.

Secondary Literature: Edward Lurie, *Louis Agassiz: A Life in Science* (Chicago: University of Chicago Press, 1960).

AMOS BRONSON ALCOTT (1799–1888)
AMERICAN AUTHOR AND EDUCATOR

Darwin and Swedenborg were both reaching the same conclusion, but Darwin was arguing from the materialist's side, while Swedenborg and others were arguing from the side of the mind. Darwin began at the bottom, and arrived at man; there he stopped. Idealists started at God and went down to nothing, which was the correct way. Conversation could not be kept up, nor can people live in communities because they were not acquainted with themselves.

Amos Bronson Alcott, *Notes of Conversations,* ed. Karen Ann English (Madison, NJ: Fairleigh Dickinson University Press, 2007), 233.

Secondary Literature: John Matteson, *Eden's Outcastes: The Story of Louisa May Alcott and Her Father* (New York: Norton, 2007).

SHOLEM ALEICHEM [SHOLEM RABINOVICH] (1859–1916)
RUSSIAN YIDDISH AUTHOR AND SATIRIST

The town of "Kasrilevke" has two theaters, presenting plays in Yiddish and Hebrew, respectively, reflecting the cultural wars between "Yiddishists" and "Hebraists." Commenting on a case where a husband and wife—both actors— were at odds, the Yiddishist newspaper editorialized.—TFG.

To Hell with Darwin!

These scholars, those so-called Darwinists, just want to talk us into something. They would have us believe that for appearance's sake there is a
force in nature known as heredity, and owing to this force a child must
resemble his parents. But this theory hasn't got a leg to stand on. We shall
prove it with facts. We do not support empty hypotheses. We like facts.
For example, here is a case that happened not so long ago, not in our own
Kasrilevke, heaven forbid, but in Yahupetz. A couple lived there, both of
whom were amateurs in the Yiddish theater. He was a fiery Yiddishist,
she, a rabid Hebraist. One day, a baby boy was born to them. Within a year
they discovered he did not look a bit like his Yiddishist father, but on the
contrary, he was the spitting image of a stranger, also an amateur, but a
Hebraist. Although this Hebraist amateur was anti-Yiddish on principle,
he did visit the Yiddishist's wife quite often. It is quite possible then—almost a sure thing—that the baby's resemblance to the amateur Hebraist
stems from the fact that the two had simply stared at each other so long.

If so, then it is further proof that the learned Darwinists who talk
about inheritance haven't got the faintest notion of who's pulling the wool
over their eyes. They only babble on at random about their silly theory and
don't even bother finding out what's happening under their very noses.

Sholem Aleichem, "Progress in Kasrilevke," in *My First Love and Other Stories,*
trans. Curt Leviant (New York: Courier Dover, 2002), 17–67, on 65–66.

Secondary Literature: Joseph Butwin and Frances Butwin, *Sholem Aleichem* (Boston: Twayne, 1977).

GRANT ALLEN (1848–1899)
English Science Writer and Novelist

Great thinkers of the elder generation, like [Alexander] Bain* and Lyell*,
felt bound to remodel their earlier conceptions by the light of the new
Darwinian hypotheses. Those who failed by congenital constitution to do
so, like [Thomas] Carlyle* and [William Benjamin] Carpenter*, were,
philosophically speaking, left hopelessly behind and utterly extinguished.
Those who only half succeeded in thus reading themselves into the new
ideas, like [George Henry] Lewes* and Max Müller*, lost ground imme-

diately before the eager onslaught of their younger competitors. "The world is to the young," says the eastern proverb; and in a world peopled throughout in the high places of thought by men almost without exception evolutionists, there was little or no place for the timid group of stranded Girondins, who still stood aloof in sullen antique scientific orthodoxy from what seemed to them the *carmagnoles* and orgies of a biological Thermidor.

Grant Allen, *Charles Darwin* (London: Longmans, Green, 1886), 198.

Secondary Literature: Peter Morton, *The Busiest Man in England: Grant Allen and the Writing Trade, 1895–1900* (New York: Palgrave Macmillan, 2005); Bernard Lightman, *Victorian Popularizers of Science: Designing Science for New Audiences* (Chicago: University of Chicago Press, 2007).

WILLIAM ALLINGHAM (1824–1889)
Irish Poet

August 11, 1868.—To Freshwater; engage bedroom over little shop, and to the Darwins. Dr. Hooker* in lower room writing away at his Address; going to put "Peter Bell's" primrose into it and wants the exact words. Upstairs Mrs. Darwin, Miss D. and Mr. Charles Darwin himself,—tall, yellow, sickly, very quiet. He has his meals at his own times, sees people or not as he chooses, has invalid's privileges in full, a great help to a studious man.

August 12.—Freshwater. Pack bag, to Farringford. Breakfast in the study, the boys pleasant; Lionel back from bathing; A. T. [Alfred Lord Tennyson*], letter from America for autograph. Mrs. T., "Lionel going to Eton." She dislikes Darwin's theory. I sit in study: A. T. teaching Hallam Latin—Catiline. Charles Darwin expected, but comes not. Has been himself called "The Missing Link." Luncheon. Then T. and I walk into croquet-ground, talking of Christianity.

William Allingham, *A Diary*, ed. H. Allingham and D. Radford (London: Macmillan, 1907), 184–85.

Secondary Literature: Alan Warner, *William Allingham* (Lewisburg, PA: Bucknell University Press, 1975).

DUKE OF ARGYLL
See GEORGE DOUGLAS CAMPBELL

MATTHEW ARNOLD (1822–1888)
ENGLISH POET AND CRITIC

You will like to see Huxley's* letters, and when the absolutely hostile attitude to Christianity of many of his friends and allies, [Alexander] Bain* of Aberdeen, [William Kingdon] Clifford*, Herbert Spencer*, etc., is considered, Huxley's adhesion, so far as it goes, is very remarkable, and was indeed much more than I expected. [John] Tyndall* has the same direction as Huxley. Old Darwin, on the other hand, though actively fierce against nothing, says that he cannot conceive what need men have either of religion or of poetry; his own nature, he says, is amply satisfied by the domestic affections and by the natural sciences.

Arnold to Miss Arnold, December 1875, in *Letters of Matthew Arnold*, ed. George W. E. Russell, 2 vols. (New York: Macmillan, 1895), 2:143.

The natural sciences do not, however, stand on the same footing with these instrument-knowledges. Experience shows us that the generality of men will find more interest in learning that, when a taper burns, the wax is converted into carbonic acid and water, or in learning the explanation of the phenomenon of dew, or in learning how the circulation of the blood is carried on, than they find in learning that the genitive plural of *pais* and *pas* does not take the circumflex on the termination. And one piece of natural knowledge is added to another, and others are added to that, and at last we come to propositions so interesting as Mr. Darwin's famous proposition that "our ancestor was a hairy quadruped furnished with a tail and pointed ears, probably arboreal in his habits." Or we come to propositions of such reach and magnitude as those which Professor Huxley* delivers, when he says that the notions of our forefathers about the beginning and the end of the world were all wrong, and that nature is the expression of a definite order with which nothing interferes.

Matthew Arnold, "Literature and Science" (1883), in his *Essays in Criticism, Second Series, Contributions to "The Pall Mall Gazette" and Discourses in America* (London: Macmillan, 1903), 334–35.

An eminent authority, Professor [John Wesley] Judd*, relates an amusing conversation he had with Matthew Arnold in 1871. "I cannot understand," said Arnold, "why you scientific people make such a fuss about Darwin. It's all in Lucretius." To which Judd replied, "Yes, Lucretius guessed what Darwin proved." Whereupon Arnold rejoined, "Ah! that only shows how much greater Lucretius really was, for he divined a truth which Darwin spent a life of labor in groping for."

Samuel Parkes Cadman, *Charles Darwin and Other English Thinkers with Reference to their Religious and Ethical Value* (Boston: Pilgrim Press, 1911), 12.

Secondary Literature: Park Honan, *Matthew Arnold: A Life* (New York: McGraw Hill, 1981); Fred Dudley, "Matthew Arnold and Science," *PMLA* 57 (1942): 275–94.

HERBERT HENRY ASQUITH (1852–1928)
English Prime Minister

One may be as skeptical about so-called Science as about so-called religion. Huxley* and his school used to be just as dogmatic about Evolution (with all its gaps and fallacies) as any of the Popes & Council & Inquisitors were about the Nicene Creed. The old proverb says that the Cowl does not make the Monk: the converse is equally true—that you may have the spirit and temper of the Monk without bearing the Cowl; and tho' Darwin himself was a modest & humble man, his disciples—until they were split up by hopeless quarrels among themselves—almost surpassed the theologians in their arrogance. Enough of this; it is rather a favorite theme of mine: and I am afraid I have inflicted it on you more than once.

Asquith to Venetia Stanley, 20 February 1915, in H. H. Asquith, *Letters to Venetia Stanley*, ed. Michael and Eleanor Brock (Oxford: Oxford University Press, 1982), 441–42.

Secondary Literature: Stephen Bates, *Asquith* (London: Haus, 2006).

EDWARD AVELING (1849–1898)
English Journalist

To the student, not unmindful of this physical and moral parallel, the mental similarity between the two men is perhaps of most moment. That which Darwin did for Biology, [Karl] Marx* has done for Economics.[†] Each of them by long and patient observation, experiment, recordal, reflection, arrived at an immense generalization—a generalization the like of which their particular branch of science had never seen; a generalization that not only revolutionized that branch, but is actually revolutionizing the whole of human thought, the whole of human life. And that the generalization of Darwin is at present much more universally accepted than that of Marx is probably due to the fact that the former affects our intellectual rather than our economic life—can, in a word, be accepted in a measure alike by the believers in the capitalistic system and by its opponents. There can be little doubt that the two names by which the nineteenth century will be known, as far as its thinking is concerned, will be those of Charles Darwin and Karl Marx.

Edward Aveling, *The Students' Marx* (London: Swan Sonnenschein, 1892), viii–ix.
[†]Aveling was the partner of Karl Marx's daughter Eleanor.

Secondary Literature: Ralph R. Colp Jr., "The Contacts of Charles Darwin with Edward Aveling and Karl Marx," *Annals of Science* 33 (1976): 387–94.

B

WALTER BAGEHOT (1826–1877)
English Political Theorist

If [Darwin] hopes finally to solve his great problem, it is by careful experiments in pigeon-fancying, and other sorts of artificial variety-making. His hero is not a self-enclosed, excited philosopher, but "that most skillful breeder, Sir John Sebright,[†] who used to say, with respect to pigeons, that he would produce any given feathers in three years, but it would take him six years to obtain a head and a beak." I am not saying that the new thought is better than the old; it is no business of mine to say anything about that; I only wish to bring home to the mind, as nothing but instances can bring it home, how matter-of-fact, how petty, as it would at first sight look, even our most ambitious science has become.

Walter Bagehot, *The English Constitution* (1867), in *The English Constitution and Other Political Essays,* rev. ed. (New York: Appleton, 1889), 319. [†]John Saunders Sebright (1767–1846), a famous breeder who wrote on various aspects of animal keeping; Darwin cited his work on pigeons.

At first some objection was raised to the principle of "natural selection" in physical science upon religious grounds; it was to be expected that so active an idea and so large a shifting of thought would seem to imperil much which men valued. But in this, as in other cases, the objection is, I think, passing away; the new principle is more and more seen to be fatal to mere outworks of religion, not to religion itself. At all events, to the sort of application here made of it, which only amounts to searching out and following up an analogy suggested by it, there is plainly no objection. Everyone now admits that human history is guided by certain laws, and all that is here aimed at is to indicate, in a more or less distinct way, an infinitesimally small portion of such laws.

Walter Bagehot, *Physics and Politics: Thoughts on the Application of the Principles of "Natural Selection" and "Inheritance" to Political Society* (1872; reprint, New York: Appleton, 1904), 44.

Secondary Literature: Alistair Buchan, *The Spare Chancellor: The Life of Walter Bagehot* (London: Chatto & Windus, 1959).

ALEXANDER BAIN (1818–1903)
Scottish Philosopher

Being repeatedly troubled with indigestion, I paid occasional visits, during the Richmond stay, to Dr. Lane's[†] hydropathic establishment at Moorpark, Farnham, and received there the greatest benefit, as well as kindness and attention. In the beginning of 1857, I was the means of recommending Charles Darwin to visit the establishment, and happened to spend a fortnight there in his company. As we generally walked together after the baths, I had opportunities of hearing of the progress of his researches, and the approaching publication of the *Origin of Species*. He found so much benefit from the treatment under Dr. Lane that he frequently returned for a fortnight's stay at a time; but I never met him again.

Alexander Bain, *Autobiography* (London: Longmans, Green, 1904), 249–50. [†]Edward Wickstead Lane (1822–91), an Edinburgh-trained London physician who ran two hydropathic establishments and wrote on hydropathy.

It has not escaped attention, that the honors paid to the illustrious Darwin, are an admission that our received Christianity is open to revision. In consequence of a few conciliatory phrases, Darwin has been credited with theism; nevertheless he has ridden rough-shod over all that is characteristic in our established creeds. Can the creeds come scathless out of the ordeal?

Alexander Bain, *Practical Essays* (New York: Appleton, 1884), 278–79.

Secondary Literature: Rick Rylance, *Victorian Psychology and British Culture, 1850–1880* (Oxford: Oxford University Press, 2000); George E. Davie, "Alexander Bain," in *Encyclopedia of Philosophy*, ed. Paul Edwards, 8 vols. (New York: Macmillan, 1967), 1:243–44.

JAMES MARK BALDWIN (1861–1934)
American Psychologist

The interest in genesis [. . .] extended itself to the great question of evolution, of which the principles are psychological no less than biological. During the years at Princeton [1885–87], I made many excursions into this territory, reviewing in various papers such topics as heredity, transmission of acquired characters, the relative importance of endowment and environment, the parallelism between individual development and racial evolution. These, with fuller discussions, were finally developed in the volume *Development and Evolution* (1902). At that date, the two great problems at issue concerned the theory of natural selection, and the possible influence of individual adaptations on the course of evolution. The Darwinians (led by [August] Weismann*) were for the moment victorious over those of the Lamarckian camp ([George John] Romanes*, [Theodore] Eimer*, [Edward Drinker] Cope*). Among the psychologists in America, Darwinism was in the ascendant, [William] James* being one of the convinced converts. The rediscovery of Mendelism had not been announced, and the question of mutation was where Darwin had left it in his description of "sports." The Darwinian theory concerned itself, as in the books of Darwin and Wallace*, with minute "accidental variations"; and the point of greatest obscurity was that of the seemingly directive or "determinate" course of evolution. The opponents argued for some vital tendency or "directive" factor, represented by the "ortho" in Eimer's theory of "orthogenesis" [straight-line evolution].

The outcome of my studies was embodied in the position known as "organic selection," printed in the *American Naturalist*, May–June 1896, and announced also, at about the same time, by H. F. [Henry Fairfield] Osborn* and Lloyd Morgan.[†] According to this point of view, natural selection operating on "spontaneous variations" is sufficient alone to produce determinate evolution (without the inheritance of acquired adaptations or modifications), since—and this is the new point—in each generation variations in the direction of, or "coincident" with, the function to be developed will favor the organisms possessing them, and their descendants will benefit by the accumulation of such variations. Thus the function will gradually come to perfection.

Carl Murchison, ed., *A History of Psychology in Autobiography,* vol. 1 (New York: Russell & Russell, 1930), 6–7. †C. Lloyd Morgan (1852–1938), an English psychologist interested in the relationship between intelligence and instinct.

Secondary Literature: John M. Broughton and D. John Freeman-Moir, eds., *The Cognitive Developmental Psychology of James Mark Baldwin: Current Theory and Research in Genetic Epistemology* (New York: Ablex, 1982); John R. Morss, *The Biologising of Childhood: Developmental Psychology and the Darwinian Myth* (London: Taylor & Francis, 1990).

ARTHUR JAMES BALFOUR (1848–1930)
English Prime Minister

It was to George Darwin that I owed my one brief glimpse of his illustrious father. George took me on a visit to their charming home near Down. My recollections of it are most pleasant, though too vague to bear particular repetition. The elder Darwins' kindness to young people was great. [...] It was alleged that during his famous voyage in the *Beagle,* perpetual tossing at sea had left behind a constitutional weakness which dry land could not wholly cure. It took the form (so ran the legend) of making him feel sick whenever he was bored. [...] Whether there had ever been any foundation for this story I know not. Certainly I perceived no confirmation of it during my visit. The kindness of the great man, his sympathy and charm [...] left a deep impression on my youthful mind. I never saw him again; and little thought, as I took leave of my kindly hosts at Down, that it would someday fall to me to take a leading part in the proceedings by which Cambridge honored the birth-centenary of one of her greatest sons.

Arthur James Balfour, *Chapters of Autobiography* (London: Cassell, 1930), 37–38.

I take it as certain that, had no such theory as Natural Selection been devised, nothing would have persuaded mankind that the organic world came into being unguided by intelligence. Chance, whatever chance may mean, would never have been accepted as a solution. [...] All this had changed, as everyone knows, by Darwin. But what exactly was it that, in this connection, Darwin did? He is justly regarded as the greatest among the founders of the doctrine of organic evolution; but there is nothing in the mere idea of organic evolution which is incongruous with design. On the contrary, it almost suggests guidance, it has all the appearance of a plan. Why, then, has Natural Selection been supposed to shake teleology to its foundation?

The reason, of course, is that though the fact of selection does not make it harder to believe in design, it makes it easier to believe in accident. [...] Before Darwin's great discovery those who denied the existence of a Contriver were hard put to it to explain the appearance of contrivance. Darwin, within certain limits and on certain suppositions, provided an explanation. [...]

I use the word "selection" as a convenient name for any non-rational process, acting through heredity, which successfully imitates contrivance. Darwin's theory, be it true or false, still provides [...] the only suggestion as to how this feat may be accomplished, and his terminology may be used without danger of misunderstanding.

Arthur James Balfour, *Theism and Humanism* (London: Hodder & Stoughton, 1915), 48–50.

Secondary Literature: Ruddock F. Mackay, *Balfour: Intellectual Statesman* (Oxford: Oxford University Press, 1985); P. L. Heath, "Arthur James Balfour," in *Encyclopedia of Philosophy,* ed. Paul Edwards, 8 vols. (New York: Macmillan, 1967), 1:246–47.

GEORGE BANCROFT (1800–1891)
AMERICAN HISTORIAN

Another significant bit is found (1860) in one of [Bancroft's] letters to Mrs. Bancroft expressing pleasure in a *London Quarterly* review in refutation of Darwin: "The faith in ideas is exactly what I approve of; and I believe 'preventing grace' precedes the formation of every living thing; as well as of every regeneration of a soul, or any event in history." With Agassiz* at his side Bancroft need not be scorned for balking at the revolutionary doctrines of Darwinism.

The Life and Letters of George Bancroft, ed. Mark DeWolfe Howe, 2 vols. (New York: Scribner's, 1908), 2:115.

Secondary Literature: Lilian Handlin, *George Bancroft: The Intellectual as Democrat* (New York: Harper & Row, 1984); Kathleen Egan Chamberlain, "George Bancroft," in *Encyclopedia of Historians and Historical Writing,* ed. Kelly Boyd, 2 vols. (London: Fitzroy Dearborn, 1999), 1:71–72.

JOHN KENDRICK BANGS (1862–1921)
AMERICAN EDITOR AND SATIRIST

"A copy of a weekly paper come to the house, with an advertisement in it of a book called the *Origin of the Species,* by a feller named Darwin, costin' two dollars and a half. That was some money in those days; but somehow or other that title sounded good and hefty, and I sent my little two-fifty by mail to the publisher, and within a week or two the *Origin of the Species* was duly received, and I went at it."

"And what did you make out of it?" I asked, my interest truly aroused.

"Nothin'—not the first damn thing at first," said the old gentleman; "except it made me wonder if I hadn't lost my mind, or something. I sat down to read the thing, and by thunder, sir, I couldn't make head nor tail out of it! I'd always thought I knew something about the English language; but this time I was stumped, and it made me mad. [. . .]

"So I went at it again. I read it, and I reread it. I wrastled with every page, paragraph, and sentence in that book. Sometimes I had to put as much as five days on one page—but by Gorry, son, *when* I got it I got it good, and when it come it come with a rush—and *now*—."

The old man paused, drew himself up very straight, and squaring his shoulders he leaned forward and put his hands on my knees.

"If there's anything you want to know about Darwin's *Origin of Species,* you ask me!"

John Kendrick Bangs, *From Pillar to Post: Leaves from a Lecturer's Notebook* (New York: Century, 1916), 58–60.

Secondary Literature: Francis Hyde Bangs, *John Kendrick Bangs, Humorist of the Nineties* (New York: Knopf, 1941).

P. T. BARNUM (1810–1891)
American Circus Impresario

[2 March 1860]
Stopped at Barnum's on my way downtown to see the much advertised non-descript, the "What-is-it." [...] The creature's [...] anatomical details are fearfully simian, and he's a great fact for Darwin.

The Diary of George Templeton Strong: The Civil War 1860–1865, ed. Allan Nevins and Milton Halsey Thomas (New York: Macmillan, 1952), 12.

[17 March 1860]
Is it a lower order of Man? Or is it a higher development of the monkey? Or is it both in combination? Nothing of the kind has ever been seen before! It is alive! And it is certainly the most marvelous creature living! [...] He possesses the countenance of a human being! He is probably ten years old; is four feet high; weighs 50 pounds; and is intelligent, docile, active and playful as a kitten.

Advertisement for "What is IT?" at Barnum's National Museum, New York, 17 March 1860, quoted in *Performance and Evolution in the Age of Darwin,* by Jane R. Goodall (London: Routledge, 2002), 54.

[Barnum] has crowded natural history into the spectacle of an evening; he has simplified for ordinary minds the Darwinian controversy by bringing all the animals concerned together and leaving them to speak for themselves.

Brooklyn Eagle, 18 April 1873, quoted in John Rickards Betts, "P. T. Barnum and the Popularization of Natural History," *Journal of the History of Ideas* 20 (1959): 360n20.

Secondary Literature: Benjamin Reiss, *The Showman and the Slave: Race, Death and Memory in Barnum's America* (Cambridge, MA: Harvard University Press, 2001).

HENRY BATES (1825–1892)
ENGLISH NATURALIST

It is really curiously satisfactory to me to see so able a man as Bates (& yourself) believing more fully in nat[ural] selection, than I think I even do myself.

Darwin to Joseph Dalton Hooker, 26 March 1862, in *The Correspondence of Charles Darwin,* ed. Frederick Burkhardt et al., 16 vols. to date (Cambridge: Cambridge University Press, 1985–), 10:135.

I recall the happy hours when, [Bates's] evening employment of beetle-sticking over, the frugal supper eaten—(like Darwin, he suffered from chronic indigestion)—the pipe lit and the talk started, he discoursed on themes evidencing his wide and varied reading. For, unlike Darwin, who tells us in the autobiography which is prefixed to his *Life and Letters* that for many years he "could not endure a line of poetry and found Shake-speare intolerably dull," even music disconcerting him and natural scenery giving him little delight, Bates reveled and rejoiced in all these ministers to the completeness of life. In fact, he was the richer of the two both in mental grasp and equipment, and such letters of Darwin to him as have survived evidence that Bates's masterly suggestiveness impressed him profoundly.

Edward Clodd, *Memories,* 3rd ed. (New York: G. P. Putnam, 1916), 66.

Secondary Literature: G. Woodcock, *Henry Walter Bates, Naturalist* (London: Faber & Faber, 1969); Graeme D. Ruxton, Thomas M. Sherratt, and Michael P. Speed, *Avoiding Attack: The Evolutionary Ecology of Crypsis, Warner Signals and Mimicry* (Oxford: Oxford University Press, 2005), chaps. 9 (Müllerian mimicry) and 10 (Batesian mimicry).

WILLIAM BATESON (1861–1926)
ENGLISH GENETICIST

As to whether my father [William Bateson] was a nice man, in spite of natural ambivalence, I incline to say yes. [. . .] He was certainly not a nice man whenever the inheritance of acquired characteristics was mentioned.

When this happened the coffee cups rattled on the table. [...] I think he always knew that there was something very wrong with orthodox Darwinian theory, but at the same time he regarded Lamarckism as a tabooed pot of jam into which he was not allowed to reach. I have his copy of *The Origin of Species,* sixth edition, in which he listed on the fly-sheet the pages on which Darwin slipped into the Lamarckian heresy.

Gregory Bateson to Arthur Koestler, 8 April 1970, quoted in *The Case of the Midwife Toad,* by Arthur Koestler (New York: Random House, 1971), 51.

Secondary Literature: Alan G. Cock and Donald R. Forsdyke, *Treasure your Exceptions: The Science and Life of William Bateson* (New York: Springer, 2008).

LYDIA ERNESTINE BECKER (1827–1890)
English Suffragist and Botanist

I do not for a moment maintain that intellectual pursuits can afford consolation in sorrow—for that we must look elsewhere; but they are undoubtedly capable of giving solace and diversion to the mind which might otherwise dwell too long on the gloomy side of things, and of beguiling the tedium of enforced solitude, or of confinement to a sick room. For an instance of this I may refer to the example of one of the most illustrious naturalists of the age. Mr. Charles Darwin had informed us that some of his most curious and interesting observations respecting the habits of climbing plants were made when he was a prisoner, night and day, to one room. [...]

How seemingly unimportant are the movements of insects, creeping in and out of flowers in search of the nectar on which they feed! If we saw a man spending his time in watching them, and in noting their flitting with curious eyes, we might be excused for imagining that he was amusing himself by idling an hour luxuriously in observing things which, though curious, were trifling. But how mistaken might we be in such an assumption! For these little winged messengers bear to the mind of the philosophical naturalist tidings of mysteries hitherto unrevealed; and as Newton saw the law of gravitation in the fall of the apple, Darwin found, in the connection between flies and flowers, some of the most important facts which support the theory he has promulgated respecting the modification of specific forms in animated beings.

Lydia Ernestine Becker, "On the Study of Science by Women," *Contemporary Review* 10 (January–April 1869): 386–404, on 389–90.

Miss Becker presents her compliments to Mr. Darwin and takes the liberty of sending him the enclosed flowers of a variety of *Lychnis dioica* common in the woods here but which she has not observed elsewhere.

Becker to Charles Darwin, 18 May 1863, in *The Correspondence of Charles Darwin*, ed. Frederick Burkhardt et al., 16 vols. to date (Cambridge: Cambridge University Press, 1985–), 11:424. This was the first in a series of letters in which Becker established herself as a correspondent and botanical informant of Darwin's.

Secondary Literature: Barbra T. Gates, *Kindred Nature: Victorian and Edwardian Women Embrace the Living World* (Chicago: University of Chicago Press, 1998).

HENRY WARD BEECHER (1813–1887)
American Clergyman

[11 March 1860]
Mr. Beecher was one of the first ministers in the Christian Church, if not the very first in this country, to advocate the doctrine of evolution as a doctrine which, so far from being inimical to the cause of Christ, was certain to prove its friend and supporter. The first reference in Mr. Beecher's public teaching which I have been able to find, that indicates an acceptance or an inclination to accept the doctrine of evolution, was an incidental reference in a Sunday morning sermon, March 11, 1860: "What word did Adam ever speak or what manly thing did he ever perform, before or after the Fall, that was worthy of record? He has a name in the Bible; that is all. The world has come uphill every single step from the day of Adam to this." But it was not until nearly twenty years after that he began the systematic advocacy of evolution as an interpretation of the divine order which is consistent with and helpful to evangelical faith.

Lyman Abbott, *Henry Ward Beecher* (Boston: Houghton, Mifflin, 1903), 317.

A vague notion exists with multitudes that science is infidel, and that Evolution in particular is revolutionary—that is, revolutionary of the doctrines of the Church. Men of such views often say, "I know that religion is true. I do not wish to hear anything that threatens to unsettle my faith." But faith that can be unsettled by the access of light and knowledge had better be unsettled. [...] Others speak of Evolution as a pseudo-science teaching that man descended from monkeys, or ascended as the case may be. They have no conception of it as the history of the divine process in the building of this world.

Henry Ward Beecher, *Evolution and Religion* (New York: Fords, Howard & Hulbert, 1885), 48.

Secondary Literature: Debby Applegate, *The Most Famous Man in America: The Biography of Henry Ward Beecher* (New York: Doubleday, 2006); Barry Werth, *Banquet at Delmonico's: Great Minds, the Gilded Age, and the Triumph of Evolution in America* (New York: Random House, 2009).

CLIVE BELL (1881–1964)
English Art Critic

By minding his own business, Darwin called into question the business of everyone else.

Clive Bell, *Art* (London: Chatto & Windus, 1914), 119.

Nevertheless [...] Proust was clumsy. He can be grossly and, what is worse, unwittingly so. He could not leave out insignificant facts, platitudinous reflections, the obvious, the well-worn, the thrice-told, all, all are set down beside what is stranger, subtler and truer than anything that has been set down in imaginative literature since Stendhal at any rate. Because he will not eliminate he is indiscriminate. He will treat facts as though he were a man of science rather than an artist. Indeed, in his way of piling instance upon instance he reminds me sometimes of Darwin; also for piling thus high he has the man of science's excuse—he accumulates that truth may prevail.

"Clive Bell on Proust" (1928), in *Marcel Proust: The Critical Heritage*, ed. Leighton Hodson (London: Routledge, 1997), 385.

Secondary Literature: William G. Bywater, *Clive Bell's Eye* (Detroit: Wayne State University Press, 1975).

SAUL BELLOW (1915–2005)
AMERICAN NOVELIST

He referred me to Darwin's autobiography (he never hesitated to keep company with the best), to those passages in which Darwin confessed that in his youth he had been moved by poetry and music while in later years these things nauseated him, and this he explained by the neglect of his responsive faculties, disused and rusting. Scientific work, immersion in insignificant detail, the noting of very small differences in organisms ruined for him the bigger emotions.

I consider Darwinian self-preservation to be a vulgar ideology. Its leading exponents are sadists who are always telling you that for the good of the species and in conformity with the law of nature they have to do in the gentle spirits they encounter on life's way.

Saul Bellow, *More Die of Heartbreak* (New York: Morrow, 1987), 50, 139.

Secondary Literature: Michael K. Glenday, *Saul Bellow and the Decline of Humanism* (Basingstoke, Hampshire, UK: Macmillan, 1990).

RUTH BENEDICT (1887–1948)
AMERICAN ANTHROPOLOGIST

The understanding we need of our own cultural processes can most economically be arrived at by a detour. When the historical relations of human beings and their immediate forbears in the animal kingdom were too involved to use in establishing the fact of biological evolution, Darwin made use instead of the structure of beetles, and the process, which in the complex physical organization is confused, in the simpler material was transparent in its cogency. It is the same in the study of cultural mechanisms. We need all the enlightenment we can obtain from the study of thought and behavior as it is organized in the less complicated groups.

Ruth Benedict, *Patterns of Culture* (1935; Boston: Mariner Books, 2005), 56.

Secondary Literature: Margaret Mary Caffrey, *Ruth Benedict: Stranger in this Land* (Austin: University of Texas Press, 1989).

HENRI BERGSON (1859–1941)
French Philosopher

It is in this quite special sense that man is the "term" and the "end" of evolution. Life, we have said, transcends finality as it transcends the other categories. It is essentially a current sent through matter, drawing from it what it can. There has not, therefore, properly speaking, been any project or plan. On the other hand, it is abundantly evident that the rest of nature is not for the sake of man: we struggle like the other species, we have struggled against other species. Moreover, if the evolution of life had encountered other accidents in its course, if, thereby, the current of life had been otherwise divided, we should have been, physically and morally, far different from what we are. For these various reasons it would be wrong to regard humanity, such as we have it before our eyes, as pre-figured in the evolutionary movement. It cannot even be said to be the outcome of the whole of evolution, for evolution has been accomplished on several divergent lines, and while the human species is at the end of one of them, other lines have been followed with other species at their end. It is in a quite different sense that we hold humanity to be the ground of evolution.

Henri Bergson, *Creative Evolution,* trans. Arthur Mitchell (New York: Henry Holt, 1913), 265–66.

Of Bergson's theory that intellect is a purely practical faculty developed in the struggle for survival, and not a source of true beliefs, we may say, first, that it is only through intellect that we know of the struggle for survival and of the biological ancestry of man: if the intellect is misleading, the whole of this merely inferred history is presumably untrue. If, on the other hand, we agree with M. Bergson in thinking that evolution took place as Darwin believed, then it is not only intellect, but all our faculties, that have been developed under the stress of practical utility. Intuition is seen at its best where it is directly useful—for example, in regard to other people's characters and dispositions. Bergson apparently holds that capacity for this kind of knowledge is less explicable by the struggle for existence than, for example, capacity for pure mathematics. Yet the savage deceived by false friendship is likely to pay for his mistake with his life; whereas even in the most civilized societies men are not put to death for mathematical incompetence. All the most striking of his instances of intuition in animals have a very direct survival value. The fact is, of course, that both intuition and intellect have been developed because they are useful, and that, speaking broadly, they are useful when they give truth and become harmful when they give falsehood. Intellect, in civilized man, like artistic capacity, has occasionally been developed beyond the point where it is useful to the individual; intuition, on the other hand, seems on the whole to diminish as civilization increases.

Bertrand Russell, *Our Knowledge of the External World as a Field for Scientific Method in Philosophy* (Chicago: Open Court, 1914), 23.

Secondary Literature: T. A. Goudge, "Henri Bergson," in *Encyclopedia of Philosophy*, ed. Paul Edwards, 8 vols. (New York: Macmillan, 1967), 1:287–95; Yvette Conry, *L'évolution créatrice d'Henri Bergson* (Paris: L'Harmattan, 2000).

ISAIAH BERLIN (1909–1997)
English Philosopher and Historian

It is very odd and interesting that all the men who wrote about the natural sciences then wrote so well. Huxley* is beautiful. Darwin (did you know that he demanded a happy ending to every story, & said that if the central figure was a pretty woman why then so much the better?), Wallace*, [John] Tyndall* all were excellent. The subject—popular science—is an abomination of the lowest order. But they did write well.

Berlin to Sheila Grant Duff, 1933, in Isaiah Berlin, *Letters, 1928–1946*, ed. Henry Hardy (Cambridge: Cambridge University Press, 2004), 76.

Secondary Literature: Michael Ignatieff, *Isaiah Berlin: A Life* (London: Chatto & Windus, 1998); Jonathan Beecher, "Isaiah Berlin," in *Encyclopedia of Historians and Historical Writing*, ed. Kelly Boyd, 2 vols. (London: Fitzroy Dearborn, 1999), 1:90–91.

ANNIE BESANT (1847–1933)
English Social Reformer

Among the witnesses we desired to subpoena [after Besant's arrest in 1877 for publishing a book on birth control] was Charles Darwin, as we needed to use passages from his works; he wrote back a most interesting letter, telling us that he disagreed with preventive checks to population on the ground that over-multiplication was useful, since it caused a struggle for existence in which only the strongest and the ablest survived, and that he doubted whether it was possible for preventive checks to serve as well as positive. He asked us to avoid calling him if we could: "I have been for many years much out of health, and have been forced to give up all society or public meetings, and it would be great suffering to me to be a witness in court. [...] If it is not asking too great a favor, I should be greatly obliged if you would inform me what you decide, as apprehension of the coming exertion would prevent the rest which I require doing me much good." Needless to add that I at once wrote to Mr. Darwin that we would not call him, but his gentle courtesy has always remained a pleasant memory to me.

Annie Besant, *Autobiographical Sketches* (London: Freethought, 1885), 136.

Secondary Literature: Anne Taylor, *Annie Besant: A Biography* (Oxford: Oxford University Press, 1992).

ELIZABETH BISHOP (1911–1979)
AMERICAN POET

I can't believe we are wholly irrational—and I do admire Darwin! But reading Darwin, one admires the beautiful and solid case being built up out of his endless heroic observations, almost unconscious or automatic—and then comes a sudden relaxation, a forgetful phrase, and one feels the strangeness of his undertaking, sees the lonely young man, his eyes fixed on facts and minute details, sinking or sliding giddily off into the unknown. What one seems to want in art, in experiencing it, is the same thing that is necessary for its creation, a self-forgetful, perfectly useless concentration.

Bishop to Anne Stevenson, 8–20 January 1964, Elisabeth Bishop Papers, Washington University, St. Louis, quoted in Zachariah Pickard, "Natural History and Epiphany: Elizabeth Bishop's Darwin Letter," *Twentieth Century Literature* 50 (2004): 121–38.

Secondary Literature: Francesco Rognoni, "Reading Darwin: On Elizabeth Bishop's Marked Copies of *The Voyage of the Beagle* and *The Autobiography of Charles Darwin*," in *Jarrell, Bishop, Lowell, and Co.*, ed. Suzanne Ferguson (Knoxville: University of Tennessee Press, 2003), 239–48.

OTTO VON BISMARCK (1815–1898)
German Statesman

Then before dinner a further walk with Rottenburg in the wood where it is cut by the road leading to Mohnsen. Lively conversation on a variety of matters serious and amusing, as for instance on Darwin and the esteem in which he is held by the Chief.

Mortiz Busch, November 1887, in *Bismark: Some Secret Pages of his History* (diary kept by Dr. Moritz Busch), 3 vols. (New York: Macmillan, 1898), 2:456.

Secondary Literature: A. J. P. Taylor, *Bismarck: The Man and the Statesman* (New York: Knopf, 1969).

FRANZ BOAS (1858–1942)
American Anthropologist

Former events [. . .] leave their stamp on the present character of a people. I consider it one of the greatest achievements of Darwinism to have brought to light this fact, and thus to have made a physical treatment of biology and psychology possible. The fact may be expressed by the words, "the physiological and psychological state of an organism at a certain moment is a function of its whole history."

Franz Boas, "The Principles of Ethnographical Classification [1887]," in *The Shaping of American Anthropology, 1883–1911*, ed. George Stocking (New York: Basic Books, 1974), 61–67.

Although the idea does not appear quite definitely expressed in Darwin's discussion of the development of mental powers, it seems quite clear that his main object has been to express his conviction that the mental faculties developed essentially without a purposive end, but that they originated as variations, and were continued by natural selection. This idea was also brought out very clearly by Wallace*, who emphasized that apparently reasonable activities of man might very well have developed without an actual appreciation of reasoning.

Franz Boas, "The Relation of Darwin to Anthropology," address delivered in a lecture series at Columbia University to commemorate the fiftieth anniversary of the *Origin of Species,* quoted in Herbert S. Lewis, "Boas, Darwin, Science, and Anthropology," *Current Anthropology* 42 (2001): 381–406, on 387.

Secondary Literature: Douglas Cole, *Franz Boas: The Early Years, 1858–1906* (Vancouver: Douglas & McIntyre, 1999).

ALICE BODINGTON (1840–1897)
English Naturalist and Science Writer

The lovers of zoology find their favorite study become ever increasingly fascinating, as the discoveries of modern paleontology more and more triumphantly vindicate the theory of evolution. Although that theory received its greatest impetus in England and on the continent from the works of Darwin, yet it is evident that the great master himself had only grasped one form of law governing evolution. He sought, at least in his earlier works, to account for all changes in animals and plants by natural selection; whereas we now see that the infinite, delicate variations in the world of organic beings are owing to the intense irritability and susceptibility to molecular changes of protoplasm, and the consequent action of the environment upon it. Natural selection evoked some unknown force vaguely of the nature of will. The action of the environment upon protoplasm requires nothing but ordinary and well-known phenomena of organic chemistry.

Alice Bodington, *Studies in Evolution and Biology* (London: Elliott Stock, 1890), quoted in *In Nature's Name: An Anthology of Women's Writing and Illustration, 1780–1930,* by Barbara T. Gates (Chicago: University of Chicago Press, 2002), 509.

Neo-Lamarckism supplies the "motif" which runs through almost every study in this little book. I had not met with the works of Lamarck when these studies were written, yet it seems to me that every advance in the physical sciences which I have endeavored to chronicle adds a fresh laurel to the fame of this most unjustly decried genius. If we, who love and honor the name of Darwin, look upon him as the Newton of evolution, we shall surely not detract from his fame if we look upon Lamarck as its Galileo.

Alice Bodington, quoted in review of *Studies in Evolution and Biology* in *American Naturalist* 25 (1891): 647–48, on 648.

Secondary Literature: Barbara T. Gates, *Kindred Nature: Victorian and Edwardian Women Embrace the Living World* (Chicago: University of Chicago Press, 1998).

WILHELM BÖLSCHE (1861–1939)
German Nature Writer

Darwin was never a handsome man from the aesthetic point of view. When he wanted to sail with FitzRoy, it was a very near question whether the splenetic captain would not reject him because he did not like his nose. His forehead had so striking a curve that [Cesare] Lombroso*, the expert, could put him down as having "the idiot-physiognomy" in his *Genius and Insanity.* At the time when he wrote the *Origin of Species* he had not the patriarchal beard that is inseparable from his image in our minds; he was bald, and his chin clean shaved. The prematurely bent form of the invalid could never have had much effect in such a place, no matter what respect was felt for him. Haeckel*, young and handsome, was an embodiment of the *mens sana in corpore sano.* He rose above the grey heads of science, as the type of the young, fresh, brilliant generation. It was an opponent at this Congress, who sharply attacked the new ideas that spoke of the "colleague in the freshness of youth" who had brought forward the subject. He brought with him the highest thing that a new idea can associate with: the breath of a new generation, of a youth that greets all new ideas with a smiling courage. Behind this was the thought of Darwin himself, a wave that swept away all dams.

Wilhelm Bölsche, *Haeckel: His Life and Work,* trans. Joseph McCabe (London: T. Fisher Unwin, 1906), 146–47.

Secondary Literature: Rosemarie Nothlich, ed., *Ernst Haeckel/Wilhelm Bölsche: Briefwechsel, 1887–1919,* 2 vols. (Berlin: Verlag für Wissenschaft und Bildung, 2002–6).

JAMES BONAR (1852–1941)
ENGLISH ECONOMIST

Darwinism is a particular form of the general theory that there is an evolution of life in the world. It differs from the Hegelian theory of development in its application to cases where preservation of identity is not possible, and where even the continuity of growth exists only *for us* and in retrospect. Darwin explains "the origin of species" [. . .] by supposing with Malthus in all living things a tendency to rapid multiplication, leading necessarily to a struggle for existence, and (second) by supposing that in each living thing produced by this power of propagation there is some slight difference from its fellows, such variations giving an advantage in the struggle to some over others, the result being a "natural selection," "a survival of the fittest." [. . .]

"Selection" implies choice of what is there, but not origination of the matter of choice. We may add that selection gives us only difference, and not development [. . .] what survives at least in the lower forms of the struggle is not the same as what went before it; there is no identity or consciousness of such. There is no pretense made by Darwinians that survival means fitness in any moral sense; the morally worse men may only be the more fit to survive in the sense that they are best able to suck advantage from their surroundings; and that is all that is claimed by the theorists. It is no more than the fitness which enables the holders of the wore [*sic*] coins to get advantage over the holders of the finer under "Gresham's law" of the currency. The worse coinage survived because it was the fitter; it was fitter because, being the worse and yet accepted, it was the more economical.

James Bonar, *Philosophy and Political Economy* (London: Frank Cass, 1893), 357–59.

Secondary Literature: Joseph Schumpeter, *History of Economic Analysis* (New York: Oxford University Press, 1954).

FRANCIS BOWEN (1811–1890)
American Philosopher

[27 March 1860]

Natural Selection can operate only upon races previously brought into being by other causes. In itself, it is powerless either to create or exterminate. In the Development Theory, its only function is, when the number of different Species is so far multiplied that they crowd upon each other, and the extinction of one or more becomes inevitable (if we can conceive of such a case), then to make the *selection,* or to determine which shall be the survivors and which the victims. As individuals of the same Species, the same Variety, and even of the same flock, certainly differ much from each other in strength, swiftness, courage, powers of endurance, and other qualities, Natural Selection has an undoubted part to play, when the struggle comes for such a flock, in determining which of its members shall succumb. But that it ever plays a corresponding part in the grand contest of Species imagined by Mr. Darwin, is a supposition resting upon no evidence whatever, but only upon the faint presumption afforded by the fact, that certain Species at widely separated times have become extinct, through what causes we know not; and therefore, for all that we know to the contrary, Natural Selection may have had something to do with their disappearance. This is to found a theory, not upon knowledge, but upon ignorance. If such reasoning be legitimate, we are entitled to affirm that the moon is inhabited by men "whose heads do grow beneath their shoulders." It may be so, for all we know to the contrary.

Francis Bowen, "Remarks on the Latest Form of the Development Theory," first published in *Memoirs of the American Academy of Arts and Sciences,* n.s., 8 (1860): 98–107, reprinted in *Darwinism Comes to America,* ed. G. Daniels (Waltham, MA: Blaisdell, 1968), 72–73.

Mr. Darwin openly and almost scornfully repudiates the whole doctrine of Final Causes. He finds no indication of design or purpose anywhere in the animate or organic world.

Instinct and structure are nicely correlated to each other, and must so be correlated, or the animal would perish. [...] But, according to the Darwinian theory, there is no ground to expect that the *variations* of structure and instinct should be even simultaneous, much less nicely correlated to each other[. . .] . The aimless and accidental character of the variations, together with their admitted infrequency as inheritable peculiarities, renders it in the highest degree improbable, not to say impossible, that an inheritable peculiarity of structure should happen to occur at the same time with a corresponding one of instinct nicely adapted to it: that, for instance, the apparatus in a bee's abdomen for secreting wax should first appear on the very day when the insect was first incited and taught to build a comb. Such a correspondence and adaptation would compel even Mr. Darwin to renounce chance and believe in design.

It is only for the advocate of Revealed Religion, drawing his premises alike from the history of human race and the fully established conclusions of physical science, to vindicate the additional truth, that the Divine action has not been limited to "the natural"—that is, to the stated repetition of uniform events—but has extended to "the supernatural," or to a break in the regularity of the succession, whenever some great purpose could be thereby directly attained. It is both unphilosophical and presumptuous for the finite to undertake to set bounds to the infinite by declaring *a priori,* that either of these modes of action has any more claim to be considered as necessary than the other.

Francis Bowen, "Darwin on the Origin of Species," *North American Review* 90 (1860): 474–506, on 475, 491–92, 504–5.

Secondary Literature: Bruce Kuklick, *The Rise of American Philosophy: Cambridge, Massachusetts, 1860–1930* (New Haven, CT: Yale University Press, 1977).

CHARLES LORING BRACE (1826–1890)
American Clergyman and Reformer

It is remarkable how the application of the law of natural selection is influencing now every department of scientific investigation. I think Mr. Darwin's name will go down for many ages with this great Law of Hypothesis. I have been amusing myself with applying it to a theory of the moral and mental development of mankind. I think it furnishes what historians and philosophers have so long sought for, a law of progress, and Darwin states the glorious point to which mankind shall eventually advance. Under this law, I hold that, in a sense, even religion may be transmitted; that is, the openness to supernatural inferences, so that ultimately a race may appear in which the highest inspiration and capacity of nobleness shall be embodied and transmitted and perpetuated. Evil seems to me destructive—good preservative. I should venture to think that the origin and influence of Christianity are out of the philosophical course of history,—that is, supernatural,—though the readiness to receive it and all other divine influence may be a part of the regular human development.

Brace to Lady Lyell, 23 December 1866, in *The Life of Charles Loring Brace,* ed. Emma Brace (New York: Scribner's, 1894), 285–86.

The study of Darwin had greatly interested Mr. Brace for some years. [...] One of his greatest recreations was to read and read again *The Origin of Species,* and we find, some years later, an allusion in one of his letters to the fact that he is reading it for the thirteenth time.

Emma Brace, ed., *The Life of Charles Loring Brace* (New York: Scribner's, 1894), 300.

I am at Mr. Darwin's with Mrs. Brace for the night. [...] Darwin was as simple and jovial as a boy, at dinner, sitting up on a cushion in a high chair, very erect, to guard his weakness. Among other things, he said "his rule in governing his children was to give them lump-sugar!" He rallied us on our vigorous movements, and professed to be dazzled at the rapidity of our operations. He says he never moves, and though he can only work an hour or two every day, by always doing that, and having no break, he accomplishes what he does. He left us for half an hour after dinner for rest, and then returned to his throne in the parlor.

We had a lively talk on the instincts of dogs (several persons being there) and on "cross-breeding," and he became animated explaining his experiments in regard to it. [...] I was telling him that the California primitive skulls were of a remarkably good type. He gave one of his lighting-up smiles, which seemed to come way out from under his shaggy eyebrows. "Yes," he said, "it is very unpleasant of these facts; they won't fit in as they ought to!" [...] He told us, with such glee, of a letter he had just got from a clergyman, saying that "he was delighted to see, from a recent photograph, that no man in England was more like the monkey he came from!" and of another from an American clergyman beginning with, "You d——m scoundrel!" and sprinkled with oaths and texts. [. . .] These things amuse him; but not a word did he say of his own success or fame. He breakfasts at half past seven, but sat by us later, as we ate, and joked and cut for us, and was as kind as could be. I never met a more simple, happy man—as merry and keen as Dr. [Asa] Gray*, whom he loves much. Both he and Lyell* think Dr. G. the soundest scientific brain in America. [. . .] "How unequally is vitality distributed," he said, as he heard what we did every day. [. . .] His parting was as of an old and dear friend. I hope this picture of the best brain in Europe will not weary you.

Brace to an unidentified friend, 12 July 1872, in *The Life of Charles Loring Brace*, ed. Emma Brace (New York: Scribner's, 1894), 319–21.

Secondary Literature: Stephen O'Connor, *Orphan Trains: The Story of Charles Loring Brace and the Children He Saved and Failed* (Boston: Houghton Mifflin, 2001).

FRANZ BRENTANO (1838–1917)
Austrian Philosopher

[Besides voluntary mental states], there are also involuntary physical changes which naturally accompany or follow certain mental states. Fright makes us turn pale, fear induces trembling, our cheeks blush red with shame, even before the expression of emotions was an object of scientific study, as Darwin recently made it once again, people had already learned a great deal about these relationships from simple custom and experience, so that the observed physical phenomena served as signs of the invisible mental phenomena. It is obvious that these signs are not themselves the things that they signify. It is not possible, therefore, as many people have quite foolishly wanted to make us believe, that this external and, as it was pretentiously called, "objective" observation of mental state could become a source of psychological knowledge, quite independently of inner "subjective" observation.

Franz Brentano, *Psychology from an Empirical Standpoint* (1874), ed. Oskar Kraus (London: Routledge, 1995), 39–40.

Secondary Literature: Dale Jacquette, ed., *The Cambridge Companion to Brentano* (Cambridge: Cambridge University Press, 2004).

ROBERT BRIDGES (1844–1930)
English Poet

I have been reading Darwin's *Life and Letters.* I don't think that he does himself justice in his autobiography. I suppose he got dull enough in his later years. His correspondence about the best of his time shows a far better man than the account written later. There is in the 2nd volume a piece of Mr. Huxley's* writing inserted in which that gentleman thinks to give an account of the reception the *Origin of Species* met with. "Instead of that," he philosophizes, and I think for rotten philosophy and poor rhetoric it would be difficult to match his performance. I mean (if you have the book) where he argues that there is no such thing as chance, and asks you to go down on the beach.

Bridges to Richard Watson Dixon, 18 April 1888, in *The Selected Letters of Robert Bridges,* ed. Donald E. Stanford (Newark: University of Delaware Press, 1983), 177.

Secondary Literature: Catherine Phillips, *Robert Bridges: A Biography* (New York: Oxford University Press, 1992).

PAUL BROCA (1824–1880)
French Anatomist and Anthropologist

[The result of population pressure on food supply results in the] fatal law of struggle of living things, struggle between species that dispute place and nourishment, struggle between individuals (of the same species) that claim a part of the common lot of their species, the universal and eternal struggle to which the weakest must succumb. This great law, long recognized by philosophers and naturalists and unpityingly constituted in human societies by the economist Malthus, Charles Darwin has used in his turn. No one before him has formulated it with such precision. No eye but his has seized it in its entirety. No mind has understood all its implications. Justly this should be called *Darwin's law.*

Paul Broca, "Sur le transformisme," *Bulletin de la Société d'Anthropologie de Paris*, 2nd ser., 5 (1870): 168–239, on 187, quoted in Joy Harvey, *"Almost a Man of Genius"; Clémence Royer, Feminism, and Nineteenth-Century Science* (New Brunswick, NJ: Rutgers University Press, 1997), 124.

Secondary Literature: Francis Schiller, *Paul Broca, Founder of French Anthropology, Explorer of the Brain* (New York: Oxford University Press, 1992).

ROBERT BROWNING (1812–1889)
English Poet

Sunday, June 10, 1888—To 29 De Vere Gardens—Upstairs, tapestry, bust of E. B. B. [Elizabeth Barrett Browning], one of R. B. [Robert Browning] (young). Enter R. B., friendly and vigorous. He asked me, Did I often come up to town? "Very seldom." On which he remarked (I think premeditatedly), that in his early life he had much secluded himself, and had often since regretted it. He sees that he lost much by it. Then he took up a book from the table, *Oannes, an Ancient Myth, as told by Berosus,* by J. Garth Wilkinson.[†] "Here's a thing has been sent to me"—and read half a page or so (without spectacles, I noticed), "to the purport that in remote times a creature, half man and half fish, came to a certain island, and taught the islanders various arts—his name was Oannes." W. A. [William Allingham]—"It might be taken as a foreshadowing of Darwinism—the origin of man in an amphibious lepidosyren." B.—"Yes, altogether different from Wilkinson's notions." W. A.—"A gentleman whom the Vicar of Wakefield met was fond of quoting Berosus." B. (smiling).—"Ay, sir, the world is in its dotage." Then we talked a little about Darwin, B. saying that, whatever his merits as investigator, his philosophy was of little or no importance. I told him of our neighbor Alfred Wallace*, and how he had arrived, as it were, at the opposite goal from Darwin in what are called "Supernatural questions"; D. at last believing almost nothing, W. almost anything!

William Allingham, *A Diary,* ed. H. Allingham and D. Radford (London: Macmillan, 1907), 372–73. [†]Reference to James J. G. Wilkinson, *Oannes according to Berosus: A Study in the Church of the Ancients* (London: James Speirs, 1888).

With George Bubb Dodington[†]

                              Thus—pelf
Smoothens the human mudlark's lodging, power
Demands some hardier wrappage to embrace
Robuster heart-beats: rock, not tree nor tower,
Contents the building eagle: rook shoves close
To brother rook on branch, while crow morose
Apart keeps balance perched on topmost bough.
No sort of bird but suits his taste somehow:
Nay, Darwin tells of such as love the bower—
His bower-birds opportunely yield us yet
The lacking instance when at loss to get
A feathered parallel to what we find
The secret motor of some mighty mind
That worked such wonders—all for vanity!
Worked them to haply figure in the eye
Of intimates as first of—doers' kind?
Actors, that work in earnest sportively,
Paid by a sourish smile. How says the Sage?
Birds born to strut prepare a platform-stage
With sparkling stones and speckled shells, all sorts
Of slimy *rubbish,* odds and ends and orts,
Whereon to pose and posture and engage
The priceless female simper.

The Poetic and Dramatic Works of Robert Browning, 6 vols. (Boston: Houghton Mifflin, 1899), 6:319–20. [†]George Bubb Dodington, Baron Melcombe (1691–1762), English politician and aristocrat.

Secondary Literature: Donald Thomas, *Robert Browning: A Life within Life* (New York: Viking, 1983).

ORESTES AUGUSTUS BROWNSON (1803–1876)
AMERICAN ESSAYIST AND EDITOR

Bacon ruined the sciences as science, when he separated them from philosophy or ideal science and made them purely empirical. Facts or one side of facts may have been examined, and the scientific men of today have no doubt, in their possession a larger mass of materials for the construction of the sciences than had the great medieval doctors and professors. St. Thomas had more science than Sir Charles Lyell*, or Professor [Richard] Owen*. The recent work of Sir Charles on the *Antiquity of Man,* as well as that of Darwin on the *Origin of Species,* shows not progress, but the deterioration of science. The same thing is shown by Agassiz* in his elaborate essay on *Classification,* and by the trouble naturalists have to settle the proper classification of man. [. . .] This must be so, because man is not a pure animal, and cannot be classed as such.

Orestes Augustus Brownson, "Science and the Sciences" (1863), in *The Works of Orestes A. Brownson,* ed. Henry F. Brownson, 20 vols. (New York: AMS, 1966), 9:254–68, on 265–66.

Secondary Literature: W. Carey Patrick, *Orestes A. Brownson: American Religious Weathervane* (Grand Rapids, MI: Eerdsman, 2004).

FERDINAND BRUNETIÈRE (1849–1906)
French Author and Critic

Do we descend from apes, or do the ape and us have a common ancestor, and is this ancestor, in turn, of some even more "animal" origin, or [. . .] do we have within us some remembrance of the forms through which we have passed until we have reached the one we have today? *Vitium hominis natura pectoris,* said Saint Augustine: "That which is vice in human beings is natural to animals." Our bad instincts are the heritage of our early ancestors [. . .] [so] anyone who thinks that morality rests on the idea of the inborn perversity of man as its indestructible foundation, has no reason to reject the "theory of descent."

Ferdinand Brunetière, *La moralité de la doctrine évolutive* (Paris: Firmin-Didot, 1896), 16–17, trans. TFG.

M. Ferdinand Brunetière proposes to apply the theory of evolution to criticism. And while the enterprise in itself appears interesting and praiseworthy, we have not forgotten the energy recently employed by the critic of the *Revue des Deux Mondes,* in subordinating science to morality, and invalidating the authority of all doctrine founded on the natural sciences. It was employed in connection with the *Disciple,* and we know whether M. Brunetière was then niggardly of his remonstrances against those who presumed to introduce theories of transformation into the province of psychology or sociology. He repelled the Darwinian ideas in the name of unchangeable morality.

He expressly states, "These ideas must be false, since they are dangerous."

And now he founds the new criticism on the hypothesis of evolution. "Our proposal," he says, "is none other than to borrow from Darwin and Haeckel* the assistance which M. [Hippolyte] Taine* obtained from Geoffroy St. Hilaire [Saint-Hilaire]*, and Cuvier."

Anatole France, *Life and Letters,* 3rd ser. (London: John Lane, 1922), viii–ix.

Secondary Literature: Giovanni Gullace, *Taine and Brunetière on Criticism* (Lawrence, KS: Coronado, 1982).

WILLIAM JENNINGS BRYAN (1860–1925)
AMERICAN POLITICIAN

Evolution is not truth; it is merely a hypothesis—it is millions of guesses strung together. It had not been proven in the days of Darwin—he expressed astonishment that with two or three million species it had been impossible to trace any species to any other species—it had not been proven in the days of Huxley*, and it has not been proven up to today. It is less than four years ago that Professor [William] Bateson* came all the way from London to Canada to tell the American scientists that every effort to trace one species to another had failed—every one. He said he still had faith in evolution but had doubts about the origin of species. But of what value is evolution if it cannot explain the origin of species? While many scientists accept evolution as if it were a fact, they all admit, when questioned, that no explanation has been found as to how one species developed into another.

Darwin suggested two laws, sexual selection and natural selection. Sexual selection has been laughed out of the classroom and natural selection is being abandoned, and no new explanation is satisfactory even to scientists. Some of the more rash advocates of evolution are wont to say that evolution is as firmly established as the law of gravitation or the Copernican theory. The absurdity of such a claim is apparent when we remember that any one can prove the law of gravitation by throwing a weight into the air and that any one can prove the roundness of the earth by going around it, while no one can prove evolution to be true in any way whatever.

William Jennings Bryan, closing statement at Scopes Trial, Dayton, Tennessee, 1925, http://www.csudh.edu/oliver/smt310-handouts/wjb-last/wjb-last.htm.

Secondary Literature: Lawrence W. Levine, *Defender of the Faith: William Jennings Bryan; The Last Decade, 1915–1925* (New York: Oxford University Press, 1968).

WILLIAM CULLEN BRYANT (1794–1878)
American Poet

It's a good while since the remark was made by an English wit that he did not like to look at monkeys, they seemed to him so much like poor relations. What was regarded at that time as a clever jest has since been taken by an eminent naturalist as the basis of an extensive system which professes to account for the origin of the human species. According to this system, man is an improved monkey, and the lowest form of animal life is found in a minute animated cell. [. . .] How does he know that the monkey is not a degenerate man, a decayed branch of the human family, fallen away from the high rank he once held. [. . .] How often do the descendants of illustrious men become the most stupid of the human race! How many are there, each of whom we call

"The tenth transmitter of a foolish face!"

—a line of [Richard] Savage, the best he ever wrote, worth all his other verses put together. [. . .]

How then can Mr. Darwin insist that if we admit the near kindred of man to the inferior animals we must believe that our progress has been upward, and that the nobler animals are the progeny of the inferior. Is not the contrary more probable? Is it not more likely that the more easy downward road has been taken, that the lower animals are derived from some degenerate branch of the human race, and that, if we do not labor to keep the rank we hold, our race may be frittered away into the meaner tribes of animals. [. . .] Then may our [Boss] Tweeds become the progenitors of those skulking thieves of the western wilds, the prairie-wolves, or swim stagnant pools in the shape of horse-leeches; our astute lawyers may be represented by foes; our great architects like colonies of beavers; our poets by clouds of mosquitoes, famished and musical.

William Cullen Bryant, "Darwin's Theory," address to Williams College alumni dinner, 28 December 1871, in *Prose Writings of William Cullen Bryant,* ed. Parke Godwin, 2 vols. (New York: Appleton, 1884), 1:291–93. Darwin's *Descent of Man* had been published on 24 February 1871.

Secondary Literature: Gilbert H. Muller, *William Cullen Bryant: Author of America* (Albany: State University of New York Press, 2008).

JAMES BRYCE (1838–1922)
English Jurist and Statesman

Five or six years later the *Origin of Species* appeared, and the impression which it produced was enormous. No book dealing with a scientific subject had ever, I suppose, been so largely read by people who were not scientific. I was an undergraduate at Oxford at the time, and I recollect very well that many of my fellow undergraduates who never opened—I will not say a scientific, but hardly even a serious book before—procured the treatise and read it with avidity. We all talked about it. We discussed it with the greatest ardor, indeed, with a positiveness which was in inverse ratio to our knowledge; and it was the same all over England. *The Origin* was not only the subject of constant comment in magazines and newspapers as well as at meetings of scientific societies, but it furnished a theme for constant jests in the comic papers, and it was an unfailing topic for conversation in all cultivated private houses. [. . .]

It was at his home that I saw him, a year and a half before his death. One could converse with him for a few minutes only because his health was so feeble that it was necessary to save all the time he could spare for the prosecution of his work, as he was only able to work for two or three hours a day, perhaps even less, and talking fatigued him. The conversation I had with him lasted less than twenty-five minutes, and at the end of that time one of his sons came in and took him away to lie down and rest. [...]

He was one of those men whose character was palpably written on his face. He had a projecting brow, with a forehead very full over the eyes, and a fine dome-shaped head. His eyes were deep set, because the brow projected so far, and were of a clear and steady blue, and he had a quiet, contemplative look, with an occasional slight smile passing over his countenance which made one feel perfectly at ease in his company. There was nothing about him to make a stranger feel constrained or timorous in his company, however deep one's reverence, because his manner was simple, natural, with nothing to indicate any consciousness of distinction.

James Bryce, "Personal Reminiscences of Charles Darwin and of the Reception of the *Origin of Species.*" *Proceedings of the American Philosophical Society* 48 (September 1909): iii–xiv, on x, xii–xiv.

Secondary Literature: John T. Seaman, *A Citizen of the World: The Life of James Bryce* (London: Tauris, 2006).

LUDWIG BÜCHNER (1824–1899)
German Materialist

Even if we had not lived during the last twenty years, through the great revolution in organic science brought about by the teachings of Darwin, the general result would yet remain certain to every philosophical mind— as it was, in fact, some scores of years since, to the minds of some naturalists, gifted with greater perspicacity than their colleagues, such as a Lamarck, a Geoffroy St. Hilaire [Saint-Hilaire]*, or most of the adepts of the so-called naturalistic school. As far back as 1855, five years before Darwin, the author of this work spoke in its *first* edition of that general result with as much certainty as was possible at that time, and represented the genesis of new species to be a *natural* process, brought about by descent, variation, and development, resting his argument upon considerations generally derived from research in paleontology, comparative anatomy and embryology. Nor did he fail to apply these considerations to the very "question of questions," and to set forth with a boldness which drew down upon him a storm of obloquy from all quarters, that "animal descent of man" about which, at present, scarcely any scientist entertains the slightest doubt.

Ludwig Büchner, *Man in the Past, Present, and Future* (New York: Peter Eckler, 1891), 164. Büchner's *Kraft und Stoff* (1845), which stressed the indestructibility of matter and force, was a cornerstone of nineteenth-century materialism.

Secondary Literature: Rollo Handy, "Ludwig Büchner," in *Encyclopedia of Philosophy,* ed. Paul Edwards, 8 vols. (New York: Macmillan, 1967), 1:411–13.

HENRY THOMAS BUCKLE (1821–1862)
English Historian

Tell your husband to read Darwin On Species, and to master it. He will find it full of thought, and of original matter.

Buckle to Mrs. Woodhead, 17 January 1860, in *The Life and Writings of Henry Thomas Buckle,* ed. Alfred Henry Huth, 2 vols. (London: Sampson Low, Marston, Searl, & Rivington, 1880), 2:28.

Secondary Literature: J. W. Burrow, *Evolution and Society: A Study in Victorian Social Theory* (Cambridge: Cambridge University Press, 1966); William T. Walker, "Henry Buckle," in *Encyclopedia of Historians and Historical Writing,* ed. Kelly Boyd, 2 vols. (London: Fitzroy Dearborn, 1999), 1:148–49.

ARABELLA BUCKLEY (1840–1929)
English Author and Science Popularizer

She had been to dine and sleep at Down [in October 1880] and I saw her in the Museum afterwards.

Samuel Butler, *The Correspondence of Samuel Butler with His Sister May,* ed. Daniel F. Howard (Berkeley: University of California Press, 1962), 91n5.

Mr. Darwin has written a whole book on the many curious and wonderful ways in which orchids tempt bees and other insects to fertilize them.

Arabella Buckley, *The Fairy-Land of Science* (New York: Appleton, 1905), 237.

But how have these red ants, which are in many ways some of the cleverest of their kind, learnt to steal young black ones, to help them in their work? Mr. Darwin suggests the answer. It is a common practice with ants to carry the cocoons of their enemies into their nests to eat them, and they tear open the cocoon to feed on the insect within. Now, nothing is more likely than that some of the black-ant cocoons, thus carried in, should be neglected, till the ants within them were perfect, and then, when they came out active and vigorous, they would be well received, as ants born in the nest generally are, and would mix with the red ones, and prove very useful. Is it too much to imagine, that thus by degrees the intelligent red ants should come to understand that it was better to have the help of the black ants than to eat them, and should learn to fetch them in numbers to help them in their work? One thing is certain, that they know their own interest now, for if by chance a female winged ant comes out of the stolen cocoons, she is killed at once by her red masters, who know that if she lived and laid eggs, these would be tended by the slaves, and the nest would soon become a black-ant city.

Then we may perhaps learn that the "struggle for existence," which has taught the ant the lesson of self-sacrifice to the community, is also able to teach that higher devotion of mother to child, and friend to friend, which ends in a tender love for every living being, since it recognizes that mutual help and sympathy are among the most powerful weapons, as they are also certainly the most noble incentives, which can be employed in fighting the battle of life.

Arabella Buckley, *Life and Her Children* (New York: Appleton, 1881), 290–91, 301.

Secondary Literature: Barbara T. Gates, *Kindred Nature: Victorian and Edwardian Women Embrace the Living World* (Chicago: University of Chicago Press, 1998); Bernard Lightman, *Victorian Popularizers of Science: Designing Science for New Audiences* (Chicago: University of Chicago Press, 2007).

EDWARD BULWER-LYTTON (1803–1873)
English Novelist and Politician

Read Darwin on the origin of species, and learn that you are fellow-Christians in an imperfect state of development.

Parliamentary debate, 13 April 1866, quoted in Thomas Archer, *William Ewart Gladstone and his Contemporaries; Fifty Years of Social and Political Progress,* vol. 3, *1852–1860* (London: Blackie & Son, 1883), 216.

But what, at the best, is man? A crude, struggling, undeveloped embryo, of whom it is the highest attribute that he feels a vague consciousness that he is only an embryo, and cannot complete himself till he ceases to be a man; that is, until he becomes another being in another form of existence. We can praise a dog as a dog, because a dog is a completed *ens,* and not an embryo. But to praise a man as man, forgetting that he is only a germ out of which a form wholly different is ultimately to spring, is equally opposed to Scriptural belief in his present crudity and imperfection, and to psychological or metaphysical examination of a mental construction evidently designed for purposes that he can never fulfill as man. That my father is an embryo not more incomplete than any present, is quite true; but that, you will see on reflection, is saying very little on his behalf. Even in the boasted physical formation of us men, you are aware that the best-shaped amongst us, according to the last scientific discoveries, is only a development of some hideous hairy animal, such as a gorilla; and the ancestral gorilla itself had its own aboriginal forefather in a small marine animal shaped like a two-necked bottle. The probability is that, some day or other, we shall be exterminated by a new development of species.

"I fear," said Kenelm, gravely, "that your change of dress betokens the neighborhood of those pretty girls of whom you spoke in an earlier meeting. According to the Darwinian doctrine of selection, fine plumage goes far in deciding the preference of Jenny Wren and her sex, only we are told that fine-feathered birds are very seldom songsters as well. It is rather unfair to rivals when you unite both attractions."

Edward Bulwer-Lytton, *Kenelm Chillingly, His Adventures and Opinions,* 2 vols. (1873; reprint, Boston: Little, Brown, 1896), 1:54–55, 263.

Secondary Literature: Leslie Mitchell, *Bulwer Lytton: The Rise and Fall of a Victorian Man of Letters* (London: Continuum, 2003).

LUTHER BURBANK (1849–1826)
American Botanist and Horticulturalist

In the same way we conceive of the evolutionary changes through which new species were evolved in the past as having been relatively sudden. I have already referred to the difficulty with which the average mind can grasp the idea that precisely the same sort of change in animal and vegetable forms is taking place to-day that has taken place in all other stages of evolution.

It was one of the great merits of Darwin's exposition of the *Origin of Species,* that he gave detailed illustrations of the struggle for existence, and brought tangibly before the minds of thoughtful people the conception that each race of beings is more or less in competition with every other race, and that the race that is adaptable enough to adjust itself to new conditions is the only one that stands any prospect of survival.

We must not forget that on occasion there may be natural methods of elimination that will single out a species and destroy it as expeditiously and as certainly as man could accomplish that end. A case in point is furnished by the chestnut, which, as we have seen in a recent chapter, has been singled out in certain regions of the Eastern United States by a fungoid blight that leaves no chestnut alive in the regions over which it spreads. Yet this blight seems powerless to affect any other species.

Here, then, we have an example of a destructive agency of an unpredicted kind that gives an example of the rapid destruction of a species, through natural selection, because that species could not rapidly enough adapt itself to a new condition.

Luther Burbank: His Methods and Discoveries and their Practical Application, ed. John Whitson, Robert John, and Henry Smith Williams, vol. 9 (New York: Luther Burbank, 1915), 205, 208–9, 211.

To would-be plant experimenters who ask my opinion of matters connected with the old versus the new interpretations of heredity, I am accustomed to say: "Read Darwin first, and gain a full comprehension of the meaning of Natural Selection. Then read the modern Mendelists in detail. But then—go back again to Darwin."

Luther Burbank, *How Plants are Trained to work For Men,* 8 vols. (New York: P. F. Collier, 1914), 1:851–52.

Secondary Literature: Peter Dreyer, *A Gardener Touched with Genius: The Life of Luther Burbank,* rev. ed. (Berkeley and Los Angeles: University of California Press, 1985).

JOHN BURROUGHS (1837–1921)
American Naturalist

Finished Darwin's *Origin of Species* last night. A true wonder-book. Few pages in modern scientific literature so noble as those last few pages of the book. Everything about Darwin indicates the master. In reading him you breathe the air of the largest and most serene mind. Every naturalist before him, and with him, he lays under contribution; every competent observer in every field. Only the greatest minds can do this as he does it. He furnishes the key to every man's knowledge. Those that oppose his theory unwittingly bring some fact or observation that fits into the scheme. His theory has such range; accounts for such a multitude of facts; easily underruns and outruns the views of all other naturalists. He is, in his way, as great and as remarkable as Shakespeare, and utilizes the knowledge of mankind in the same way. His power of organization is prodigious. He has the candor, the tranquility, the sincerity, the singleness of purpose that go with, and are a promise of, the highest achievement. He is the father of a new generation of naturalists. He is the first to open the doors into Nature's secret Senate chambers. His theory confronts, and is as grand as, the nebular hypothesis, and is in the same line of creative energy.

John Burroughs diary, 26 September 1883, in *The Heart of John Burrough's Journals,* ed. Clara Barrus (Boston: Houghton Mifflin, 1928), 100–101.

Darwin cared nothing for religion, so called, because his mind and his conscience were enlisted in his science. He was serving God disinterestedly.

John Burroughs, "The Decadence of Theology," *North American Review* 156 (1893): 585.

Secondary Literature: Edward J. Renehan, *John Burroughs: An American Naturalist* (Post Mills, VT: Chelsea Green, 1992).

J. B. BURY (1861–1927)
ENGLISH HISTORIAN

From the more general influence of Darwinism on the place of history in the system of human knowledge, we may turn to the influence of the principles and methods by which Darwin explained development. It had been recognized even by ancient writers (such as Aristotle and Polybius) that physical circumstances (geography, climate) were factors conditioning the character and history of a race or society. In the sixteenth century Bodin emphasized these factors, and many subsequent writers took them into account. The investigations of Darwin, which brought them into the foreground, naturally promoted attempts to discover in them the chief key to the growth of civilization. Comte[†] had expressly denounced the notion that the biological methods of Lamarck could be applied to social man. [Henry Thomas] Buckle[*] had taken account of natural influences, but had relegated them to a secondary plane, compared with psychological factors. But the Darwinian theory made it tempting to explain the development of civilization in terms of "adaptation to environment," "struggle for existence," "natural selection," "survival of the fittest," etc. (Recently O. Seeck[‡] has applied these principles to the decline of Graeco-Roman civilization in his *Untergang der antiken Welt*, 2 volumes, Berlin, 1895, 1901.)

J. B. Bury, "Darwinism and History" (1909), in *Selected Essays of J. B. Bury*, ed. Harold Temperely (Cambridge: Cambridge University Press, 1930), 34. [†]Auguste Comte (1798–1857), French philosopher and mathematician who founded a school of positivist philosophy. [‡]Otto Seeck (1850–1921), Geman historian of antiquity.

Secondary Literature: D. S. Goldstein, "J. B. Bury's Philosophy of History: A Reappraisal," *American Historical Review* 82 (1977): 896–919; Roger Collins, "J. B. Bury," in *Encyclopedia of Historians and Historical Writing,* ed. Kelly Boyd, 2 vols. (London: Fitzroy Dearborn, 1999), 1:155–56.

SAMUEL BUTLER (1835–1902)
English Novelist

We have used the words "mechanical life," "mechanical kingdom," "the mechanical world" and so forth, and we have done so advisedly, for as the vegetable kingdom was slowly developed from the mineral, and as in like manner the animal supervened upon the vegetable, so now in these last few ages an entirely new kingdom has sprung up, of which we as yet have only seen what will one day be considered the antediluvian prototypes of the race.

We regret deeply that our knowledge both of natural history and of machinery is too small to enable us to undertake the gigantic task of classifying machines into the genera and sub-genera, species, varieties and sub-varieties, and so forth, of tracing the connecting links between machines of widely different characters, of pointing out how subservience to the use of man has played that part among machines which natural selection has performed in the animal and vegetable kingdoms, of pointing out rudimentary organs which exist in some few machines, feebly developed and perfectly useless, yet serving to mark descent from some ancestral type which has either perished or been modified into some new phase of mechanical existence. We can only point out this field for investigation; it must be followed by others whose education and talents have been of a much higher order than any which we can lay claim to.

Some few hints we have determined to venture upon, though we do so with the profoundest diffidence. Firstly, we would remark that as some of the lowest of the vertebrata attained a far greater size than has descended to their more highly organized living representatives, so a diminution in the size of machines has often attended their development and progress. Take the watch for instance. Examine the beautiful structure of the little animal, watch the intelligent play of the minute members which compose it; yet this little creature is but a development of the cumbrous clocks of the thirteenth century—it is no deterioration from them. The day may come when clocks, which certainly at the present day are not diminishing in bulk, may be entirely superseded by the universal use of watches, in which case clocks will become extinct like the earlier saurians, while the watch (whose tendency has for some years been rather to decrease in size than the contrary) will remain the only existing type of an extinct race.

Samuel Butler, "Darwin among the Machines (13 June 1863)," in *A First Year in Canterbury Settlement with Other Early Essays* (London: A. C. Fifield, 1914), 179–85, on 180–82.

Of course he read Mr. Darwin's books as fast as they came out and adopted evolution as an article of faith. "It seems to me," he said once, "that I am like one of those caterpillars which, if they have been interrupted in making their hammock, must begin again from the beginning. So long as I went back a long way down in the social scale I got on all right, and should have made money but for Ellen; when I try to take up the work at a higher stage I fail completely." I do not know whether the analogy holds good or not, but I am sure Ernest's instinct was right in telling him that after a heavy fall he had better begin life again at a very low stage, and as I have just said, I would have let him go back to his shop if I had not known what I did.

Samuel Butler, *The Way of All Flesh* (1903; reprint, New York: Dutton, 1917), 412.

I see they are talking of making the old schools into a memorial of Darwin. I am not surprised, for he is most certainly the most widely know Shrewsbury man, but I feel very sure that he will not keep the high reputation he has at present. Nevertheless he has been such a prominent figure during the last twenty years and more that any notes of his boyhood would be prized, a hundred years hence, as much, or more, perhaps than now.

The Family Letters of Samuel Butler 1841–1886, ed. Arnold Silver (Stanford, CA: Stanford University Press, 1962), 207.

I regret that reviewers have in some cases been inclined to treat the chapters on Machines as an attempt to reduce Mr. Darwin's theory to an absurdity. Nothing could be further from my intention, and few things would be more distasteful to me than any attempt to laugh at Mr. Darwin; but I must own that I have myself to thank for the misconception, for I felt sure that my intention would be missed, but preferred not to weaken the chapters by explanation, and knew very well that Mr. Darwin's theory would take no harm. The only question in my mind was how far I could afford to be misrepresented as laughing at that for which I have the most profound admiration. I am surprised, however, that the book at which such an example of the specious misuse of analogy would seem most naturally levelled should have occurred to no reviewer; neither shall I mention the name of the book here, though I should fancy that the hint given will suffice.

Samuel Butler, *Erewhon* (1901; reprint, Harmondsworth: Penguin, 1970), 30.

It is obvious that the having fatally impaired the theory of his predecessors could not warrant Mr. Darwin in claiming, as he most fatuously did, the theory of evolution. That he is still generally believed to have been the originator of this theory is due to the fact that he claimed it, and that a powerful literary backing at once came forward to support him.

Samuel Butler, "The Deadlock in Darwinism," in *Essays on Life, Art and Science* (London: Fifield, 1908), 243, first published in *Universal Review*, April–June 1890.

I remember when I was at Down we were talking of what it is that sells a book. Mr. Darwin said he did not believe it was reviews or advertisements, but simply "being talked about" that sold a book.

I believe he is quite right here, but surely a good flaming review helps to get a book talked about. I have often inquired at my publishers' after a review and I never found one that made any perceptible increase or decrease of sale, and the same with advertisements. I think, however, that the review of *Erewhon* in the *Spectator* did sell a few copies of *Erewhon*, but then it was such a very strong one and the anonymousness of the book stimulated curiosity. A perception of the value of a review, whether friendly or hostile, is as old as St. Paul's Epistle to the Philippians.

The Note-Books of Samuel Butler, ed. Henry Festing Jones (New York: Dutton, 1917), 161.

Secondary Literature: Samuel Butler: Victorian against the Grain, ed. James G. Paradis (Toronto: University of Toronto Press, 2007).

THOMAS BUTLER (1806–1886)
English Clergyman

Lady Powis was staying at Barmouth and used to ask me to dinner every now and then and the Darwins hated that kith and kin because Lord Powis[†] was almost the only great man of the neighborhood who employed [the physician Dr. Thomas] DuGard, so Darwin used to jeer me about going to Lady Powis', and I took offence not seeing why I should not go when she made her little dinners pleasant. I never saw him again after that summer [1828] till he came back wasted to a shadow after his return from the Beagle expedition, when I traveled with him and [Robert] Southey in a stage coach from Birmingham to Shrewsbury. After that I never saw him again. [. . .]

I remember one day at Barmouth he had gone alone to shoot sea birds [. . .] and there was a beast of a black bull on the opposite side of the Barmouth river that made it really unsafe for foot passengers. Darwin however was on a pony and had a gun and a guide and on returning with a headache so that he could hardly sit his horse. The guide exclaimed, If you please Sir I think there is the bull! [. . .] He put a bullet into his gun but said he could not have seen to shoot owing to his headache. Did the bull understand his coming greatness?

Thomas Butler to his son Samuel, 9 May 1882, in *The Family Letters of Samuel Butler, 1841–1886,* ed. Arnold Silver (Stanford, CA: Stanford University Press, 1962), 209. [†]Edward Herbert (1785–1848), second Earl of Powis.

Secondary Literature: Peter Raby, *Samuel Butler: A Biography* (Iowa City: University of Iowa Press, 1991).

C

ELIZABETH DWIGHT CABOT (1830–1901)
American Social Reformer

[March 1860]

[James] Elliot [Cabot] has just finished Darwin and sees no answer to be made to it, and does not at all agree with Agassiz* that there is anything "dangerous" about the book, though he has discussed it with him. He thinks just what you say, that there is no less design in one plan of creation than the other.

Cabot to Mrs. Edward Twistleton, 4 March 1860, in *Letters of Elizabeth Cabot*, 2 vols. (Boston: privately printed, 1905), 1:235. Ellen Twistleton was a member of the Parkman family of Boston.

Secondary Literature: Mary Saracino Zboray, *Everyday Ideas: Socioliterary Experience among Antebellum New Englanders* (Knoxville: University of Tennessee Press, 2006); Sylvia D. Hoffert, *Private Maters: American Attitudes toward Childbearing and Infant Nurture in the Urban North, 1800–1860* (Champaign: University of Illinois Press, 1989).

GEORGE DOUGLAS CAMPBELL, DUKE OF ARGYLL (1823–1900)
Scottish Politician and Author

I was a close reader of all [Richard] Owen's* books, and he seldom wrote any important scientific paper without sending me an early copy. His book on the *Nature of Limbs* was an education in itself to me. I had been accustomed from childhood to look at the wings of a bird as in the nature of an elaborate apparatus for the accomplishment of flight. I saw that this explanation of them was in no way superseded, but that another and larger question was raised by Owen—namely, What were wings in relation to the arms of men and to the legs of dogs, even to the fins of fishes, and to the limbs of all other creatures having a vertebrate skeleton?

When Professor Owen rose above the horizon, I soon made acquaintance with him, and I read with intense interest his celebrated book *On the Homologies of the Vertebrate Skeleton*. This dealt with the same question in still larger and more fundamental aspects. By this course of reading, which included, of course, all the relative papers in the scientific journals, I felt myself comparatively well furnished with data, both of facts and of inferences from them, to deal with the memorable book of Darwin when it was published in 1859.

George Douglas Campbell, Eighth Duke of Argyll, *Autobiography and Memoirs*, ed. Duchess of Argyll, 2 vols. (New York: Dutton, 1906), 1:411.

[February 1860]
I wish I had been able to talk over Darwin's book. It is a most delightful one, suggesting endless subjects for discussion and inquiry. I think he fails fundamentally in these two cardinal points: First, in showing that in the existing or contemporary world breeding does effect any changes such as tend to the formation of new species. Second, he fails to show that in the past worlds there is any proof or clear evidence of such gradations of change as his theory requires. I am thoroughly dissatisfied, too, with the explanations by which the latter difficulty is met.

Argyll to Charles Lyell, 29 February 1860, in George Douglas Campbell, Eighth Duke of Argyll, *Autobiography and Memoirs*, ed. Duchess of Argyll, 2 vols. (New York: Dutton, 1906), 2:482.

The truth is that the phrase "natural selection," and the group of ideas which hide under it, is so elastic that there is nothing in heaven or on earth that by a little ingenuity may not be brought under its pretended explanation. Darwin in 1859–1860 wondered "how variously" his phrase had been "misunderstood." The explanation is simple: it was because of those vague and loose analogies which are so often captivating. It is the same now, after thirty-six years of copious argument and exposition. Darwin ridiculed the idea which some entertained that natural selection was set up as an active power of the deity; yet this is the very conception of it which is at this moment set up by one of the most faithful worshippers in the Darwinian cult. Professor [Edward Bagnall] Poulton*, of Oxford, gives to natural selection the title of a "motive power" first discovered by Darwin.

George Douglas Campbell Argyll, *Organic Evolution Cross-Examined; Or Some Suggestions on the Great Secret of Biology* (New York: Little, Brown, 1898), 88.

The scientific world, in the Duke's opinion, has been for some time bowing down to the idol of Darwin and the theory of evolution, which is the fundamental dogma of that cult. Like a prophet of old he raises a warning voice, and points out that the feet of the golden image are in part composed of clay. In the North has been hewn the stone which shall shatter those fragile supports and lay the idol prone in the dust!

T. G. Bonney, "A Conspiracy of Silence," *Nature* 37 (1887–88): 25–26.

Secondary Literature: Robert Alun Jones, *The Secret of the Totem* (New York: Columbia University Press, 2005).

ALPHONSE DE CANDOLLE (1806–1893)
Swiss Botanist

I longed to converse once more with Darwin, whom I had seen in 1839, and with whom I kept up a most interesting correspondence. It was on a fine autumn morning in 1880 that I arrived at Orpington station, where my illustrious friend's break met me. I will not here speak of the kind reception given to me at Down, and of the pleasure I felt in chatting familiarly with Mr. and Mrs. Darwin and their son Francis. I note only that Darwin at seventy was more animated and appeared happier than when I had seen him forty-one years before. His eye was bright; and his expression cheerful, whilst his photographs show rather the shape of his head, like that of an ancient philosopher. His varied, frank, gracious conversation, entirely that of a gentleman, reminded me of that of Oxford and Cambridge *savants*. The general tone was like his books, as is the case with sincere men, devoid of every trace of charlatanism. He expressed himself in English easily understood by a foreigner, more like that of [Edward] Bulwer[-Lytton]* or [Thomas Babington] Macaulay*, than that of

[Charles] Dickens* or [Thomas] Carlyle*. I asked him for news of the committee, of which he was a member, for reforming English spelling, and when I said that moderate changes would be best received by the public, he laughingly said, "As for myself, *of course*, I am for the most radical changes." We were more in accord on another point, that a man of science, even up to advanced age, ought to take an interest in new ideas, and to accept them, if he finds them true. "That was very strongly the opinion of my friend Lyell*," he said; "but he pushed it so far as sometimes to yield to the first objection, and I was then obliged to defend him against himself." Darwin had more firmness in his opinions, whether from temperament, or because he had published nothing without prolonged reflection. [...]

I looked for the greenhouse in which such beautiful experiments on hybrid plants had been made. It contained only a vine. One thing struck me, although it is not rare in England, where animals are loved. A heifer and a colt were feeding close to us with the tranquility which tells of good masters, and I heard the joyful barking of dogs. "Truly," I said to myself, "the history of the variations of animals was written here, and observations must be going on, for Darwin is never idle." I did not suspect that I was walking above the dwellings of those lowly beings called earthworms, the subject of his last work, in which Darwin showed once more how little causes in the long run produce great effects. He had been studying them for thirty years, but I did not know it.

Returning to the house, Darwin showed me his library, a large room on the ground floor, very convenient for a studious man; many books on the shelves; windows on two sides; a writing-table and another for apparatus for his experiments. Those on the movements of stems and roots were still in progress. The hours passed like minutes. I had to leave. Precious memories of that visit remain.

Alphonse de Candolle, "Darwin considéré au point de vue des causes de son succès et de l'importance de ses travaux," Archives des Sciences Physiques et Naturelles (Geneva), 3rd ser., 7 (1882), The Complete Works of Darwin Online CUL-DAR134.11; English translation in George Thomas Bettany, *Life of Charles Darwin* (London: Walter Scott, 1887), 148–50.

Secondary Literature: P. E. Pilet, "Alphonse de Candolle," in *Dictionary of Scientific Biography*, ed. Charles Coulston Gillispie, 16 vols. (New York: Scribner's, 1970–80), 3:42–43.

JANE WELSH CARLYLE (1801–1886)
English Woman of Letters

[January 1860]

But even when Darwin, in a book that all the scientific world is in ecstasy over, proved the other day that we are all come from shell-fish, it didn't move me to the slightest curiosity whether we are or not—I did not feel that the slightest light would be thrown on my practical life for me, by having it ever so logically made out that my first ancestor, millions of millions of ages back, had been, or even had not been, an oyster. It remained a plain fact that I was no oyster, nor had any grandfather an oyster within my knowledge; and for the rest, there was nothing to be gained, for this world, or the next, by going into the oyster-question, till all more pressing questions were exhausted! So—if I can't read Darwin, it may be feared I shall break down in Mrs. Duncan. Thanks to you, however, for the book, which will be welcome to several of my acquaintances. There is quite a mania for geology at present, in the female mind.

Jane Carlyle to Mary Russell, 28 January 1860, in *Letters and Memorials of Jane Welsh Carlyle,* ed. James Anthony Froude, 2 vols. (New York: Charles Scribner's, 1883), 2:155–56.

Secondary Literature: Philip J. Waller, *Writers, Readers, and Reputations: Literary Life in Britain, 1870–1918* (Oxford: Oxford University Press, 2006).

THOMAS CARLYLE (1795–1881)
English Author

To John A. Carlyle, Moffat: 10 March, 1853. [...] The Ashburtons lately have done unexpectedly a really handsome thing to me. Lord Ashburton is on the committee of the Athenaeum Club; he said once, Shall I propose you, this spring, for immediate election? I answered grumblingly, vaguely. [...] But now the other day, comes news that I am elected, the money all paid, entrance money and subscription in a lump; and that I have only to go in when I like and stay out when I like! [...] Lord Ashburton took me the other night to my first dinner and entrance there: I do not much believe I shall go often; but that will be seen. Old Crabbe [*sic*] Robinson,[†] visible in the reading room, inquired after you that night: very old, and clattery. Darwin, [Richard] Owen*, etc., were also visible: plenty of loungers there, if one wanted lounging!

Carlyle to John A. Carlyle, 10 March 1853, in *New Letters of Thomas Carlyle*, ed. Andrew Carlyle, 2 vols. (London: John Lane, 1904), 2:145–46. [†]Henry Crabb Robinson (1775–1867), English journalist, lawyer, and traveler.

[1875]
Here a personal recollection comes into view which, as it throws a pleasant light on the relations of Carlyle and Darwin, may be worth recording. Like many other noble ladies, Lady Derby was a warm friend of Carlyle;[†] and once, during an entire summer, Keston Lodge was placed by Lord Derby at Carlyle's disposal. From the seat of our common friend, Sir John Lubbock*, where we had been staying, the much-mourned William Spottiswoode[‡] and myself once walked over to the Lodge to see Carlyle. He was absent; but as we returned we met him and his niece, the present Mrs. Alexander Carlyle, driving home in a pony-carriage. I had often expressed to him the wish that he and Darwin might meet; for it could not be doubted that the nobly candid character of the great naturalist would make its due impression. The wish was fulfilled. He met us with the exclamation: "Well, I have been to see Darwin." He paused, and I expressed my delight. "Yes," he added, "I have been to see him, and a more charming man I have never met in my life."

John Tyndall, *New Fragments* (New York: Appleton, 1898), 388. [†]Lady Constance Villier (1840–1922), wife of the Conservative politician Frederick Arthur Stanley, Earl of Derby (1841–1905), promoted the idea of a knighthood for Carlyle. [‡]William Spottiswoode (1825–1883), English mathematician and physicist, president of the Royal Society from 1878 to 1883.

Tuesday, October 5, 1875—Back from Eastbourne—at C.'s 3.20. Carlyle and Mary back from Keston Lodge. C. and I go out, rain, shelter in public-house, allowed with scant civility, and then go back for his coat. Carlyle never carried an umbrella; in wet weather a waterproof coat. He very seldom caught cold. C.—We had a successful stay at Keston. Saw Sir John Lubbock*; Charles Darwin several times. I had not seen him for twenty years. He is a pleasant jolly-minded man (I thought this a very curious phrase), with much observation and a clear way of expressing it. Has long been an invalid. I asked him if he thought there was a possibility of men turning into apes again. He laughed much at this, and came back to it over and over again.

William Allingham, *A Diary*, ed. H. Allingham and D. Radford (London: Macmillan, 1907), 239.

February 22, 1879.—Drove with Carlyle.—Darwin and Haeckel*. C.: "For Darwin personally I have great respect; but all that of *Origin of Species*, etc., is of little interest to me. What we desire to know is, who is the Maker? and what is to come to us when we have shuffled off this mortal coil. Whoever looks into himself must be aware that at the centre of things is a mysterious Demiurgus—who is God, and who cannot in the least be adequately spoken of in any human words."

William Allingham, *A Diary*, ed. H. Allingham and D. Radford (London: Macmillan, 1907), 274.

Secondary Literature: Fred Kaplan, *Thomas Carlyle: A Biography* (Ithaca, NY: Cornell University Press, 1983); Philip J. Waller, *Writers, Readers, and Reputations: Literary Life in Britain, 1870–1918* (Oxford: Oxford University Press, 2006).

WILLIAM BENJAMIN CARPENTER (1813–1885)
English Naturalist and Physiologist

I have had a letter from Carpenter this morning. He reviews me in the *National* [*National Review*]. He is a convert, but does not go quite so far as I, but quite far enough, for he admits that all birds are from one progenitor, and probably all fishes and reptiles from another parent. But the last mouthful chokes him. He can hardly admit all vertebrates from one parent. He will surely come to this from Homology and Embryology. I look at it as grand having brought round a great physiologist, for great I think he certainly is in that line. How curious I shall be to know what line [Richard] Owen* will take; dead against us, I fear; but he wrote me a most liberal note on the reception of my book, and said he was quite prepared to consider fairly and without prejudice my line of argument.

Darwin to Charles Lyell, 5 December 1859, in *The Life and Letters of Charles Darwin*, ed. Francis Darwin, 2 vols. (New York: Appleton, 1887), 2:35.

[January 1860]

To such as look upon this question from the purely scientific point of view, any theological objection, even to Mr. Darwin's rather startling conclusion, much more to his very modest premises, seems simply absurd. We never heard of anybody who thought that a religious question was involved in the inquiry whether our breeds of dog are derived from one or from several ancestral stocks; nor should we suppose that the stoutest believers in the Mosaic cosmogony would be much dismayed if it could be shown that the dog is really a derivation from the wolf. [...]

Why, then, should Mr. Darwin be attacked (as he most assuredly will be) for venturing to carry the same method of inquiry a step further: and be accused [...] of superseding the functions of the Creator, of blotting out his Attributes from the page of Nature, and of reducing Him to the level of a mere Physical Agency? To our apprehension, the Creator did not finish his labors with the creation of the protoplasts of each species: his work is always in progress; the origin and development of each new being that comes into life, is a new manifestation of his creative power.

Naturalists have gone on quite long enough on the doctrine of the "permanence of species." Their catalogues are becoming more and more encumbered with these hypothetical "distinct creations." And the difficulty of distinguishing between true species and varieties increases, instead of diminishing, with the extent of their researches. The doctrine of progressive modification by Natural Selection propounded by Mr. Darwin, will give a new direction to inquiry into the real genetic relationship of species, existing and extinct; and it has a claim to respectful consideration, not merely on account of the high value of Mr. Darwin's previous contributions to zoological science, and the thoroughly philosophical spirit in which it is put forth, but also because it brings into mutual reconciliation the antagonistic doctrines of two great schools—that of Unity of Type, as put forward by [Étienne] Geoffroy St Hilaire [Saint-Hilaire] and his followers of the Morphological School, and that of Adaptation to conditions of existence, which has been the leading principle of Cuvier and the Teleologists.

William Benjamin Carpenter, "Darwin on the Origin of Species," *National Review* 10 (1860): 188–214, reprinted in *Darwin and his Critics*, by David Hull (Cambridge, MA: Harvard University Press, 1973), 88–114, on 93, 114.

My Dear Carpenter: I have just read your excellent article in the *National*. It will do great good; especially if it becomes known as your production. It seems to me to give an excellently clear account of Mr. Wallace's* and my views. How capitally you turn the flanks of the theological opposers by opposing to them such men as [George] Bentham and the more philosophical of the systematists! I thank you sincerely for the extremely honorable manner in which you mention me. I should have liked to have seen some criticisms or remarks on embryology, on which subject you are so well instructed. I do not think any candid person can read your article without being much impressed with it. The old doctrine of immutability of specific forms will surely but slowly die away. It is a shame to give you trouble, but I should be very much obliged if you could tell me where differently colored eggs in individuals of the cuckoo have been described, and their laying in twenty-seven kinds of nests. Also do you know from your own observation that the limbs of sheep imported into the West Indies change color? I have had detailed information about the loss of wool; but my accounts made the change slower than you describe.

Darwin to Carpenter, 6 January 1860, in *The Life and Letters of Charles Darwin*, ed. Francis Darwin, 2 vols. (New York: Appleton, 1887), 2:57.

Secondary Literature: Vance M. D. Hall, "The Contribution of the Physiologist William Benjamin Carpenter (1813–1885) to the Development of the Principles of the Correlation of Forces and the Conservation of Energy," *Medical History* 23 (1979): 129–55.

LEWIS CARROLL (1832–1898)
English Author

Three other unseen correspondents may be recorded here. [...] Thirdly, Mr. C. Darwin whose book *The Expression of the Emotions in Man and Animals* I am reading, and to whom I have given a print of "No Lessons today."

The Diaries of Lewis Carroll, ed. R. Green, 2 vols. (New York: Oxford University Press, 1954), 2:315–16.

Secondary Literature: Philip J. Waller, *Writers, Readers, and Reputations: Literary Life in Britain, 1870–1918* (Oxford: Oxford University Press, 2006).

EMILIO CASTELAR (1832–1899)
Spanish Politician and President

If it is necessary to subject science to the religion of the State, how can a professor who teaches the doctrine of Lyell* explicate Geology? Natural History, a professor who supports the ideas of Darwin or Wallace*; Moral Law, a follower of Kant; [. . .] Philosophy or History, an instructor who professes the doctrine of Hegel?

From Castelar's explanation in 1875 of why he resigned his chair in history, quoted in *La Institución Libre de Enseñanza y su ambiente: Los orígines,* by A. Jiménez Landi (Madrid: Taurus, 1973), 663, trans. TFG.

Secondary Literature: C. A. M. Hennessy, *The Federal Republic in Spain: Pi y Margall and the Federal Republican Movement, 1868–74* (Oxford: Clarendon, 1962).

WILLA CATHER (1873–1947)
American Novelist

If Darwin had wished to study further the part played by environment in the differentiation [of] species, he could have taken no better subjects than the canal people. Originally the boatmen were Englishmen, with all the earmarks of the British working man. They have become a solitary and peculiar people who have not their like in the world, an Englishman only in speech. He is sort of half-land, half-water gypsy.

Willa Cather, "The Canal Folk of England," in the *World and the Parish: Articles and Reviews, 1893–1902,* 2 vols. (Lincoln: University of Nebraska Press, 1970), 2:902 (19 August 1902).

Secondary Literature: James Woodress, *Willa Cather: A Literary Life* (Lincoln: University of Nebraska Press, 1989).

JOHN W. CHADWICK (1840–1914)
AMERICAN UNITARIAN MINISTER

Darwin was not one of those who cannot see the forest for the trees, who,

> Viewing all things intermittently,
> In disconnection dull and spiritless
> Break down all grandeur.

The parts did not obscure for him the whole. He did not murder to dissect. The healthy vision of the natural man enjoyed the lovely synthesis of outward things, unspoiled by any peeping or analysis that was essential to his scientific search. A worshiper he must have been, and was, a wonderer, for it is truly written, "The more thou searchest the more thou shalt wonder." In the popular theology he made no investment. He came of Unitarian stock, and he went forward and not backward from his inherited opinions. His favorite religious journal was our own Boston *Index*. He wrote with perfect frankness, over his own name, "I do not believe that any revelation has ever been made." Since it became certain that his doctrine was to become established science, the orthodox have done their best to capture him. But they have only had their labor for their pains.

Rev. John W. Chadwick, "Charles Robert Darwin," in *Evolution: Popular Lectures and Discussions before the Brooklyn Ethical Association* (Boston: James H. West, 1889), 44.

Secondary Literature: Charles D. Cashdollar, *The Transformation of Theology, 1830–1890: Positivism and Protestant Thought in Britain and America* (Princeton, NJ: Princeton University Press, 1989).

ANTON CHEKHOV (1860–1904)
RUSSIAN AUTHOR

I am reading Darwin. Magnificent! I simply love him.

Chekhov to V. Bilibin, [11 March 1886], in *Letters of Anton Chekhov*, ed. Avraham Yamolinksy (New York: Viking, 1973), 30.

If you stayed the night with him, he would put [Dmitry] Pisarev* or Darwin on your bedroom table; if you said you had read it, he would go and bring Dobrolubov. [. . .] In the district this was called free-thinking, and many people looked upon this free-thinking as an innocent and harmless eccentricity.

Anton Chekhov, "Neighbours," in *The Duel and other Stories,* trans. Constance Garnett (New York: Macmillan, 1916), 244.

Ivanoff. [. . .] Do you actually think it is worse to be the wife of a strong man than to nurse some whimpering invalid?
     Sasha. Yes, it is worse.
     Ivanoff. Why do you think so? *[Laughing loudly]* It is a good thing Darwin can't hear what you are saying! He would be furious with you for degrading the human race.

Anton Chekhov, "Ivanoff," in *Plays by Anton Chekhov,* trans. Marian Fell (New York: Charles Scribner's Sons, 1912), 131.

The kitten lay awake thinking. Of what? Unacquainted with real life, having no store of accumulated impressions, his mental processes could only be instinctive, and he could but picture life in accordance with the conceptions that he had inherited, together with his flesh and blood, from his ancestors, the tigers (*vide* Darwin).

Anton Chekhov, "Who was to blame?" in *The Cook's Wedding and other Stories* (New York: Macmillan, 1922), 228.

Secondary Literature: Donald Rayfield, *Anton Chekhov: A Life* (London: Harper-Collins, 1997).

G. K. CHESTERTON (1874–1936)
ENGLISH AUTHOR

One could not better sum up Christianity than by calling a small white insignificant flower "The Star of Bethlehem." But then, again, one could not better sum up the philosophy deduced from Darwinism than in the one verbal picture of "having your monkey up."

G. K. Chesterton, *Alarms and Discursions* (1910; reprint, Teddington, UK: Echo Library, 2006), 23–24.

I have just read a story in a magazine about Java and how modern white inhabitants of that island are prevailed on to misbehave themselves by the personal influence of poor old *Pithecanthropus.* That the modern inhabitants of Java misbehave themselves I can very readily believe; but I do not imagine that they need any encouragement from the discovery of a few highly doubtful bones. Anyhow, those bones are far too few and fragmentary and dubious to fill up the whole of the vast void that does in reason and in reality lie between man and his bestial ancestors, if they were his ancestors. On the assumption of that evolutionary connection (a connection which I am not in the least concerned to deny), the really arresting and remarkable fact is the comparative absence of any such remains recording that connection at that point. The sincerity of Darwin really admitted this; and that is how we came to use such a term as the Missing Link. But the dogmatism of Darwinism has been too strong for the agnosticism of Darwin; and men have fallen into turning this entirely negative term into a positive image. They talk of searching for the habits and habitat of the Missing Link; as if one were to talk of being on friendly terms with the gap in a narrative or the hole in an argument, of taking a walk with a nonsequitur or dining with an undistributed middle. In this sketch, therefore, of man in his relation to certain religious and historical problems, I shall waste no further space on these speculations on the nature of man before he became man. His body may have been evolved from the brutes; but we know nothing of any such transition that throws the smallest light upon his soul as it has shown itself in history.

G. K. Chesterton, *The Everlasting Man* (New York: Dodd, Mead, 1925), pt. 1, chap. 2, "Professors and Prehistoric Men."

Secondary Literature: Stephen R. L. Clark, *G. K. Chesterton: Thinking Backward, Looking Forward* (West Conshohocken, PA: Templeton Foundation Press, 2006), chap. 13, "Darwinism, Scientific and Social."

WINSTON CHURCHILL (1874–1965)
English Prime Minister

From November to May I read for four or five hours every day history and philosophy. Plato's *Republic* [...] Malthus *On Population,* Darwin's *Origin of Species:* all interspersed with other books of lesser standing, It was a curious education.

Winston Churchill, *My Early Life: 1874–1904* (New York: Scribner, 1996), 112.

Secondary Literature: Chris Wrigley, *Winston Churchill: A Biographical Companion* (Santa Barbara, CA: ABC-CLIO, 2002).

CARL CLAUS (1835–1899)
Austrian Zoologist

Darwin has been unjustly reproached with having left chance to play a considerable part in his attempt to account for the origin of varieties, with having accounted for everything by the struggle for existence, and with having given too little prominence to the direct influence of physical action on the mutation of forms. This reproach seems to arise from a misapprehension. Darwin says himself that the expression "chance," which he often uses to explain the presence of any small alteration, is a totally incorrect expression, and is only used to express our complete ignorance of the physical reasons for such particular variation.

When the fundamental arguments of the Darwinian theory of selection and the transmutation theory founded upon it are submitted to criticism, it is soon apparent that direct proof by investigation is now, and perhaps always will be, impossible; for the theory is founded upon postulates which cannot be submitted to direct inquiry. Periods of time which cannot be brought within the limits of human observation are required for the alteration of forms under natural conditions of life; and the extremely complicated interactions, which in the natural state under the form of natural selection are tending to change plants and animals, can only be grasped in a general sense, while in their details they are practically unknown to us.

C. Claus, *Elementary Text-Book of Zoology,* trans. Adam Sedgwick, 2 vols. (London: A. Swan Sonnenschein, 1884), 1:147–48, 150.

Secondary Literature: Lynn K. Nyhart, *Biology Takes Form* (Chicago: University of Chicago Press, 1995).

GEORGES CLEMENCEAU (1841–1929)
French Statesman

Analogies of structure and function in the entire hierarchy of the organic world were one day perceived, and Lamarck and Darwin drew from these their well-known conclusions, to the confusion of biblical tradition. Comparative anatomy and comparative physiology are now flourishing sciences of which academicians find it less easy to assimilate the results than to proclaim the failure. At the point we have reached in the knowledge of vital manifestations all along the scale of living creatures, unlimited material is day by day accumulating for the science of comparative psychology which will soon be established.

Georges Clemenceau, *Surprises of Life,* trans. Grace Hall (Garden City, NY: Doubleday, Page, 1920), 197–98.

In his early youth [Clemenceau] was deeply impressed with the teaching of J. S. [John Stuart] Mill*, and in later years he was manifestly under the successive sway of Sir Charles Lyell* and of Herbert Spencer*. But by the time he collected his essays in *La Melée Sociale* [1918], he was completely infatuated by the system of Darwin. He had long been familiar with *The Origin of Species* and *The Descent of Man;* the death of Darwin in 1882 had deprived him of a master and, as it seemed, a friend, while the publication of *Life and Letters* in 1887 had given a coherency and, we might say, an atmosphere, to his conception of the illustrious English savant. When, therefore, M. Clemenceau put together the material of *La Melée Sociale,* he did so in the quality of an advanced Darwinian, and he produced his first book almost as a tribute of affection to the memory of the greatest exponent of the tragedy of natural selection. But the habit of his mind, and no doubt the conditions of his own fortunes, led him into a field more tragical than any haunted by the spirit of the placid philosopher of Down.

Charles Darwin refrained from pushing his observations to such sinister conclusions as this: "La mort, partout la mort. Les continents et les mers gémissent de l'effroyable offrande de massacre."

Edmund Gosse, *Aspects and Impressions* (London: Cassell, 1922), 232–33.

Secondary Literature: Linda L. Clark, *Social Darwinism in France* (University: University of Alabama Press, 1984).

WILLIAM KINGDON CLIFFORD (1845–1879)
English Mathematician and Philosopher

When he was at Cambridge he was an ardent High Churchman, but from that creed he speedily broke away to become the enthusiastic propounder and defender of the faith as it is in Darwin.

Edward Clodd, *Memories* (New York: G. P. Putnam's Sons, 1916), 39.

[ca. 1863–66]
Meanwhile [Clifford] was eagerly assimilating the ideas which had been established as an assured possession of biological science by Mr. Darwin, and the kindred ones already at an earlier time applied and still being applied to the framing of a constructive science of psychology, and to the systematic grouping and gathering together of human knowledge, by Mr. Herbert Spencer*; who had, in Clifford's own words, "formed the conception of evolution as the subject of general propositions applicable to all natural processes." Clifford was not content with merely giving his assent to the doctrine of evolution: he seized on it as a living spring of action, a principle to be worked out, practiced upon, used to win victories over nature, and to put new vigor into speculation. For two or three years the knot of Cambridge friends of whom Clifford was the leading spirit were carried away by a wave of Darwinian enthusiasm: we seemed to ride triumphant on an ocean of new life and boundless possibilities. Natural Selection was to be the master-key of the universe; we expected it to solve all riddles and reconcile all contradictions.

Frederick Pollock, "Biographical Introduction," in William Kingdon Clifford, *Lectures and Essays,* ed. Leslie Stephen and Frederick Pollock, 2nd ed. (London: Macmillan, 1886), 1–32, on 24.

Secondary Literature: Gowan Dawson, "The Refashioning of William Kingdon Clifford's Posthumous Reputation," in *Darwinism, Literature and Victorian Respectability* (Cambridge: Cambridge University Press, 2007), 162–89.

EDWARD CLODD (1840–1930)
English Banker and Author

As for the reception of the book abroad, the French savants were somewhat coy, but the Germans, with Haeckel* at their head, were enthusiastic. Darwin had, like all prophets, more honor in other countries than in his own, Evolution being rechristened *Darwinismus*. Translation after translation of the *Origin* followed apace, and the personal interest that gathered round the central idea led to the perusal of the book by people who had never before opened a scientific treatise. *Punch* seized on it as subject of caricature; and writers of light verse found welcome material for "chaff" which the winds of oblivion have blown away, a stanza here and there surviving, as in Mr. [William John] Courthope's Aristophanic lines:

> Eggs were laid as before, but each time more and more varieties
>   struggled and bred,
> Till one end of the scale dropped its ancestor's tail, and the other got rid
>   of his head.
> From the bill, in brief words, were developed the Birds, unless our tame
>   pigeons and ducks lie;
> From the tail and hind legs, in the second-laid eggs, the apes—and
>   Professor Huxley!

Edward Clodd, *Pioneers of Evolution from Thales to Huxley* (New York: Appleton, 1897), 164.

But the remnant who care to know through what tribulation the fighters in the sixties entered the kingdom of the free may be told that the battle was the fiercer by reason of divisions in the camp of science, whereas the theologians were a solid phalanx. True it is that one of the earliest converts to Darwinism was a clerical ornithologist, Canon [Henry Baker] Tristram* (still with us), who applied the theory of natural selection to explanation of the colors of birds of the Sahara. Charles Kingsley*, too, was sympathetic; but these were as men "born out of due time." [Richard]

Owen's* malignant attitude has had reference; Sir John Herschel* said that natural selection was "the law of higgledy-piggledy," the exact meaning of which, Darwin confessed, puzzled him, as well it might; Adam Sedgwick* read parts of the book with "absolute sorrow, as false and grievously mischievous." But he hoped to meet Darwin "in heaven." [William] Whewell's* opposition took the form of refusing the *Origin of Species* a place in the library of Trinity College; Lyell* at first, and [William Benjamin] Carpenter*, with others, throughout, accepted with reservations; while the tone of the more intellectual organs was reflected in the *Athenaeum,* for long years an anti-Darwinian journal. Touching on the theological issues involved, it committed Darwin "to the tender mercies of the Divinity Hall, the College, the Lecture-room, and the Museum."

On both sides of the Atlantic the drum ecclesiastic was beaten in pulpits where, needless to say, vituperative rhetoric did duty for argument; preachers in cathedrals and little Bethels were at one in condemnation of a "brutal philosophy," whose success meant the denial of Scripture and the dethronement of God; while Episcopacy voiced itself through the Bishop of Oxford's [Samuel Wilberforce*] philippic in the *Quarterly Review,* which, albeit inspired by Owen, exhibited "preposterous incapacity" in dealing with elementary biology.

Edward Clodd, *Thomas Henry Huxley* (New York: Dodd, Mead, 1902), 106–7.

Secondary Literature: Joseph McCabe, *Edward Clodd: A Memoir* (London: John Lane, 1932); Philip J. Waller, *Writers, Readers, and Reputations: Literary Life in London, 1870–1918* (Oxford: Oxford University Press, 2006).

FRANCES POWER COBBE (1822–1904)
IRISH SOCIAL REFORMER AND ANIMAL-RIGHTS ACTIVIST

That the doctrine of the descent of man from the lower animals, of which Mr. Darwin has been the great teacher, should be looked on as well-nigh impious by men not mentally chained to the Hebrew cosmogony has always appeared to me surprising.

Mr. Darwin's theories have hitherto chiefly invaded the precincts of traditional theology. We have now to regard him as crowning the edifice of utilitarian ethics by certain doctrines respecting the nature and origin of the moral sense, which, if permanently allowed to rest upon it, will, we fear, go far to crush the idea of duty level with the least hallowed of natural instincts. It is needless to say that Mr. Darwin puts forth his views on this, as on all other topics, with perfect moderation and simplicity, and that the reader of his book has no difficulty whatever in comprehending the full bearing of the facts he cites, and the conclusions he draws from them.

In the [*Descent of Man*], he has followed out to their results certain hints given in his *Origin of Species* and *Animals under Domestication,* and has, as it seems, given Mr. Herbert Spencer's* abstract view of the origin of the moral sense its concrete application. Mr. Spencer broached the doctrine that our moral sense is nothing but the "experiences of utility organized and consolidated through all past generations." Mr. Darwin has afforded a sketch of how such experiences of utility, beginning in the ape, might (as he thinks) consolidate into the virtue of a saint; and adds some important and quite harmonious remarks, tending to show that the virtue so learned is somewhat accidental, and might perhaps have been what we now call vice.

Frances Power Cobbe, *Darwinism in Morals* (1872; reprint, Boston: George H. Ellis, 1883), 6, 9.

With the great naturalist who has revolutionized modern science I had rather frequent intercourse till the same sad barrier of a great difference of moral opinion arose between us. Mr. Charles Darwin's brother-in-law, Mr. Hensleigh Wedgwood, was, for a time tenant here at Hengwrt [*sic*]; and afterwards took a house named Caer-Deon in this neighborhood, where Mr. and Mrs. Charles Darwin and their boys also spent part of the summer. As it chanced, we also took a cottage that summer close by Caer-Deon and naturally saw our neighbors daily. I had known Mr. Darwin previously, in London, and had also met his most amiable brother, Mr. Erasmus Darwin, at the house of my kind old friend Mrs. Reid, the foundress of Bedford Square College. The first thing we heard concerning the illustrious arrivals was the report, that one of the sons had had "a fall off a Philosopher"; a word substituted by the ingenious Welsh mind for "velocipede" (as bicycles were then called) under an easily understood confusion between the rider and the machine he rode!

Next, the Welsh parson of the little church close by, having fondly calculated that Mr. Darwin would certainly hasten to attend his services, prepared for him a sermon which should slay this scientific Goliath and spread dismay through the ranks of the skeptical host. He told his congregation that there were in these days persons, puffed up by science, falsely-so-called, and deluded by the pride of reason, who had actually been so audacious as to question the story of the six days of Creation as detailed in Sacred Scripture. But let them note how idle were these skeptical questionings! Did they not see that the events recorded happened before there was any man existing to record them, and that, therefore, Moses must have learned them from God himself, since there was no one else to tell him?

Alas! the philosopher, I fear, never went to be converted (as he surely must have been) by this ingenious Welsh parson, and we were for a long time merry over his logic. Mr. Darwin was never in good health, I believe, after his Beagle experience of sea sickness, and he was glad to use a peaceful and beautiful old pony of my friend's, yclept† Geraint, which she placed at his disposal. His gentleness to this beast and incessant efforts to keep off the flies from his head, and his fondness for his dog Polly (concerning whose cleverness and breeding he indulged in delusions which Matthew Arnold's* better dog-lore would have swiftly dissipated), were very pleasing traits in his character.

In writing at this time to a friend I said:

"I am glad you like [John Stuart] Mill's* book. Mr. Charles Darwin, with whom I am enchanted, is greatly excited about it, but says that Mill could learn some things from physical science; and that it is in the struggle for existence and (especially) for the possession of women that men acquire their vigor and courage. Also he intensely agrees with what I say in my review of Mill about inherited qualities being more important than education, on which alone Mill insists. All this the philosopher told me yesterday, standing on a path 60 feet above me and carrying on an animated dialogue from our respective standpoints."

Mr. Darwin was walking on the footpath down from Caer-Deon among the purple heather which clothes our mountains so royally; and impenetrable brambles lay between him above and me on the road below; so we exchanged our remarks at the top of our voices, being too eager to think of the absurdity of the situation, till my friend coming along the road heard with amazement words flying in the air which assuredly those "valleys and rocks never heard" before, or since. When we drive past that spot, as we often do now, we sigh as we look at the "Philosopher's Path," and wish (o, *how* one wishes!) that he could come back and tell us what he has learned since!

At this time Mr. Darwin was writing his *Descent of Man,* and he told me that he was going to introduce some new view of the nature of the Moral Sense. I said: "Of course you have studied Kant's *Grundlegung der Sitten?*"‡ No, he had not read Kant, and did not care to do so. I ventured to urge him to study him, and observed that one could hardly see one's way in ethical speculation without some understanding of his philosophy. My own knowledge of it was too imperfect to talk of it to him, but I could lend him a very good translation. He declined my book, but I nevertheless packed it up with the next parcel I sent him.

On returning the volume he wrote to me:

"It was very good of you to send me *nolens uolens* Kant, together with the other book. I have been extremely glad to look through the former. It has interested me much to see how differently two men may look at the same points. Though I fully feel how presumptuous it sounds to put myself even for a moment in the same bracket with Kant—the one man a great philosopher looking exclusively into his own mind, the other a degraded wretch looking from the outside through apes and savages at the moral sense of mankind."

In a letter to a friend (Nov., 1869) I say:

"We lunched with Mr. Charles Darwin at Mr. Erasmus D's house on Sunday. He told us that a German man of science, (I think Carl Vogt*), the other day gave a lecture, in which he treated the Mass as the last relic of that *Cannibalism* which gradually took to eating only the heart, or eyes of a man to acquire his courage. Whereupon the whole audience rose and cheered the lecturer enthusiastically! Mr. Darwin remarked how much more *decency* there was in speaking on such subjects in England."

This pleasant intercourse with an illustrious man was, like many other pleasant things, brought to a close for me in 1875 by the beginning of the Anti-vivisection crusade. Mr. Darwin eventually became the centre of an adoring clique of vivisectors who (as his Biography shows) plied him incessantly with encouragement to uphold their practice, till the deplorable spectacle was exhibited of a man who would not allow a fly to bite a pony's neck, standing forth before all Europe (in his celebrated letter to Prof. [Frithiof] Holmgren of Sweden) as the advocate of Vivisection.

Life of Frances Power Cobbe, as Told by Herself (London: Swann Sonnenschein, 1904), 485–88, 490–91. [†]Yclept, "named, called" (Old English). [‡]*Grundlegung zur Metaphysik der Sitten* (Fundamental Principles of the Metaphysics of Morals), of 1785, Kant's statement on moral philosophy.

Secondary Literature: Sally Mitchell, *Frances Power Cobbe: Victorian Feminist, Journalist, Reformer* (Charlottesville: University of Virginia Press, 2004).

FERDINAND COHN (1828–1898)
German Biologist

Our sojourn in England was nearing its end when we received an invitation from the great naturalist Charles Darwin; since we had heard that he had barely survived an illness and there were deaths in the family, we had hardly dared to introduce ourselves in person; now he himself sent an invitation to us in a most gracious twist, saying that we were not allowed to leave England until we had visited him in his home; with English precision the train station and the time of departure that we were supposed to use for our journey were stipulated for us. So we left London from Charing Cross Station, in the west of the city and, once we had crossed the Thames, we delighted in the splendid panoramic view from the Houses of Parliament up to the lofty cupola of St. Paul's Cathedral, enthroned on its hill; then we went through Southwark, where Shakespeare's Globe Theatre once stood. Finally the endless rows of houses with their yellow and green tiled roofs, over which rows of high quaint smokestacks arose, were left behind, making way for the open countryside.

The train, turning to the southeast, took us through Kent, the charming garden of England. The terrain became hilly; on the hills were newly laid out colonies of red cottages lined up one after the other with their dainty balconies and little blooming gardens in front, where the populace, always fleeing farther out of the metropolis, sought the fresh air. On one of these hills, peering forth out of old trees, stood Camden Place and Chislehurst, once the refuge and death place of the overthrown Napoleon [III]. The scenery became ever more charming as the train climbed up the height of the tableland and soon arrived at Orpington. Here we left the train; there was already an open buggy awaiting us, which Charles Darwin had sent. With precautionary hospitality, plaid rugs, coats and umbrellas were supplied for us. The countryside was truly English. The fields were meticulously groomed, bordered by hawthorn hedges; in the distance were countless villages, hidden behind tall trees, over which only the church spires jutted out. The splendid road was bordered on both sides with an avenue of proud elm trees, such as one cannot find on the Continent; heaps of flint, with which the whole area was strewn, point to the fact that we were in the English chalk region, which slopes steeply to the Sea in the white cliffs of the South coast. Innumerable flights of pigeons flew around us; they reminded us already of that great friend of the pigeons, who was able to draw such far-reaching conclusions from the observations he made of their variety; therefore they seemed to us no less venerable than those from St. Mark's in Venice. Thus the delightful drive lasted one hour in the fresh fall air, past many estates with stately manors. Then we arrived at the top of the dune, which had given the village of Down its name. Here Charles Darwin had taken up residence for forty years in a modest country estate, in order to, like those wise men of antiquity, in solitary sequestration, devote his work filled life solely and alone to the investigation of the truth, far from the noise of the city. A gateway in a garden wall opened off the village street. The carriage drove into a courtyard bedecked with equipment and stopped to the left at the doorway to the residence. Here a tall man met us; he wore the comfortable jacket of a country gentleman; it was Charles Darwin himself, who greeted us graciously. His head was reminiscent of that of Socrates, yet it was noticeably lengthened; truly Socratic was the unusually broad and high forehead, which continued up to the high arch of his bald skull, surrounded on both sides on the temples by gray hair. Especially distinctive were the broad protruding, flush arches of the eyebrows, which shaded his honest looking eyes like a protective roof. His mouth and nose exhibited strong features; a full gray beard covered the entire lower part of his face, which, thinner on the cheeks, was rather long on the strong prognathous chin. If the less features of his face were not pretty enough to remark, they animated themselves during speaking in a marvelous fashion; when listening those same features exposed a soft, almost sad expression, the like of which I had once seen on Alexander von Humboldt.[†]

The house in which Darwin lived resembled the numerous country manors, such as had been scattered across all of England for hundreds of years. Up a few stairs from the front door opened up a hall, in which hats and coats were left and, in the background, steps led to the upper floor. On the ground floor lay a large room—half parlor, half library—into which we were first led; here we were greeted by Mrs. Darwin, a friendly old gentlewoman in mourning, who took a seat alongside her daughter and niece, across from the fireplace, on a *causeuse* [love seat] surrounded by wingbacks and chairs.

The most beautiful room decorations were the library of sumptuously-bound books that ran all around the walls. This room opened on the veranda, upon which rocking chairs and armchairs were set up. Climbing plants twined around the pillars and roof of the veranda; they were the very vines on which Darwin had made his wonderful observations about twining and climbing plants. The window opened up on a pleasant view of the garden which, like that of all English country manors, was dominated by a single large lawn, where colorful borders of flowers and here and there a clump of trees interrupted the splendid, fragrant velvet green; the path, which was shaded on either side by tall trees, ran around the lawn. Seemingly without borders, the garden petered out into the park-like open countryside, which was inhabited by grazing cattle.

Next to the parlor was Darwin's study. In its center stood the desk from which came all his epoch-making works, which have influenced our overall worldview more strongly in the last quarter century than those of his contemporaries. One corner led to a round bay, which was segmented by windows on all sides, in order to let in light from all parts of the sky; this was taken up to a great extent by a work table, on which a Hartnack microscope (Paris and Potsdam) stood. Around it lay needles and scalpels, bells, glasses and vials of strange shapes, the whole ménage of an experimenter and naturalist. On the opposite side of the hall—the corridor, as one would say in our parts—was the dining room. Family portraits decorated its walls, painted in oil, held in stately frames. They pictured in part members of Darwin's family, including the extraordinary predecessor of our sage, Erasmus Darwin. Some portrayed the Wedgwood family, famous for their ceramics, to which his wife belongs. The entire installation bore witness to genuine prosperity. The meal too was perhaps the finest that we enjoyed in England, especially the table wine, a fine red Bordeaux, unlike the customary Claret spoiled by the addition of spirits. Darwin didn't fail to call to our attention that he had served us roast venison, a great rarity in England.

But if the meal, like those classical symphonies, has stayed in never to be forgotten memory, it was the conversations with the great scientist that bestowed a blessing without compare.

Darwin was a man of compelling kindness. His entire character was imbued with a sincere sense of verity, with meticulous scrupulousness. He was reluctant to pass off rash conjectures as proven facts, but that, which after thorough scrutiny emerges as true, he declared without reservations or trimming. Ungrudging and unselfish was the recognition that he proffered with respect to the research of others, even that of novices, even when they came out in direct contradiction to his pet ideas; also stirring was the humility with which he judged his own accomplishments, as if he had no idea of their phenomenal significance. He showed the warmest admiration for German science; for even though he had only very imperfect control of our language, he nevertheless studied with dogged persistence each new piece of German scientific literature, and his writings demonstrated a thorough and comfortable knowledge of that body of work, a feat of which likely only a very few of our colleagues could boast.

In the beautiful language of the Greeks, philosophy means the love of wisdom. By wisdom, however, they meant not merely the knowledge of the truth, but also the pursuit of the good, and the cultivation of a noble character, in all circumstances equally tried and tested. Charles Darwin has been the first philosopher of our time in the realm of science. For this the evidence lies exposed to all the world in his writings, of which almost every year for more than two decades a sturdy octavo in a green canvas cover has always surprised and delighted the friends of science anew. But he was also the one of the most perfect personages, who, like Socrates, Spinoza and Goethe, shine before mankind as ideal role models, and this until now only those could know who had the luck to be able to encounter Darwin in conversation in person or even in correspondence.

Ferdinand Cohn, "Ein Besuch bei Charles Darwin [1876]," *Breslauer Zeitung,* 23 April 1882, trans. Allison Palm. †Alexander von Humboldt (1769–1859), German naturalist whose writings on biogeography influenced Darwin.

Secondary Literature: Gerhart Drewsa, "The Roots of Microbiology and the Influence of Ferdinand Cohn in Microbiology of the 19th Century," *FEMS Microbiology Reviews* 24 (2000): 225–49.

EDWIN GRANT CONKLIN (1863–1952)
AMERICAN BIOLOGIST AND EUGENICIST

In order to estimate the significance of any biological process or tendency, it is necessary to get the evolutionary point of view. Only one who can, like Darwin, see such processes in four dimensions—length, breadth, depth, and *duration*—only such a one can estimate the probable effects of our past and present immigration upon the future of America and of the world; and even a Darwin might well have declined to hazard any prophecy regarding a situation in which the factors involved are so numerous and complex. [. . .] There was once the supreme chance of breeding here the finest race and nation in the whole history of mankind. Here was an almost unoccupied continent to which came at first only the bravest, most hardy, most independent peoples of western Europe. The English Puritans thought that "God had sifted the whole nation that he might send choice grain into the wilderness," and this sifting gave us the sturdy, liberty-loving, New England stock of our early history. If God had only continued to sift the nations for our benefit, or if our fathers had exercised only reasonable caution in sifting out those who were to form the American nation, we might have had here only the choicest blood and the highest types of culture of all lands, we might have replaced the slow and wasteful methods of natural selection by intelligent selection and thus have enormously advanced and hastened human evolution. That chance has gone forever.

Edwin Grant Conklin, "Some Biological Aspects of Immigration," *Scribner's Magazine,* January 1921, 352–59, on 356–58.

Secondary Literature: Garland E. Allen, "Edwin Grant Conklin," *Dictionary of Scientific Biography*, ed. Charles Coulston Gillispie, 16 vols. (New York: Scribner's, 1970–80), 3:389–91.

MONCURE D. CONWAY (1832–1907)
American Clergyman

Freethought is a kind of applied science. Charles Darwin, whom I used to know, regarded the damage done to dogmas by science as incidental and unintentional; the scientific men by their method of exactness, by their demand for the most thorough evidence, were unconsciously criticizing the vague and untrustworthy evidence on which Christianity rested. I have known personally the leading scientific men in England and America in my time, and though their writings and lectures undermined orthodox dogmas, they were tender and cautious in their relations with individuals and their sentiments.

Moncure D. Conway, "Dogma and Science (1904)," in *Addresses and Reprints, 1850–1907* (Boston: Houghton Mifflin, 1909), 354–67, on 359–60.

The determined repudiation by Agassiz* of the discovery of Darwin caused something like dismay in scientific circles throughout Europe as well as in America. Concerning this I have some memories that may interest men of science. When I belonged to the class of Agassiz (1853–54), he repeatedly referred to the hypothesis of continuous development of species in a way which has suggested to me a possibility that he may have had some private information of what was to come from Charles Darwin. In his Introduction (1859) Darwin speaks of having submitted a sketch of his work to Sir Charles Lyell* and Sir Joseph Hooker*,—"the latter having seen my sketch of 1844." Either of these, or Darwin himself, might have consulted Agassiz. Most of us knew about such a theory only through the popular *Vestiges of Creation*, to which he paid little attention. [...] At any rate, during 1854 especially his mind was much occupied with the subject. I also remember well that during this time he often dwelt upon what he called the "ideal connection" between the different forms of life, describing with drawings the embryonic changes; in that progress no unbridged chasm after the dawn of organic life. At the end of every week a portion of the afternoon was given for our putting questions to Agassiz, the occasion often giving rise to earnest discussion. These repeatedly raised the theory of development in *The Vestiges of Creation*. [...]

There was at Concord a course of lectures every year, one of which was given by Agassiz. His coming was an important event. He was always a guest of the Emersons,† where the literary people of the village were able to meet him. On one such occasion I remember listening to a curious conversation between Agassiz and A. [Amos] Bronson Alcott*,—who lived and moved in a waking dream. After delighting Agassiz by repudiating the theory of the development of man from animals, he filled the professor with dismay by equally decrying the notion that God could ever have created ferocious and poisonous beasts. When Agassiz asked who could have created them, Alcott said they were the various forms of human sin. Man was the first being created. And the horrible creatures were originated by his lusts and animalisms. When Agassiz, bewildered, urged that geology proved that the animals existed before man, Alcott suggested that man might have originated them before his appearance in his present form. Agassiz having given a signal of distress, Emerson came to the rescue with some reconciling discourse on the development of life and thought, with which the professor had to be content, although there was a *soupçon* of Evolutionism in every word our host uttered. There was a good deal of suspicion in America that the refusal of Agassiz to accept Darwin's discovery was due to the influence of religious leaders in Boston, and particularly to that of his father-in-law, Thomas Gary, who had so freely devoted his wealth to the professor's researches. Some long intimacy with those families convinced me that there was no such influence exerted by the excellent Mr. Gary, but that it was the old Swiss pastor, his father, surviving in him. He had, indeed, departed far from the paternal creed; he repudiated all miracles at a time when Mr. Gary and other Unitarians upheld them tenaciously. He threw a bomb into the missionary camp by his assertion of racial diversity of origin. His utterances against Darwinism were evidently deistic, and had nothing whatever to do with any personal interest, except that he had a horror of being called an atheist. [...] [Henry Wadsworth] Longfellow* told me that Agassiz was entreating him to write a poem on the primeval world.

Moncure Daniel Conway, *Autobiography: Memories and Experiences,* 2 vols. (Boston: Houghton Mifflin, 1905), 1:151–53. [†]That is, the Ralph Waldo Emersons.

The subjects of my [lectures at Sheffield, 1867] were advertised, and one on "The Pre-Darwinite and the Post-Darwinite World" attracted Darwin. I was told that he listened to it; but he rarely came to London, and probably the discourse was reported to him. I soon after received an invitation to visit him at Down, his house near Bromley. I went to Bromley with the Wedgwoods. Hensleigh Wedgwood was a very interesting gentleman, but inclined to put some faith in "occultism." Mrs. Wedgwood told me anecdotes about her brother (Darwin), one of which is quaint. Darwin could never realize the world-wide impression made by his discovery, nor his own fame. [William] Gladstone*, then Prime Minister, being in the neighborhood of Down, had called. When he had gone Darwin said, "To think of such a great man coming to see me!" The other guests at Down, besides the Wedgwoods and myself, were my friends Charles [Eliot] Norton* and his sister of Cambridge, Massachusetts. A sister of Mrs. Charles Norton married a son of Darwin. Darwin was not in perfect health, and his wife and daughters took care that he should retire early. My opportunity for conversing with him came next day. In the soft spring morning before sunrise I looked out of my bedroom window and saw Darwin in his garden, inspecting his flowers. His grey head was bent to each bush as if bidding it good-morning. And what a head! All that the phrenologists had written was feeble compared with a look at that big head with its wonderful dome, and the lobes above each luminous eye. All the forms of organic nature had contributed something to represent them visibly in the constitution of the head able to interpret them. I was soon with Darwin in the garden, which was in floral glory. He expressed satisfaction that I had been able to derive from evolution the hopeful religion set forth in my discourse, but I remember that he did not express agreement with it. He spoke pleasantly of W. J. Fox M. P., my predecessor at South Place (whom he well knew), and asked me about [Ralph Waldo] Emerson*, whose writings interested him. But he had not been aware of the extent of Emerson's poetic anticipations of his discovery many years before it was published. [. . .] When we were presently at breakfast the post came,—a pile of letters which the daughters began to open, separating from those of friends the large number from strangers in all parts of the world. A few of these were read aloud for our amusement, letters from crude people reporting to Darwin observations which they believed important. One American farmer wrote about the marvelous intelligence of his dog, who always knew when he was about to take a walk, dancing about so soon as he touched his cane. One had some commonplaces to tell about his new variety of beans, another something about his pigeons. The rest of us laughed, but Darwin said, "Let them all be pleasantly answered. It is something to have people observing the things in their gardens and barnyards." Adjoining the house was the conservatory in which Darwin carried on his experiments. Into this he invited me, overcoming my hesitation by saying that he particularly desired it. I felt indeed that it was right, because I was minister of the chief rationalistic congregation and was endeavoring to transfer the religious sentiment from a supernatural to a scientific basis. He took pains to show me everything. There was the enclosure in which he and Sir John Lubbock*, who resided near him, conducted their experiments with ants. But Darwin was at that time chiefly occupied with the earthworm, his volume on which impresses me as next to his *Origin of Species* in value.

Darwin and Emerson died at nearly the same time (April 20 and April 27, 1882). The relation of these two minds to each other and to their time is striking. In the year (1836) when Darwin abandoned theology to study nature, Emerson, having also abandoned theology, published his first book, *Nature,* whose theme is Evolution. It was a notable circumstance that on the death of these two men who have done away with supernaturalism, no voice of *odium theologicum* broke the homage of England and America. The scene in Westminster Abbey at the burial of Darwin was impressive. From the chapel of St. Faith the body of the great man was borne by the procession along the remote cloisters. We who had long been in our appointed seats in the Abbey presently heard a faint melodious strain; nearer the dirge of the invisible choir approached; and when at length the great door of the Abbey opened, and the choristers appeared, and the coffin laden with wreaths from all parts of Europe, a stir of emotion passed through the waiting company. There were following that coffin more than a hundred of the first men in England and some from other countries. On many faces the grief was visible. Huxley*, [John] Tyndall*, Francis Galton*, Sir John Lubbock*, Sir Joseph Hooker*, could with difficulty control their grief. It was dark in the Abbey and the lights but feebly struggled with the gloom. There was something almost spectral in the slow moving of the procession with noiseless tread. Around in every direction the throng of marble statues were discernible, as if a cloud of witnesses gathered to receive the new-comer in their Valhalla, But it was an earthly Valhalla. The darkness of the Abbey, only made visible by occasional lamps, might have been regarded by saints of the still radiant windows as emblematic of the curtain drawn by knowledge beyond the grave. To me the gloom deepened when the service thanked God for removing such a man out of this wicked world, but lifted a little when the white-robed choristers gathered around the three graves—those of Newton, [John] Herschel*, and Darwin—and sang a new anthem, "Happy is the man that findeth wisdom!" Amid the universal homage to Darwin one adverse sentiment is widely noted and rebuked. *L'Univers,* the Roman Catholic organ in Paris, said, "When hypothesis tends to nothing less than the destruction of faith, the shutting out of God from the heart of man, and the diffusion of the filthy leprosy of Materialism, the savant who invents and propagates them is either a criminal or a fool. *Voilà ce que nous avons a dire du Darwin des singes.*"

Moncure Daniel Conway, *Autobiography: Memories and Experiences,* 2 vols. (Boston: Houghton Mifflin, 1905), 2:356–64.

Secondary Literature: John D'Entremont, *Moncure Conway, the American Years, 1832–1865* (New York: Oxford University Press, 1987).

CALVIN COOLIDGE (1872–1933)
AMERICAN PRESIDENT

I see [Oliver Wendell] Holmes [Sr.]* is dead, the Autocrat of the Breakfast table on whom the years sat so lightly and who had only just declared that he was 85 years young. No one but [William] Gladstone* is left of those great men that were born in 1809. Darwin is gone, the great expounder of evolution, a scientist equal to Newton. Our own [Abraham] Lincoln* finished his life's work when he struck the shackles from four millions of slaves and saw the surrender of General Lee.

Coolidge to his father, 2 November 1894, in *Your Son, Calvin Coolidge: A Selection of Letters from Calvin Coolidge to his Father,* ed. Edward Connery Lathem (Montpellier: Vermont Historical Society, 1968), 61.

Secondary Literature: Robert Sobel, *Coolidge: An American Enigma* (Washington, DC: Regnery, 1998).

EDWARD DRINKER COPE (1840–1897)
AMERICAN PALEONTOLOGIST

In endeavoring to assign a cause for the existence of the peculiar structures which define the divisions among animals, Messrs. Wallace* and Darwin have proposed the now well-known law of natural selection. This states, that, inasmuch as slight variations appear continually in all species, it is evident that some will be more beneficial to the animal than others, in its exertions to supply itself with food, protect itself from enemies, the weather, etc. It then asserts that those whose peculiarities are beneficial will excel those less favored, in the successful use of their powers, and hence will live better, grow better, and increase more rapidly. That by the force of numbers, if not by direct conflict, they will ultimately supersede the weaker and destroy or drive them away.

Then, as there are many fields of action and possibilities of obtaining support in the world, that the weaker will first be driven to adopt such of these as their peculiarities may adapt them for, or not exclude them from. Thus all the positions in the world's economy are filled and the surplus destroyed. This is styled by Spencer* the "survival of the fittest"; an expression both comprehensive and exact. This doctrine is no doubt a true one, and has regulated the preservation of the variations of species, and assigned them their locations in the economy of nature. It was natural that this great law should have been brought out by such men as Darwin and Wallace, who are by nature much more of observers of life in the field, or out-door physiologists, than they are (or were) anatomists and embryologists. Their writings in their chosen field of the mutual relations of living beings in their search and struggle for means of existence are admirable, and almost unique, especially some of those of Darwin. It is to be observed, however, that they both (especially Darwin) start with the variations observed. This is assumed at the outset, and necessarily so, for "selection" requires alternatives, and these are the product of variation. Great obscurity has arisen from the supposition that natural selection can originate anything, and the obscurity has not been lessened by the assertion often made that these variations are due to inheritance! What is inheritance but repetition of characters possessed by some (no matter what) ancestor; and if so, where did that ancestor obtain the peculiarity? The origin of variation is thus only thrown upon an earlier period.

Edward Drinker Cope, *The Origin of the Fittest* (New York: Appleton, 1887), 14–15.

Secondary Literature: Jane P. Davidson, *The Bone Sharp: The Life of Edward Drinker Cope* (Philadelphia: Academy of Natural Sciences, 1997); David Rains Wallace, *The Bonehunters' Revenge: Dinosaurs, Greed, and the Greatest Scientific Feud of the Gilded Age* (New York: Houghton Mifflin Harcourt, 1999).

MANDELL CREIGHTON (1843–1901)
ENGLISH HISTORIAN AND PRELATE

You have the advantage of me in reading Darwin: I am afraid I don't take sufficient interest in the subject of his speculations. I gather from scientific men that his view cannot ever claim to be more than a hypothesis, and they all estimate differently the amount of probability they attach to it; and no one has yet told me of any course of inquiry which could be adopted to convert the hypothesis into an ascertained fact. Consequently the whole matter seems to me to be very ingenious and amusing, but I have not time for it, and would rather read some Italian history.

Creighton to unidentified recipient, 1 May 1871, in *Life and Letters of Mandell Creighton*, ed. Louise Creighton, 2 vols. (London: Longmans, Green, 1904), 1:93.

I quite agree with you about evolution. It is quite established by quiet acceptance: the thing is to see what it comes to, how much it explains and how much it does not. Has it ever struck you that evolution has been the working theory of historians long before Darwin examined it in reference to species? Hegel's *Philosophie der Geschichte* contained its metaphysical basis, and Ranke's *Weltgeschichtliche Bewegung* set forth the "survival of the fittest" in human affairs. I hope your book will meet with the attention it deserves.

Creighton to William Samuel Lilly, 5 August 1889, in *Life and Letters of Mandell Creighton,* ed. Louise Creighton, 2 vols. (London: Longmans, Green, 1904), 1: 410–11.

Secondary Literature: Owen Chadwick, *The Victorian Church,* 2 vols. (New York: Oxford University Press, 1966–70).

JOÃO CRUZ E SOUSA (1861–1898)
Brazilian Poet

You come straight out of Darwin, from the common ancestral form of organized beings: I can clearly see your cranial protuberances—just like those of the orangutan—your lustful body language, the predatory, animalistic air of the ape.

João Cruz e Sousa, "Psicologia do feio," in *Obra completa,* ed. Alexei Bueno (Rio de Janeiro: Nova Aguilar, 1995), 473, trans. TFG.

It is the Century of revolt, of high transformism,
of Darwin, of Littré,[†] of Spencer[*], of Lafitte—
To speak, to make laws, is blood-drenched nihilism
that pays its dividend in dynamite.

João Cruz e Sousa, "À Revolta," in *Obra completa,* ed. Alexei Bueno (Rio de Janeiro: Nova Aguilar, 1995), 235, trans. TFG. [†]Émile Littré (1801–81), French positivist philosopher and lexicographer.

Secondary Literature: David T. Haberly, *Three Sad Races: Racial Identity and Natural Consciousness in Brazilian Literature* (Cambridge: Cambridge University Press, 1983).

JAMES DWIGHT DANA (1813–1895)
American Geologist

Under such a system of evolution—evolution by regional progress—the causes of variation mentioned by Darwin are all real causes. But they act directly, after the Lamarckian method, without dependence for success on the principle of natural selection. Use and disuse, labor, strife, physical changes or conditions, and organic influences act as such and have their direct effects. The plants that migrated in the Tertiary from the arctic regions southward over Japan and North America, and became new species on the way, simply changed. That is the sum of knowledge on the subject. [...]

The survival of the fittest is a fact; and the fact accounts in part for the geographical distribution of the races of men now existing and still in progress; but not the existence of the fittest, or for the power that has determined survival.

It is of no avail to speak of *chance* variation. The use of the word *chance* indicates personal ignorance. *Chance* has no place in nature's laws and can have none in nature-science.

Whatever the results of further research, we may feel assured, in accord with Wallace*, who shares with Darwin in the authorship of the theory of Natural Selection, that the intervention of a Power above Nature was at the basis of Man's development. Believing that Nature exists through the will and ever-acting power of the Divine Being [...] or, in the words nearly of Wallace, that the whole Universe is not merely dependent on, but actually is, the Will of one Supreme Intelligence, Nature, with Man as its culminant species, is no longer a mystery.

James D. Dana, *Manual of Geology*, 4th ed. (New York: American Book, 1896), 1034, 1046.

Secondary Literature: Daniel Coit Gilman, *The Life of James Dwight Dana* (New York: Harper, 1899); Michael Laurent Prendergast, *James Dwight Dana: The Life and Thought of an American Scientist* (Ann Arbor: University Microfilms, 1981).

RICHARD HENRY DANA (1815–1882)
AMERICAN AUTHOR AND LAWYER

In philosophy, he studied metaphysics and moral and intellectual philosophy in college, and planned at one time to pursue these studies in a postgraduate course with Professor James Marsh of Burlington,[†] so great was his interest in them. Besides reading those philosophical works I have previously mentioned, he in later days discussed Spencer*, [Auguste] Comte, Darwin, [John] Tyndall*, Huxley*, [Ralph Waldo] Emerson*, and some of the modern German philosophers, with his father and brother Edmund, who, with their leisure, had time to read in full what he, with his quick mind, took in only talks or from essays, short passages and criticism.

[Dana:] "While in college I took several courses in philosophy. [...] I was surprised, in talking with my father, to find how much he knew of what I thought was beyond his reading." Later in life, he became more tolerant of some of the new philosophical ideas as they became better understood, and accepted them. Mr. Dana did not, like the late Philip Henry Gosse, F.R.S., in his *Omphalos,* believe that the earth, with its fossils and glacial marks, was created just as it is by one catastrophic act in a day of twenty-four hours.[‡] He admitted long periods of development and evolution, but he believed it was evolution by successive creations, not by natural selection alone. Whatever the ultimate conclusion, there are at least sudden developments, now called "mutations," which are preserved, modified or lost by natural selection, but which are not caused by it.

As to Mr. Dana's attitude towards Darwinism, we must remember that Agassiz* had not accepted that when he died, and Mr. Dana survived Agassiz less than eight years. [...] As to not seeking out Darwin in 1856, Mr. Dana would not have had sufficient sympathy with Darwin's views to look him up, even had he been visiting England later. But, though somewhat known among scientists, Darwin had not made his great popular reputation in 1856.

Richard Henry Dana Jr., *Speeches in Stirring Times and Letters to a Son,* ed. Richard Henry Dana 3rd (Boston: Houghton Mifflin, 1910), 19–20, 23–24, 28. †James Marsh (1794–1842), an influential theologian and educator. ‡Gosse (1810–1888) attempted to reconcile biblical chronology with that of Charles Lyell in a famous book titled *Omphalos* (1857).

Secondary Literature: Samuel Shapiro, *Richard Henry Dana* (East Lansing: Michigan State University Press, 1961).

NIKOLAI DANILEVSKY (1822–1885)
RUSSIAN NATURALIST

Darwinism was a purely English doctrine, with all the particularities of orientation of the English mind, and all the qualities of the English spirit. Practical use and competitive struggle—here are two characteristics [...] that give direction to English life and also to English science. On usefulness and utilitarianism is founded Benthamite ethics, and essentially Spencer's* also; on the war of all against all, now termed the struggle for existence—Hobbes's theory of politics; on competition—the economic theory of Adam Smith and all of that primarily English science, political economy. Malthus applied the very same principle to the problem of population. [...] Darwin extended both Malthus's partial theory and the general theory of the political economists to the organic world.

N. Ia. Danilevskii, *Darvinizm: Kriticheskoe issledovanie,* 2 vols. (St. Petersburg: Komarov, 1885), 1:478, trans. and quoted in *Darwin without Malthus: The Struggle for Existence of Russian Evolutionary Thought,* by Daniel P. Todes (New York: Oxford University Press, 1989), 41–42.

Danilevsky's *Darwinism,* as stated by the author, is intended for educated persons without special training in biology. In this category of works, devoted primarily to the diffusion of scientific knowledge, particularly important is the scientific accuracy of presented scientific material. [...] In my judgment, Danilevsky's work has not obeyed this rule; throughout the work the foundations of Darwin's theory have been presented incorrectly. No other person has made the social influence of Darwin's work appear so pernicious and destructive, and none has been so distraught and bitter about it. Spreading such views about the significance of Darwin's work—views that contradict the judgment of all experts without exception—is, in my opinion, most regrettable, undesirable, and harmful.

A. S. Famintsyn, "N. Ia. Danilevskii i darvinizm," *Vestnik Evropy*, no. 2 (1889): 616–43, on 643, quoted in *Darwin in Russian Thought*, by Alexander Vucinich (Berkeley and Los Angeles: University of California Press, 1988), 139.

The polemic [evoked by Danilevsky's *Darwinism*] mirrored the full field of our recent cultural history. It expressed the oppressive atmosphere of the eighties, when the triumphant forces of reaction thought it possible to make a decisive break with the legacy of the recent past and, acting on behalf of a primordial, national-Byzantine principle, to declare war on the whole of Western culture. Danilevsky's campaign against Darwinism was one of the characteristic episodes of that struggle. He received help from all reactionary spheres. Ministers, influential St. Petersburg circles, obliging capitalists (without whose help Danilevsky's thick volumes would not have been published in the first place), literature [. . .] leading dailies, philosophers, and official science [. . .] all were on the side of Danilevsky.

Kliment Timiriazev, *Sochineniia*, 10 vols. (Moscow, 1937–40), 7:28; quoted in *Darwin in Russian Thought*, by Alexander Vucinich (Berkeley and Los Angeles: University of California Press, 1988), 150.

Secondary Literature: Alexander Vucinich, *Darwin in Russian Thought* (Berkeley and Los Angeles: University of California Press, 1988).

JOHN WILLIAM DAWSON (1820–1899)
Canadian Geologist and Paleobotanist

These calculations would give us, say, eighty-six millions of years since the earth began to have a solid crust, which would, like Lord Kelvin's earlier estimate, give us nearly fifty millions of years for the geological time since the introduction of life. [. . .] This reduced estimate of geological time would still give scope enough for the distribution of animals and plants, but it will scarcely give that required by certain prevalent theories of evolution. When Darwin says, "If the theory (of natural selection) be true, it is indisputable that before the lowest Cambrian stratum was deposited long periods elapsed, as long as, or probably far longer than, the whole interval from the Cambrian to the present day," he makes a demand which geology cannot supply; for independently of our ignorance of any formations or fossils, except those included in the Archaean, to represent this vast succession of life, the time required would push us back into a molten state of the planet. This difficulty is akin to that which meets us with reference to the introduction of many and highly specialized mammals in the Eocene, or of the forests of modern type in the Cretaceous. To account for the origin of these by slow and gradual evolution requires us to push these forms of life so far back into formations which afford no trace of them, but, on the contrary, contain other creatures that appear to be exclusive of them, that our faith in the theory fails. The only theory of evolution which seems to meet this difficulty is that advanced by [St. George Jackson] Mivart*, [Joseph] Leconte*, and Saporta,[†] of "critical periods," or periods of rapid introduction of new species alternating with others of comparative inaction. This would much better accord with the apparently rapid introduction of many new forms of life over wide regions at the same period. It would also approach somewhat near, in its manner of stating the problem to be solved, to the theory of "creation by law" as held by the Duke of Argyll*, or to what may be regarded as "mediate creation," proceeding in a regular and definite manner, but under laws and forces as yet very imperfectly known, throughout geological time. It seems singular, in view of the facts of palaeontology, that evolutionists of the Darwinian school are so wedded to the idea of one introduction only of each form of life, and its subsequent division by variation into different species, as it progressively spreads itself over the globe, or is subjected to different external conditions.

John William Dawson, *Some Salient Points in the Science of the Earth* (New York: Harper, 1894), 417–18. [†]Gaston Saporta (1823–95), French paleontologist.

Secondary Literature: Susan Sheets-Pyenson, *John William Dawson: Faith, Hope, Science* (Montreal: McGill-Queens University Press, 1996).

HUGO DEVRIES (1848–1935)
Dutch Botanist

Before Darwin, little was known concerning the phenomena of variability. The fact, that hardly two leaves on a tree were exactly the same, could not escape observation: small deviations of the same kind were met with everywhere, among individuals as well as among the organs of the same plant. Larger aberrations, spoken of as monstrosities, were for a long time regarded as lying outside the range of ordinary phenomena. A special branch of inquiry, that of Teratology, was devoted to them, but it constituted a science by itself, sometimes connected with morphology, but having scarcely any bearing on the processes of evolution and heredity.

Darwin was the first to take a broad survey of the whole range of variations in the animal and vegetable kingdoms. His theory of Natural Selection is based on the fact of variability. In order that this foundation should be as strong as possible he collected all the facts, scattered in the literature of his time, and tried to arrange them in a scientific way. He succeeded in showing that variations may be grouped along a line of almost continuous gradations, beginning with simple differences in size and ending with monstrosities. He was struck by the fact that, as a rule, the smaller the deviations, the more frequently they appear, very abrupt breaks in characters being of rare occurrence. [...]

Darwin was led to regard small deviations as the source from which natural selection derives material upon which to act. But even these are not all of the same type, and Darwin was well aware of the fact.

It should here be pointed out that in order to be selected, a change must first have been produced. This proposition, which now seems self-evident, has, however, been a source of much difference of opinion among Darwin's followers. The opinion that natural selection produces changes in useful directions has prevailed for a long time. In other words, it was assumed that natural selection, by the simple means of singling out, could induce small and useful changes to increase and to reach any desired degree of deviation from the original type. In my opinion this view was never actually held by Darwin. It is in contradiction with the acknowledged aim of all his work—the explanation of the origin of species by means of natural forces and phenomena only. Natural selection acts as a sieve; it does not single out the best variations, but it simply destroys the larger number of those which are, from some cause or another, unfit for their present environment. In this way it keeps the strains up to the required standard, and, in special circumstances, may even improve them.

Hugo DeVries, "Variation," in *Essays in Commemoration of the Centenary of the Birth of Charles Darwin and of the Fiftieth Anniversary of the Publication of the Origin of Species,* ed. A. C. Seward (Cambridge: Cambridge University Press, 1909), 66–84.

Secondary Literature: Garland E. Allen, "Hugo de Vries and the Reception of the Mutation Theory," *Journal of the History of Biology* 2 (1969): 55–87.

JOHN DEWEY (1859–1952)
AMERICAN PHILOSOPHER

No wonder, then, that the publication of Darwin's book, a half century ago, precipitated a crisis. The true nature of the controversy is easily concealed from us, however, by the theological clamor that attended it. The vivid and popular features of the anti-Darwinian row tended to leave the impression that the issue was between science on one side and theology on the other. Such was not the case—the issue lay primarily within science itself, as Darwin himself early recognized. The theological outcry he discounted from the start, hardly noticing it save as it bore upon the "feelings of his female relatives." But for two decades before final publication he contemplated the possibility of being put down by his scientific peers as a fool or as crazy; and he set, as the measure of his success, the degree in which he should affect three men of science: Lyell* in geology, Hooker* in botany, and Huxley* in zoology.

Darwin was not, of course, the first to question the classic philosophy of nature and of knowledge. The beginnings of the revolution are in the physical sciences of the sixteenth and seventeenth centuries. When Galileo said: "It is my opinion that the earth is very noble and admirable by reason of so many and so different alterations and generations which are incessantly made therein," he expressed the changed temper that was coming over the world, the transfer of interest from the permanent to the changing. When Descartes said: "The nature of physical things is much more easily conceived when they are beheld coming gradually into existence, than when they are only considered as produced at once in a finished and perfect state," the modern world became self-conscious of the logic that was henceforth to control it, the logic of which Darwin's *Origin of Species* is the latest scientific achievement. Without the methods of Copernicus, Kepler, Galileo, and their successors in astronomy, physics, and chemistry, Darwin would have been helpless in the organic sciences. But prior to Darwin the impact of the new scientific method upon life, mind, and politics, had been arrested, because between these ideal or moral interests and the inorganic world intervened the kingdom of plants and animals. The gates of the garden of life were barred to the new ideas; and only through this garden was there access to mind and politics. The influence of Darwin upon philosophy resides in his having conquered the phenomena of life for the principle of transition, and thereby freed the new logic for application to mind and morals and life. When he said of species what Galileo had said of the earth, *e pur se muove,* he emancipated, once and for all, genetic and experimental ideas as an organon of asking questions and looking for explanations.

John Dewey, *The Influence of Darwinism on Philosophy* (New York: Henry Holt, 1910), 2, 7–9.

Secondary Literature: Jerome A. Popp, *Evolution's First Philosopher: John Dewey and the Continuity of Nature* (Albany: State University of New York Press, 2007); Richard J. Bernstein, "John Dewey," in *Encyclopedia of Philosophy,* ed. Paul Edwards, 8 vols. (New York: Macmillan, 1967), 2:380–85.

CHARLES DICKENS (1812–1870)
English Novelist

It is well for Mr. Charles Darwin, and a comfort to his friends, that he is living now, instead of having lived in the sixteenth century; it is even well that he is a British subject, and not a native of Austria, Naples, or Rome. Men have been kept for long years in durance, and even put to the rack or the stake, for the commission of offenses minor to the publication of ideas less in opposition to the notions held by the powers that be.

But we have come upon more tolerant times. If a man can calmly support his heresy by reasons, the heresy will be listened to; and, in the end, will either be received or refuted, or simply neglected and forgotten. Mr. Darwin also enjoys the benefit of the bygone heresies of previous heretics: one heresy prepares the way for, and weakens, the shock occasioned by another. Astronomical and geological innovations render possible the acceptance of doctrines that would have made people's hair stand on end three centuries ago. [...] Mr. Darwin is already supported by a small party of disciples and fellow-laborers, who put faith in his inspiration; while the great majority shrink back in alarm at the boldness of his conclusions, and at the illimitable lapse of time which it unfolds before their wondering and bewildered gaze.

If Mr. Darwin's theory be true, nothing can prevent its ultimate and general reception, however much pain and shock those to whom it is propounded for the first time. If it be merely a clever hypothesis, an ingenious hallucination, to which a very industrious and able man has devoted the greater and best part of his life, its failure will be nothing new in the history of science.

"Natural Selection," in *All Year Round* 7 (July 1860): 293–99, on 293, 299. As editor of the magazine, Dickens is presumed to have approved the contents of the article.

To five little stone lozenges, each about a foot and a half long, which were arranged in a neat row beside their grave, and were sacred to the memory of five little brothers of mine—who gave up trying to get a living exceedingly early in that universal struggle[†]—I am indebted for a belief I religiously entertained that they had all been born on their backs with their hands in their trousers-pockets, and had never taken them out in this state of existence.

Charles Dickens, *Great Expectations* (New York: Bantam, 2003), 2. [†]There seems to be a consensus in Dickens scholarship that the two words "universal struggle" in *Great Expectations* constitute the only unambiguous Darwinian phrase in all of Dickens's work.

Secondary Literature: Charles Dickens, *Great Expectations,* ed. Graham Law and Adrian James Pinnington (Peterborough, ON: Broadview, 1997), 505: "Probably referring to Darwin's theories of evolution by natural selection, recently summarized in *On the Origin of Species* [. . .] and discussed three times around this period in Dickens's family journal *All the Year Round* (June 2 & July 7 1860, & March 29 1861." Cf. *On the Origin of Species: A Facsimile of the First Edition* (Cambridge, MA: Harvard University Press, 1964), 62: "Nothing is easier than to admit in words the truth of the universal struggle for life, or more difficult at least I have found it so than constantly to bear this conclusion in mind."

EMILY DICKINSON (1830–1886)
AMERICAN POET

I'm forced to the Darwinian conclusion
That here's a masterpiece of evolution.
From the first skiff of sutured skins or bark
To the three-decker with its thundering guns,—
From Jason's classic junk, or Noah's ark,
To the grand steamship of five thousand tons,—
The thing developed: just as Man was once—
Well, not a monkey; that he never was—
But something less, evolved through Nature's laws.

Allah il Allah! great is Evolution,
And Darwin eminently is its Prophet!

.　.　.　.　.　.　.　.　.　.　.　.　.　.

All which I do most potently believe,
Taking large stock in Natural Selection.
But, gentlemen, I cannot quite conceive—
Since centuries of plotting and reflection
Have brought to pass the steamboat's last perfection
What power, without intelligence or plan,
Evolved the wonders of the World and Man.

[Emily Dickinson], *A Masque of Poets including Guy Vernon, a Novelette in Verse* (Boston: Roberts Brothers, 1878), 215–16.

Why the Thief ingredient accommodates all Sweetness Darwin does not tell us.

Dickinson to Elizabeth Holland, 1871, quoted in *Emily Dickinson and the Art of Belief,* by Roger Lundin (Grand Rapids, MI: Eerdmans, 2004), 210.

Secondary Literature: James R. Guthrie, "Darwinian Dickinson: The Scandalous Rise and Fall of the Common Clover," *Emily Dickinson Journal* 16, no. 1 (2007): 73–91; Brenda Wineapple, *White Heat: The Friendship of Emily Dickinson and Thomas Wentworth Higginson* (New York: Knopf, 2008).

BENJAMIN DISRAELI (1804–1881)
English Prime Minister

It was in reference to dissension within the Church of England that Mr. Disraeli made a remarkable speech at a meeting of the Oxford Diocesan Society for the endowment of small benefices. It would not be easy to decide what some of those who were present made of his peculiar declarations, and though it has been represented that they were intended to denote the policy and convictions of the Conservative party, it would perhaps have been difficult for any member of that party, who had himself deeply considered the position of the establishment, and the manner in which it might have to meet the dangers by which it was supposed to be assailed, to gather from the speech any practical suggestion for preventing or repulsing them, except by excommunication alike for the theories of Darwin and the speculations of the authors of *Essays and Reviews.*

"Instead of believing," said Mr. Disraeli, "that the age of faith has passed, when I observe what is passing around me, what is taking place in this country, and not only in this country but on the Continent, in other countries and in other hemispheres, instead of believing that the age of faith has passed I hold that the characteristic of the present age is a craving credulity. [. . .] I hold that the highest function of science is the interpretation of nature, and the interpretation of the highest nature is the highest science. What is the highest nature? Man is the highest nature. But I must say that when I compare the interpretations of the highest nature with the most advanced, the most fashionable and modern school of modern science; when I compare that with older teachings with which we are familiar, I am not prepared to say that the lecture-room is more scientific than the church. What is the question which is now placed before society with the glib assurance which to me is most astounding. That question is this, is man an ape or an angel? I, My lord, I am on the side of the angels. I repudiate with indignation and abhorrence those new-fangled theories. I believe they are foreign to the conscience of humanity; and I say more that, even in the strictest intellectual point of view, I believe the severest metaphysical analysis is opposed to such conclusions.

Benjamin Disraeli, speech at the Oxford Diocesan Conference, 25 November 1864, quoted in *William Ewart Gladstone and his Contemporaries,* by Thomas Archer, 4 vols. (London: Blackie & Son, 1883), 3:200–201.

Secondary Literature: T. A. Jenkins, *Disraeli and Victorian Conservatism* (New York: St. Martin's, 1996).

ANTON DOHRN (1840–1909)
German Zoologist

[1870]
I sent a telegram and received the most kind invitation to visit Darwin the following day, Monday September 26. One will understand that this day has left in me a long lasting impression, which it is truly superfluous to justify. At noon when I arrived at Down, Darwin was still out for a ride. Mrs. Darwin greeted me like an old acquaintance which made me believe that having had friendly relations with the Huxley* family for several years, meant that for Darwin too I was no longer someone unknown. At half past 12 Darwin entered the room in the company of another gentleman whom he was accompanying to the door. He gave me his hand in a friendly manner and asked me to excuse his keeping me waiting. He returned at once, after having bidden farewell to the other gentleman and said smiling:

"You will think me respectable: this gentleman was our clergyman!"— indicating that he was not yet a complete outcast and still had friendly relations with individual clergymen. We then moved to his private room. I must confess, Darwin's personal appearance surprised me very much. I had expected to find a sick-looking man; instead I saw before me a tall, strong, grey bearded stature, full of life and cheerfulness and heart-winning amiability. As may easily be understood the conversation started with the immense historical events that had just taken place, and Darwin showed the greatest sympathy with the turn of events that had caused the fall of Napoleon III and had so very severely punished French haughtiness and frivolity. [...]

Darwin took the most vivid interest in my Neapolitan plans, looking forward to seeing the [Marine Biology] Station well equipped and then conversed with me about his own scientific works.

During this conversation I asked him, "Excuse me, if I ask a perhaps very stupid question: how do you begin your studies at all?"

"I'll tell you: I begin always with *a priori* solutions. If anything happens to impress me, I have hundreds of hypotheses before I know the facts. I apply one after the other, till I find the one that covers the whole ground. But I am exceedingly careful and slow in printing."

We talked for an hour and a half. Then Mrs. Darwin entered and took him with her, because he should never talk longer than that. After having rested for half an hour he returned. We went for lunch. After that I took leave, not without having to promise to return as soon as possible.

Anton Dohrn, "Memories," autobiographical notes, 1895–1909, reproduced in Charles Darwin and Anton Dohrn, *Correspondence,* ed. Christine Groeben (Naples: Macchiaroli, 1982), 93–94.

Dohrn was here yesterday, and was remarking that no one stood higher in the public estimation of Germany than Lyell.

Darwin to [Joseph Dalton Hooker], 27 September 1870, in *More Letters of Charles Darwin,* ed. Francis Darwin, 2 vols. (New York: Appleton, 1903), 1:396.

Secondary Literature: Christiane Groeben, "Anton Dohrn—The Statesman of Darwinism," *Biological Bulletin* 168 (June 1985): 4–25.

IGNATIUS DONNELLY (1831–1901)
American Novelist and Politician

I could spend many hours, my dear brother, telling you of the splendor of this hotel, called The Darwin, in honor of the great English philosopher of the last century. It occupied an entire block from Fifth Avenue to Madison Avenue, and from Forty-sixth Street to Forty-Seventh. The whole structure consists of an infinite series of cunning adjustments, for the delight and gratification of the human creature.

I said to Max:

"What will those millions do to-morrow?"

"Starve," he said.

"What will they do next week?"

"Devour each other," he replied.

There was silence for a time.

"Will not civil government arise again out of this ruin?" I asked.

"Not for a long time," he replied. "Ignorance, passion, suspicion, brutality, criminality, will be the lions in the path. [...] After about three-fourths of the human family have died of hunger, or been killed, the remainder, constituting, by the law of the survival of the fittest, the most powerful and brutal, will find it necessary, for self-defense against each other, to form squads or gangs. The greatest fighter in each of these will become chief, as among all savages."

Ignatius Donnelly, *Caesar's Column: A Story of the Twentieth Century* (Chicago: F. J. Schulte, 1890; Forgotten Books, 2008), 7, 252–53.

Secondary Literature: Martin Ridge, *Ignatius Donnelly: The Portrait of a Politician* (Chicago: University of Chicago Press, 1962).

FYODOR DOSTOEVSKY (1821–1881)
Russian Novelist

[1864–73]

The [young female] nihilist explains herself in stupid terms, although completely in accordance with Darwin.

Fragment titled "The Struggle of Nihilism with Honesty," quoted in Michael R. Katz, "Dostoevsky and the Natural Sciences," *Dostoevsky Studies* 9 (1988): 63–76.

[1866]

He had tried expounding to him the system of Fourier and the Darwinian theory, but of late Pyotr Petrovich began to listen too sarcastically and even to be rude.

Fyodor Dostoevsky, *Crime and Punishment* (Ware, Hertfordshire, UK: Wordsworth, 2000), 308.

[1876]

By the way: remember the contemporary theories of Darwin and others concerning the descent of man from monkeys. Without engaging in any theories, Christ explicitly declares that in man, in addition to an animal world, there is also a spiritual world. And what of it? What difference does it make where man is descended from [. . .] , God still breathed the breath of life into him.

Dostoevsky to V. A. Alekseev, June 1876, quoted in Michael R. Katz, "Dostoevsky and the Natural Sciences," *Dostoevsky Studies* 9 (1988).

Secondary Literature: Joseph Frank, *Dostoevsky: The Mantle of the Prophet, 1871–1881* (Princeton, NJ: Princeton University Press, 2002).

ARTHUR CONAN DOYLE (1859–1930)
ENGLISH AUTHOR

"It was magnificent," [Sherlock Holmes] said, as he took his seat. "Do you remember what Darwin says about music? He claims that the power of producing and appreciating it existed among the human race long before the power of speech was arrived at. Perhaps that is why we are so subtly influenced by it. There are vague memories in our souls of those misty centuries when the world was in its childhood."

Arthur Conan Doyle, *A Study in Scarlet* (1887; reprint, Ware, Hertfordshire, UK: Wordsworth, 2001), 60.

Half an hour later I was seated in the newspaper office with a huge tome in front of me, which had been opened at the article, "[August] Weismann* versus Darwin," with the sub heading, "Spirited Protest at Vienna. Likely Proceedings." My scientific education having been somewhat neglected, I was unable to follow the whole argument, but it was evident that the English professor had handled his subject in a very aggressive fashion, and had thoroughly annoyed his Continental colleagues. "Protests," "Uproar," and "General appeal to the Chairman" were three of the first brackets which caught my eye. Most of the matter might have been written in Chinese for any definite meaning that it conveyed to my brain.

Arthur Conan Doyle, *The Lost World* (1912; reprint, Fairfield, IA: First World, 2004), 22.

Secondary Literature: E. J. Wagner, *The Science of Sherlock Holmes* (New York: John Wiley, 2007).

SIMON DUBNOW (1860–1941)
Russian Jewish Historian

At that time, progressive Jewish society was not overly concerned about such harbingers of [nationalist] reaction: it was still very much fascinated by the idea of emancipation that still prevailed among the nobler Russian intelligentsia. [...] Enthusiastic about a dignified human life among a free Russian people, young Jewish men streamed from the Pale of Settlement to the high schools and the universities that received them with open arms. By means of Russian literature, they drew near to the noblest European ideals. Chernishevsky, Dobrutov, [Dmitry] Pisarev*, [Henry Thomas] Buckle*, Darwin, [John Stuart] Mill*, Spencer*—the rulers of the mind of that generation—became the idols of the Jewish youth. Jewish minds, which but yesterday had been engrossed in the Talmud in the *khederim* [traditional elementary schools] and *yeshivot* [Talmudical academies], turned to new ideas of positivism, evolutionism and social reforms.

Simon Dubnov, *History of the Jews*, vol. 5, *From the Congress of Vienna to the Emergence of Hitler* (New York: Thomas Yoseloff, 1973), 342.

Secondary Literature: Simon Rabinovitch, "The Dawn of a New Diaspora: Simon Dubnov's Autonomism, from St. Petersburg to Berlin," *Yearbook of the Leo Baeck Institute* 50 (2005): 267–88; John Radzilowski, "Simon Dubnov," in *Encyclopedia of Historians and Historical Writing*, ed. Kelly Boyd, 2 vols. (London: Fitzroy Dearborn, 1999), 1:324–25.

EUGÈNE DUBOIS (1858–1940)
Dutch Anatomist and Paleontologist

I'm sorry to be so discouraging, but I think your chances of success are slight. Logic is all very well, but [. . .] well, let me just say that I think you have let yourself be carried away by that crazy book of Darwin's. There is little truth in it, you know, and much speculation. Evolution is not a fact, it is a theory. I regret, Dr. Dubois, that there is nothing I can do to help you. I suggest you forget all about the missing link and concentrate on your career at the university.

Secretary-General, Colonial Office, The Hague, quoted in "Prof. Dubois blikt," *De Telegraaf,* 1938, quoted in *The Man Who Found the Missing Link: Eugène Dubois and His Lifelong Quest to Prove Darwin Right,* by Pat Shipman (New York: Simon & Schuster, 2001), 67.

For some time now, the military surgeon Eug. Dubois busies himself with Paleontological investigations and, because of his predilection for such studies, works with exceptional diligence and assiduity. He is totally absorbed by his work, so to speak. As a firm Darwinist, he dreams of making a discovery which the great master of evolution will greet with joy. Namely, the discovery of the until recently missing link between the animal world and man.

Should this be taken amiss? I do not believe so. At present Darwinism is the backbone of the education of most high school graduates. The heavy facts that are brought up against Darwin's theory by the most competent authorities—these leave them cold. Examine their libraries and, ten to one, you will not find a single paper in which Darwin's theory is opposed. [. . .] It is old-fashioned to think differently, their teachers have told them.

[P. A. Daum?], "Palaeontologische onderzoeking op Java," *Bataviaasch Nieuwsblad,* 6 February 1893, quoted in *The Man Who Found the Missing Link: Eugène Dubois and His Lifelong Quest to Prove Darwin Right,* by Pat Shipman (New York: Simon & Schuster, 2001), 176.

Secondary Literature: Bert Theunissen, *Eugène Dubois and the Ape-Man from Java: The History of the First "Missing Link" and its Discoverer* (Dordrecht, Netherlands: Springer, 1988).

EMIL DU BOIS-REYMOND (1818–1896)
German Physiologist

ALBERT WIGAND (1821–1886)
German Botanist

Darwin *versus Gallani, and Teleology or Chance.* These are two pamphlets—one by [Emil] Du Bois Reymond, the other by Professor [Albert] Wigand—the latter a sharp criticism of the former. Du Bois-Reymond is a rabid Darwinian; Wigand, a moderate one. The former thinks that Darwin has given teleology its death-blow; the latter thinks that reason is compelled by its very nature to seek for *causae finales.* We certainly agree with him in this opinion. Indeed, the present opponents of teleology are unable to avoid teleology. They are constantly finding "purpose," "design" etc., as soon as they forget their *a priori* theories. Wigand makes a severe criticism of Du Bois-Reymond, and well he deserves it. Wigand advocates also the assumption of a causative first principle independent of nature and its laws, maintaining that without it we cannot account for the existence of the world such as we have it before our eyes.

"Literary Intelligence and Reviews," *Dickinson's Theological Quarterly* 4 (1878): 315.

Secondary Literature: K. E. Rothschuh, "Emil Heinrich Du Bois-Reymond," in *Dictionary of Scientific Biography,* ed. Charles Coulston Gillispie, 16 vols. (New York: Scribner's, 1970–80), 4:200–205; P. Smit, "Albert Wigand," ibid., 14:350.

PIERRE DUHEM (1861–1916)
French Philosopher of Science

English philosophers are almost wholly concerned with applications of philosophy: psychology, ethics, social science. They have little liking for more abstract research and do it poorly. They proceed less by abstract argument than by accumulation of examples. Instead of connoting deductions, they accumulate facts. Darwin and Spencer* do not engage in learned fencing of discussion with their adversaries; they crush them by stoning them.

Pierre Duhem, "The English School and Physical Theories (1893)," in *Essays in the History and Philosophy of Science* (Chicago: Hackett, 1966), 50–74, on 52.

Secondary Literature: R. N. D. Martin, "Darwin and Duhem," *History of Science* 20 (1982): 64–74; Peter Alexander, "Pierre Duhem," in *Encyclopedia of Philosophy,* ed. Paul Edwards, 8 vols. (New York: Macmillan, 1967), 2:423–25.

GEORGE DU MAURIER (1834–1896)
ENGLISH NOVELIST

And as he took a seat in a second-class carriage (it would be third in these democratic days), south corner, back to the engine, with *Silas Marner,* and Darwin's *Origin of Species* (which he was reading for the third time), and *Punch,* and other literature of a lighter kind, to beguile him on his journey, he felt rather bitterly how happy he could be if the little spot, or knot, or blot, or clot which paralyzed that convolution of his brain where he kept his affections could but be conjured away!

And thus, amicably, they entered a small wooded hollow. Then the vicar, turning of a sudden his full blue gaze on the painter, asked, sternly:
    "What book's that you've got in your hand, Willie?"
    "A-a—its the *Origin of Species,* by Charles Darwin. I'm very f-f-fond of it. I'm reading it for the third time. [. . .] It's very g-g-good. It *accounts* for things, you know."

George Du Maurier, *Trilby, A Novel* (1894; New York: International Book, 1899), 262, 285–86.

Secondary Literature: Richard Kelly, *George Du Maurier* (Boston: Twayne, 1983).

ÉMILE DURKHEIM (1858–1917)
FRENCH SOCIOLOGIST

Men submit to the same law. In the same city, different occupations can co-exist without being obliged mutually to destroy one another, for they pursue different objects. [. . .] [I]t is easy to understand that all condensation of the social mass, especially if it is accompanied by an increase in population, necessarily determines advances in the division of labor. [...] The division of labor is, then, a result of the struggle for existence, but it is a mellowed *dénouement*. Thanks to it, opponents are not obliged to fight to a finish, but can exist one beside the other. [...] It is this progressive disjunction that Darwin called the law of the divergence of character.

Émile Durkheim, *The Division of Labor in Society,* trans. George Simpson (1933; reprint, New York: Free Press, 1964), 266–76.

Secondary Literature: Camille Limoges, "Milne-Edwards, Darwin, Durkheim and the Division of Labour: A Case Study in Reciprocal Conceptual Exchanges between the Social and the Natural Sciences," in *The Natural and the Social Sciences,* ed. I. B. Cohen (Dordrecht, Netherlands: Kluwer, 1994), 317–43.

THEODORE EIMER (1843–1898)
GERMAN ZOOLOGIST

Definitely directed evolution, orthogenesis, is a universally valid law. It disproves definitively [August] Weismann's* contention of the omnipotence of natural selection—a mere exaggeration of Darwinism and implicitly involving the other view which Weismann has heretofore upheld unconditionally and which Darwin too had once advocated, that all existing characters of animals have some utility. Orthogenesis shows that organisms develop in definite directions without the least regard for utility through purely physiological causes as the result of *organic growth,* as I term the process. No absolutely injurious character could in the nature of the case continue to exist, but neither could natural selection which Weismann assumes to be the only determining factor in transformation have any efficacy unless something previously existed which from being already useful could be taken hold of by natural selection and so made to serve its purposes.

Th. Eimer, *On Orthogenesis and the Impotence of Natural Selection in Species-Formation* (Chicago: Open Court, 1898), 2.

Secondary Literature: Stephen Jay Gould, *The Structure of Evolutionary Theory* (Cambridge, MA: Harvard University Press, 2002).

ALBERT EINSTEIN (1879–1955)
GERMAN-AMERICAN PHYSICIST

In my opinion, a great era of atomic science cannot be assured by organizing science in the way large corporations are organized. One can organize the application of a discovery already made, but one cannot organize the discovery itself. Only a free individual can make a discovery. However, there can be a kind of organization wherein the scientist is assured freedom and proper conditions of work. Professors of science in American universities, for instance, should be relieved of some of their teaching so as to have more time for research. Can you imagine an organization of scientists making the discoveries of Charles Darwin?

Albert Einstein, "On the Atomic Bomb," *Atlantic Monthly,* November 1945, 43–45, reprinted in *Einstein on Politics,* ed. David E. Rowe and Robert Schulmann (Princeton, NJ: Princeton University Press, 2007), 376.

Darwin's theory of the struggle for existence and the selectivity connected with it has by many people been cited as authorization of the spirit of competition. Some people also in such a way have tried to prove pseudo-scientifically the necessity of the destructive economic struggle of competition between individuals. But this is wrong, because man owes his strength in the struggle for existence to the fact that he is a socially living animal. As little as a battle between single ants of a hill is essential for survival, just so little is this the case with the individual members of a human community.

Albert Einstein, *Out of My Later Years* (New York: Citadel, 1995), 34.

Secondary Literature: Walter Isaacson, *Einstein: His Life and Universe* (New York: Simon & Schuster, 2007).

GEORGE ELIOT (1819–1880)
ENGLISH NOVELIST

[5 December 1859]
We have been reading Darwin's book on the *Origin of Species* just now: it makes an epoch, as the expression of his thorough adhesion, after long years of study, to the Doctrine of Development—and not the adhesion of anonyms like the author of the *Vestiges,* but of a long-celebrated naturalist. The book is sadly wanting in illustrative facts—of which he has collected a vast number, but reserves them for a future book, of which this smaller one is the *avant-coureur.* This will prevent the work from becoming popular as the *Vestiges* did, but it will have a great effect on the scientific world,

causing a thorough and open discussion of a question about which people have hitherto felt timid. So the world gets on step by step towards brave clearness and honesty! But to me the Development Theory and all other explanations of processes by which things came to be, produce a feeble impression compared with the mystery that lies under the processes.

Eliot to Madame Bodichon, in *George Eliot's Life as related in her Letters and Journals,* ed. J. W. Cross, 3 vols. (Boston: Houghton Mifflin, 1909), 2:164–65.

I rejoice to see how many valuable articles you get for the Magazine [*Cornhill*]. I don't read the fiction, but if that is at all in keeping with the good matter I often find in other contributions it will be a public benefit if "natural selection" turns out to be in favor of the Cornhill. But natural selection is not always good, and depends (see Darwin) on many caprices of very foolish animals.

Eliot to George Smith (editor of *Cornhill Magazine*), 26 July 1867, in *The George Eliot Letters,* ed. Gordon S. Haight, 9 vols. (New Haven, CT: Yale University Press, 1954–55), 4:377.

As soon as I received your letter, Mr. [George Henry] Lewes* wrote to Darwin on your behalf, and I enclose two photographs of him together with his letter, which I think you will be glad to have because of the feeling which it expresses towards your husband.

Eliot to Frau Karl von Siebold, June 1871, in *The Letters of George Henry Lewes,* ed. William Baker, 2 vols. (Victoria, BC: University of Victoria, 1995), 2:175n1.

Secondary Literature: Gillian Beer, *Darwin's Plots: Evolutionary Narrative in Darwin, George Eliot, and Nineteenth-Century Fiction* (New York: Cambridge University Press, 2000).

T. S. ELIOT (1888–1965)
English Poet

The eighteenth and nineteenth centuries were the ages of logical science: not in the sense that this science actually made more progress than the others, but in the sense that it was biology that influenced the imagination of non-scientific people. Darwin is the representative of those years, as Newton of the seventeenth, and Einstein perhaps of ours. Creative evolution is a phrase that has lost both its stimulant and sedative virtues. It is possible that an exasperated generation may find comfort in admiring, even if without understanding, mathematics, may suspect that precision and profundity are not incompatible, may find maturity as interesting as adolescence, and permanence more interesting than change. It must at all events be either much more demoralized intellectually than the last age, or very much more disciplined.

T. S. Eliot, "London Letter," *Dial* 71 (October 1921): 455.

Secondary Literature: Lois A. Cuddy, *T. S. Eliot and the Poetics of Evolution* (Lewisburg, PA: Bucknell University Press, 2000); Manju Jain, *T. S. Eliot and American Philosophy: The Harvard Years* (Cambridge: Cambridge University Press, 1993).

HAVELOCK ELLIS (1859–1939)
English Sociologist

It has sometimes been maintained—never more energetically than to-day, especially among the nations which most eagerly entered the present conflict—that war is a biological necessity. War, we are told, is a manifestation of the "Struggle for Life"; it is the inevitable application to mankind of the Darwinian "law" of natural selection. There are, however, two capital and final objections to this view. On the one hand it is not supported by anything that Darwin himself said, and on the other hand it is denied as a fact by those authorities on natural history who speak with most knowledge. That Darwin regarded war as an insignificant or even non-existent part of natural selection must be clear to all who have read his books. He was careful to state that he used the term "struggle for existence" in a "metaphorical sense," and the dominant factors in the struggle for existence, as Darwin understood it, were natural suitability to the organic and inorganic environment and the capacity for adaptation to circumstances; one species flourishes while a less efficient species living alongside it languishes, yet they may never come in actual contact and there is nothing in the least approaching human warfare. The conditions much more resemble what, among ourselves, we may see in business, where the better equipped species, that is to say, the big capitalist, flourishes, while the less well equipped species, the small capitalist, succumbs. Mr. [Peter] Chalmers Mitchell*, Secretary of the London Zoological Society and familiar with the habits of animals, has lately emphasized the contention of Darwin and shown that even the most widely current notions of the extermination of one species by another have no foundation in fact.

Havelock Ellis, *Essays in Wartime: Further Studies in the Task of Social Hygiene* (Boston: Houghton Mifflin, 1917), 16–17.

Secondary Literature: Phyllis Grosskurth, *Havelock Ellis: A Biography* (New York: Knopf, 1980).

RALPH WALDO EMERSON (1803–1882)
American Philosopher

Consolation for Readers of Darwin

Why is Man the head of Creation the power of all, but because his education had been so longaeval and immense? He has taken all the degrees; he began in the beginning, & has passed through all the steps through radiate, articulate, mollusk, articulate vertebrate, through all the forms to the mammal, & now holds in essence the virtues or powers of all, though in him all the exaggeration of each subdued into harmony [. . .] through the antagonism of their opposites, & is at last the harmony, the flower & top of all their being—their honored representative. Lion. Many individuals have taken their Master's degree prematurely, & will yet have mortifications, & perhaps drop into one of the lower classes for a time.

The Topical Notebooks of Ralph Waldo Emerson, ed. Ronald A. Bosco, 2 vols. (Columbia: University of Missouri Press, 1993), 2:336.

Secondary Literature: Laura Dassow Walls, *Emerson's Life in Science: The Culture of Truth* (Ithaca, NY: Cornell University Press, 2003); Lee Rust Brown, *The Emerson Museum: Practical Romanticism and the Pursuit of the Whole* (Cambridge, MA: Harvard University Press, 1997).

WILLIAM EMPSON (1906–1984)
English Critic and Poet

[1928]

Invitation to Juno

Lucretius could not credit centaurs;
Such bicycle he deemed asynchronous.
"Man superannuates the horse;
Horse pulses will not gear with ours."

Johnson could see no bicycle would go;
"You bear yourself, and the machine as well."
Gennets for germans sprung not from Othello,
And Ixion rides upon a single wheel.

Courage. Weren't strips of heart culture seen
Of late mating two periodicities?
Did not at one time even Darwin
Graft annual upon perennial trees?

William Empson, in *The Atlantic Book of British and American Poetry*, ed. Dame
Edith Sitwell (Boston: Little, Brown, 1958), 977.

Somebody had Julian Bell's great gushing *Life and Letters;* it would have
offended him rather I think. But my word how did he write letters. And
what is this business of keeping letters? Do *you* keep letters? It must be a
widespread practice judging from biographies. [. . .] Darwin I remember,
though, a different kind of man, said it had been a great help to him never
to destroy a single letter. [. . .] By the way, what on earth was the disease
of Darwin? Was it simply neurotic? They never say, and he seems a very
unneurotic man otherwise.

Empson to John Hayward, 4 May 1939, in *Selected Letters of William Empson,* ed.
John Haffenden (London: Oxford University Press, 2006), 120–21.

Secondary Literature: John Haffenden, *William Empson,* 2 vols. (New York: Oxford University Press, 2005–6).

FRIEDRICH ENGELS (1820–1895)
German Political Theorist

Of the Darwinian doctrine I accept the theory of evolution, but Darwin's method of proof (struggle for life, natural selection) I consider only a first, provisional, imperfect expression of a newly discovered fact. Until Darwin's time the very people who now see everywhere only struggle for existence ([Carl] Vogt*, [Ludwig] Büchner*, Moleschott,[†] etc.) emphasized precisely cooperation in organic nature, the fact that the vegetable kingdom supplies oxygen and nutriment to the animal kingdom and conversely the animal kingdom supplies plants with carbonic acid and manure, which was particularly stressed by [Justus] Liebig. Both conceptions are justified within certain limits, but the one is as one-sided and narrowminded as the other. The interaction of bodies in nature—inanimate as well as animate—includes both harmony and collision, struggle and cooperation. When therefore a self-styled natural scientist takes the liberty of reducing the whole of historical development with all its wealth and variety to the one-sided and meager phrase "struggle for existence," a phrase which even in the sphere of nature can be accepted only *cum grano salis,* such a procedure really contains its own condemnation.

Engels to Pyotr Lavrov, November 1875, in *Collected Works,* 50 vols. (New York: International Publishers, 1975–2008), 45:106–9. [†]Jacob Moleschott (1822–93), Dutch physiologist and theorist of materialism.

The main reproach levelled against Darwin is that he transferred the Malthusian population theory from political economy to natural science, that he was held captive by the ideas of the animal breeder, that in his theory of the struggle for existence he pursued unscientific semi-poetry, and that the whole of Darwinism, after subtracting what had been borrowed from Lamarck, is a piece of brutality directed against humanity.

Darwin brought back from his scientific travels the view that plant and animal species are not constant but subject to variation. In order to follow up this idea after his return home, there was no better field available than that of the breeding of animals and plants. It is precisely in this field that England is the classical country; the achievements of other countries, for example Germany, fall far short of what England has achieved in this connection. Moreover, most of these successes have been won during the last hundred years, so that there is little difficulty in establishing the facts. Now Darwin found that this breeding artificially produced differences among animals and plants of the same species greater than those occurring in what are generally recognized as different species. Thus there was established the variability of species up to a certain point, on the one hand, and the possibility of a common ancestry for organisms with different specific characteristics, on the other. Darwin then investigated whether there were not possibly causes in nature which—without the conscious intention of the breeder—would nevertheless necessarily produce in living organisms over the long run changes similar to those produced by artificial breeding. [. . .]

Against this Darwinian theory Herr Dühring† now says that the origin of the idea of the struggle for existence, as, he claims, Darwin himself admitted, has to be sought in a generalization of the views of the economist and population theorist, Malthus, and that the idea therefore suffers from all the defects inherent in Malthus' clerical views on overpopulation.

Now Darwin would not dream of saying that the *origin* of the idea of the struggle for existence is to be found in Malthus. He only says that his theory of the struggle for existence is the theory of Malthus applied to the animal and plant world as a whole. However great Darwin's blunder in accepting the Malthusian theory so naively and uncritically, anyone can see at the first glance that no Malthusian spectacles are required to perceive the struggle for existence in nature—the contradiction between the countless host of embryonic germs nature so lavishly produces and the small number of those which can ever reach maturity, a contradiction which in fact finds its solution for the most part in a struggle for existence—often of extreme cruelty. And just as the law of wages has retained its validity even after the Malthusian arguments on which [David] Ricardo based it have long been consigned to oblivion, so the struggle for existence can take place in nature, even without any Malthusian interpretation. For that matter, the organisms of nature also have their laws of population which have been left practically uninvestigated, although their establishment would be of decisive importance for the theory of the evolution of species. But who was it that lent the decisive impetus to work in this direction too? None other than Darwin.

Friedrich Engels, *Anti-Dühring (Herr Eugen Dühring's Revolution in Science)* (Peking: Foreign Languages Press, 1976), 84–86. [†]Eugen Dühring (1833–1921), an anti-Marxian German philosopher and economist.

Just as Darwin discovered the law of development of organic nature, so Marx[*] discovered the law of development of human history: the simple fact, hitherto concealed by an overgrowth of ideology, that mankind must first of all eat, drink, have shelter and clothing, before it can pursue politics, science, art, religion, etc.; that therefore the production of the immediate material means of subsistence and consequently the degree of economic development attained by a given people or during a given epoch form the foundation upon which the state institutions, the legal conceptions, art, and even the ideas on religion, of the people concerned have been evolved, and in the light of which they must, therefore, be explained, instead of *vice versa,* as had hitherto been the case.

Friedrich Engels, "Speech at the Graveside of Karl Marx," in *On Marx* (Peking: Foreign Languages Press, 1975), 16–17.

Secondary Literature: D. A. Stack, "The First Darwinian Left: Radical and Socialist Response to Darwin, 1859–1914," *History of Political Thought* 21 (2000): 682–710; William A. Pelz, "Friedrich Engels," in *Encyclopedia of Historians and Historical Writing,* ed. Kelly Boyd, 2 vols. (London: Fitzroy Dearborn, 1999), 1:357–58.

REGINALD BALIOL BRETT ESHER (1852–1930)
ENGLISH POLITICIAN AND HISTORIAN

Odo Russell[†] from Versailles went to see one of the villas which was much exposed. He found there an officer, who had not had his boots off for two months, reading Darwin.

Reginald Baliol Brett Esher, 30 August 1873, in *Extracts from Journals, 1872–1881* ([Whitefish, MT?]: Kessinger, 2007), 34. [†]Russell (1829–84) was the British ambassador to Germany.

I spent four hours this afternoon in the society of George Eliot[*]. She talks like the best parts of her books, the parts where she analyzes without dissecting, the parts out of which compilers get her "wise, witty, and tender" sayings. She shut up George [Henry] Lewes* when he tried to talk about her. She does not seem vain. She adores Charles Darwin, because of his humility. I suppose it is an event to have spent the day in her company.

Reginald Baliol Brett Esher, 1 June 1876, in *Extracts from Journals, 1872–1881* ([Whitefish, MT?]: Kessinger, 2007), 145.

Secondary Literature: James Lees-Milne, *The Enigmatic Edwardian: The Life of Reginald, 2nd Viscount Esher* (London: Sidgwick & Jackson, 1986).

JEAN HENRI FABRE (1823–1910)
FRENCH ENTOMOLOGIST

Favier [Fabre's gardener] [...] told me that, when people want to move a Cat from one farm to another at some distance, they place the animal in a bag, which they twirl rapidly at the moment of starting, thus preventing the animal from returning to the house which it has quitted. Many others, besides Favier, described the same practice to me. According to them, this twirling round in a bag was an infallible expedient: the bewildered Cat never returned. I communicated what I had learnt to England, I wrote to the sage of Down and told him how the peasant had anticipated the researches of science. Charles Darwin was amazed; so was I; and we both of us almost reckoned on a success.

Jean Henri Fabre, *The Mason-Bees,* trans. Alexander Teixerra de Mattos (New York: Dodd, Mead, 1914), 76–77.

Fabre is one of the most astute students of Natural History the world has known. He accepts no one's opinion, considers theorizing a waste of time, and believes what he sees his insects do, not what someone else says they do. He spurns Darwinism, finds no evidence for the mechanical explanation of life, and says that the facts do not warrant the ascribing of intelligence to the mental life of the insect. Fabre is in good company here, as his well-known Catholic fellow-biologist, Erich Wasmann, S. J.*, has brilliantly maintained these positions for some years.

Bramble Bees and Others shows Fabre to be a most remarkable observer, not surpassed in this respect even by Darwin; a clean, careful reasoner who confines himself to facts with much better success than Darwin; and a humble, patient scientist, who is never embarrassed when admitting his ignorance.

"Bramble Bees and Others," *Catholic World* 101 (1915): 540–41, on 540.

Secondary Literature: Alex Delage, *Henri Fabre: L'observateur incomparable; Biographie* (Rodez, France: Rouerge, 2005).

MICHAEL FARADAY (1791–1867)
ENGLISH PHYSICIST

I am exceedingly obliged by your kindness in sending me Mr. Darwin's remarkable book. I have received it at Brighton where it arrived before me & shall read it with great attention.

Faraday to John Murray, 2 December 1859 (letter 3689), in *The Correspondence of Michael Faraday, Volume 5, 1855–60,* ed. Frank A. J. L. James (London: Institution of Electrical Engineers, 2008), 607.

Secondary Literature: James Hamilton, *Faraday: The Life* (London: HarperCollins, 2002).

FREDERIC WILLIAM FARRAR (1831–1903)
ENGLISH PRELATE

Acknowledging his gift of the *Descent of Man,* I said that one insuperable difficulty in the acceptance of his theories was, that from all I had ever read about Anthropology, and from all my studies in Comparative Philology, it seemed to me indisputable that different germs of language and different types of race were traceable from the farthest prehistoric days. The argument has, since then, been indefinitely strengthened by the discovery of the earliest known skulls and remains of primeval races, which show that, even in those immeasurably distant days, there were higher and lower types of humanity. Mr. Darwin admitted the fact, but made this very striking answer: "You are arguing from the last page of a volume of many thousands of pages!"

When Darwin died, I happened to see Professor Huxley[*] and Mr W. [William] Spottiswoode[†] in deep and earnest conversation at the Athenaeum. I asked them why no memorial had been sent to the Dean of Westminster, requesting that one who had been an honor to his age should be buried in the great historic Abbey. "There is nothing which we should like so much," said Professor Huxley. "Nothing would be more fitting; it is the subject on which we were talking. But we did not mean to make the request, for we felt sure it would be refused." I replied, with a smile, "that we clergy were not all so bigoted as he supposed"; and that, though I had no authority to answer for the Dean, I felt no doubt that, if a memorial were sent to him, the permission would be accorded. I said that I would consult the Dean, and let them know at once. Leave was given. I was asked to be one of the pall-bearers, with nine men of much greater distinction—Sir J. [John] Lubbock*, Professor Huxley, Mr. J. R. [James Russell] Lowell*, Mr. A. R. Wallace*, the Dukes of Devonshire and Argyll*, the late Earl of Derby, Sir J. Hooker*, and Mr. W. Spottiswoode; and on the Sunday evening I preached at the Nave Service the funeral sermon of the great author of the Darwinian hypothesis. Ecclesiasticism was offended; but if what God requires of us is "to do justly, and to love mercy, and to walk humbly with Him," I would rather take my chance in the future life with such a man as Charles Darwin, than with many thousands who, saying, "Lord, Lord," and wearing the broadest of phylacteries, show very faint conceptions of honor, kindness, or the love of truth, and are sadly to seek in the most elementary Christian virtues.

The Life of Frederic William Farrar, ed. Reginald Farrar (New York: Thomas Y. Crowell, 1904), 108–9. †See Thomas Carlyle entry.

In his funeral sermon he [Farrar] thus spoke of Darwin: "This man, on whom for years bigotry and ignorance poured out their scorn, has been called a materialist. I do not see in all his writings one trace of materialism. I read in every line the healthy, noble, well-balanced wonder of a spirit profoundly reverent, kindled into deepest admiration for the works of God. [. . .] Calm in the consciousness of integrity; happy in sweetness of home life; profoundly modest; utterly unselfish; exquisitely genial; manifesting, as his friend has said of him, 'an intense and passionate honesty, by which all his thoughts and actions were irradiated as by a central fire'—Charles Darwin will take his place, side by side, with [John] Ray and Linnaeus; with Newton and Pascal; with [John] Herschel* and [Michael] Faraday*—among those who have not only served humanity by their genius, but have also brightened its ideal by holy lives. [. . .] And because these false antagonisms have been infinitely dangerous to faith, over Darwin's grave let us once more assure the students of science that, for us, the spirit of mediaeval ecclesiasticism is dead. We desire the light. We believe in the light. We press forward into the light. If need be, let us perish in the light. But we know that in the light we shall never perish. For to us God is light; and Christ is, and will be, to the end, 'the Light of the World.'"

The Life of Frederic William Farrar, ed. Reginald Farrar (New York: Thomas Y. Crowell, 1904), 109–10.

Secondary Literature: Owen Chadwick, *The Victorian Church*, 2 vols. (New York: Oxford University Press, 1966–70).

SANDOR FERENCZI (1873–1933)
Hungarian Psychoanalyst

The active courtship which precedes actual mating has become attenuated in man in the course of the development of civilization, often to the point of being quite unrecognizable, in such a way that again we can recognize its meaning only through the observation of animals. We have already mentioned that according to our assumption the central striving to return to the maternal womb dominates both sexes equally; courting can accordingly have as its purpose nothing else than that the female sex, in giving up or restricting her own gratification, be made amenable to tolerating the sex act on the part of the male. We should like to cite in support of this assertion two pronouncements of Charles Darwin, who can surely speak with authority on the question. "As appearances more than once lead us to believe," he says somewhere, "the female does not take the male that seems most attractive to her but the one that is least repulsive to her." [...] In another place Darwin states that sexual variation in the sense of a sexual dimorphism always originates in the male sex, even if it is later adopted in part by the female also. All this agrees strikingly with Freud's assertion that all libido is *per se* "masculine," even when, as in the female, it seeks goals of gratification of a passive type.

Sandor Ferenczi, *Thalassa: A Theory of Genitality* (New York: Psychoanalytic Quarterly, 1938), 30–31.

Secondary Literature: Arnold L. Rachman, *Sandor Ferenczi* (Northvale, NJ: J. Aronson, 1997).

JOHN FISKE (1842–1901)
AMERICAN SOCIAL PHILOSOPHER

In America, this discussion was centered in a measure around Harvard College by reason of the fact that two of the leading scientists in this country engaged in this controversy and representing the opposing sides were professors in the Lawrence Scientific School and instructors in the college—Professor Agassiz*, one of the world's great zoologists, and Professor Asa Gray*, one of the world's great botanists and the firm supporter of the views of Mr. Darwin. The points of difference between these two eminent teachers as to origins of organic life were apparent in their instruction, while the larger scientific implications of their views as to "origins" were set forth in their public discussions.

It is not [the] place here to enter into the full details of the Darwinian discussion. But inasmuch as it was an active element in the Harvard thought of the time, and inasmuch as the labors of Mr. Darwin were a very important contribution to the doctrine of Evolution, in the setting-forth of which Fiske was to take a conspicuous part in subsequent years, and particularly as in years to come we are to see Fiske in close friendly relations with Mr. Darwin growing out of their respective labors in behalf of Evolution, a brief presentation of the origin of the discussion is appropriate here.

The first half of the last century was a period of great scientific activity, and it was specially marked by searching inquiries into the phenomena of organic life as revealed in the past and present condition of the globe. To this end the departments of geology, palaeontology, embryology, zoology, ethnology, physiology, and botany were interrogated by able observers intent upon getting at the fundamental facts conditioning organic life, both in its particulars and in its widest generalities.

The Life and Letters of John Fiske, ed. John Spencer Clark, 2 vols. (Boston: Houghton Mifflin, 1917), 1:180–81. Fiske had read Agassiz's *Essay on Classification,* Gray's *Structural and Systematic Botany,* and Darwin's *Origin of Species* consecutively in March 1860, just before entering Harvard College (ibid., 106, 107n).

Mr. Darwin has invited me to visit him, at his place in Kent: am I likely to be able to accomplish all these things by reaching London about October 15th and remaining there till Christmas?

I read [George Henry] Lewes's* book *(Problems of Life and Mind)* in the sheets, and I consider his treatment of Kant one of the most masterly pieces of philosophical criticism I ever read. I told Darwin about it, and found that he has a great admiration for Lewes's straightforward and clean-cut mind. I have made up my mind that Lewes will have a permanent place in history as the critic of Kant, to say nothing of the other things he has done. What a comical old fellow he is! At *the* dinner the other day [Spencer's* dinner in honor of Fiske] I was saying that very soon we should see Evolution taken up by the orthodox. "To be sure," says Lewes, "for don't you see that Evolution requires an Evolver?" Huxley* was telling about something I said in my Agassiz* article, when Spencer blandly interrupted with "What will Agassiz say to all that?" "O," said Lewes, "he will say what Louis XIV said after the battle of Ramillies— *Dieu m'a abandonné; et après tout ce que j'ai fait pour Lui!!!*"

Fiske to Mrs. Fiske, 18 November 1872, in *The Life and Letters of John Fiske,* ed. John Spencer Clark, 2 vols. (Boston: Houghton Mifflin, 1917), 1:481–82.

To-day, I lunched with Darwin and Mrs. Darwin, Mrs. Litchfield (Darwin's daughter), Frank Darwin (whom I saw in Boston two years ago) and Miss Bessie Darwin, and Dr. Hooker*, the greatest living botanist, and Mrs. Hooker. [. . .] Darwin is the dearest, sweetest, loveliest old Grandpa that ever was. And on the whole he impresses me with his strength more than any man I have yet seen. There is a charming kind of quiet strength about him and about everything he does. He is not burning and eager like Huxley*. He has a mild blue eye, and is the gentlest of gentle old fellows. I think he would make a noble picture after the style of mother's picture which I call "Galileo." His long white hair and enormous beard make him very picturesque. And what is so delightful to see, as that perfect frankness and guileless simplicity of manner, which comes from a man having devoted his whole life to some great idea, without a thought of self, and without ever having become a "man of the world"? I had a warm greeting from the dear old man, and I am afraid I shall never see him again, for his health is very bad, and he had to make a special effort to see me to-day. Of all my days in England, I prize to-day the most; and what I pity you most of all for, my dear, is that you haven't seen our dear grand old Darwin! I think we both felt it might be the last time. He came to the door with me and gave me a warm grip of the hand and best wishes, and watched me down the road till I turned the corner, when I took off my hat and bowed good-bye.

Fiske to Mrs. Fiske, 13 November 1873, in *The Life and Letters of John Fiske,* ed. John Spencer Clark, 2 vols. (Boston: Houghton Mifflin, 1917), 1:477.

Wednesday, June 18, 1879. [. . .] After lecture went down by cars to Orpington in Kent and found Darwin's carriage awaiting me at the station. Drove four miles through exquisite English lanes (the air heavily scented with blossoms) to Darwin's house. Jolly place with lots of garden. George and Horace were there, and Mrs. Litchfield, and two or three Wedgwoods. Nice dinner and smoke on verandah, and Miss Carrie Wedgwood played considerable Bach, Scarlatti, Schumann, and Schubert on the grand piano. Afterwards grandpa and Hezzy got into a very abstruse discussion, and, when the clock said ten, up came Mrs. Darwin and pointed with warning finger to the clock, and so grandpa said he must obey orders and trotted off to bed. I stayed up till eleven and smoked another cigar with the boys. At ten Darwin was to sit for his portrait in his red Doctor-of-Laws gown, for the university of Cambridge. He put the gown on after breakfast, to the great glee of the little grandchildren, and the merriment of all, as he stepped up on a chair to get a full view of himself in the mirror.

Excerpt from Fiske's journal, in *The Life and Letters of John Fiske,* ed. John Spencer Clark, 2 vols. (New York: Houghton Mifflin, 1917), 2:133–34.

Secondary Literature: Milton Berman, *John Fiske: The Evolution of a Popularizer* (Cambridge, MA: Harvard University Press, 1961).

EDWARD FITZGERALD (1809–1883)
English Author and Translator

[2 March 1860]
Darwin's *Species* brings one back to the old *Vestiges of Creation:* which I always had a leaning to, though [Adam] Sedgwick* and the Big Wigs *don'd* it down as impious: which it may be.

Fitzgerald to Stephen Spring Rice, in *The Letters of Edward Fitzgerald*, ed. A. M. Terhune and A. B. Terhune, 4 vols. (Princeton, NJ: Princeton University Press, 1980), 2:355.

Secondary Literature: Philip J. Waller, *Writers, Readers, and Reputations: Literary Life in Britain, 1870–1918* (Oxford: Oxford University Press, 2006).

F. SCOTT FITZGERALD (1896–1940)
AMERICAN NOVELIST

Somehow Amory's dissatisfaction with his lack of enthusiasm culminated in an attempt to put the blame for the whole war on the ancestors of his generation [. . .] all the people who cheered for Germany in 1870. [. . .] All the materialists rampant, all the idolizers of German science and efficiency. So he sat one day in an English lecture and heard "Locksley Hall" quoted and fell into a brown study with contempt for [Alfred Lord] Tennyson* and all he stood for, for he took him as a representative of the Victorians.

"Victorians, Victorians, who never learned to weep
Who sowed the bitter harvest that your children go to reap"

scribbled Amory in his note-book. The lecturer was saying something about Tennyson's solidity and fifty heads were bent to take notes. Amory turned over to a fresh page and began scrawling again.

"They shuddered when they found what Mr. Darwin was about,
They shuddered when the waltz came in and Newman* hurried out"

F. Scott Fitzgerald, *This Side of Paradise* (1920; reprint, New York: Dover, 1996).

Secondary Literature: Andrew Hook, *F. Scott Fitzgerald: A Literary Life* (New York: Palgrave Macmillan, 2002).

GUSTAVE FLAUBERT (1821–1880)
FRENCH NOVELIST

Darwin. The fellow who says we're sprung from monkeys.

Gustave Flaubert, in *Dictionary of Accepted Ideas*, trans. Jacques Barzun (New York: New Directions, 1968), 29.

I have just read *la Creation naturelle* by Haeckel*, a pretty book, pretty book! Darwinism seems to me to be better expounded there than in the books of Darwin himself.

Flaubert to George Sand, 3 July 1874, in *The George Sand–Gustave Flaubert Letters*, trans. Aimee L. McKenzie (New York: Boni & Liveright, 1921), 320.

Secondary Literature: Allen Thiher, *Fiction Rivals Science: The French Novel from Balzac to Proust* (Columbia: University of Missouri Press, 2001), chap. 3, "Flaubert and the Ambiguous Victory of Positivism."

BENJAMIN ORANGE FLOWER (1858–1918)
AMERICAN JOURNALIST

I remember on one occasion, several years ago, being in the presence of some very zealous members of a well known Christian fellowship. The names of Charles Darwin and Dr. Alfred Russel Wallace* were mentioned, when one lady, who probably had never read a line by either of these great thinkers, exclaimed, "I don't want to hear anything about those monkey-men"—a sentiment with which her companions seemed in full accord.

Benjamin Orange Flower, *Progressive Men and Women of the Past Twenty-five Years* (Boston: New Arena, 1914), 20.

Secondary Literature: Roy P. Fairfield, "Benjamin Orange Flower: Father of the Muckrakers," *American Literature* 22 (November 1950): 27–82.

ANTONIO FOGAZZARO (1842–1911)
ITALIAN NOVELIST AND SENATOR

In February, 1891, I had the honor of speaking to a grave and learned audience on the subject of the relation which the famous doctrine that generally goes by the name of Charles Darwin bears to the Catholic doctrine of the Creation. My object then was to establish the liberty of the Catholic conscience with regard to an hypothesis, according to which it is maintained that living organisms did not appear on the earth at intervals by virtue of distinct acts of the Creator, but were modified and developed from generation to generation from one single originative form up to the immense present variety. I even went a step farther, and pronounced which of the two theories seemed to me to correspond most closely to the truth and to the religious ideal. [. . .]

Till the end of 1860 German scientists, with one exception, breathed no word either for or against it. [. . .] Readers went on increasing. If official science had not yet given Darwin its suffrage, at least the sweet-smelling smoke of celebrity was being offered up to him from every quarter. But he was wrong in asking from the general public a verdict on the value of his theories. Granted the nature of the subject and the reticence of scientific men, it was idle to expect a precise and explicit judgment on the theory of Natural Selection from the unscientific world. By making the man and his book famous, the public did indeed substantially pronounce itself in favor of some rational method to prove that species, like individuals, came into the world naturally. [. . .] And this blind majority is growing perpetually wider as fresh generations slowly attain to the possession of culture and popular prejudices. I do not, however, intend here to trace the steps of Darwin's fame; it attained a diffusion not surpassed even by the names of Newton, Copernicus, and Galileo. While [Darwin] was still alive, a German psychological society went to the length of discussing the shape of his brain, and concluded that his bump of respect for authority was big enough for ten priests! [. . .] Darwin became, in the eyes of the multitude, the legitimate father of the Transformist hypothesis, and it was popularly called Darwinism after him, whereas he had only conceived a practical method of setting it on its feet. [. . .]

Darwin gets lost just because he cannot free himself from the idea that, according to the advocates of a divine plan of the universe, everything in nature must have a separate and visible aim. It does not suit him, for instance, to think that the peacock's plumage was so richly painted merely to please the eye of man. Yet at the same time he cannot persuade himself that humanity is a product of chance. So he concludes that for man to meditate on the plan of the universe is like a dog meditating on the mind of Newton. On the other hand, Huxley*, his most faithful disciple, has confessed that in the place of the old dead teleology a new and grander one may arise, with this same fundamental idea of evolution as its basis. [. . .]

It was always Fogazzaro's hope that a reconciliation between the Church and science would be effected, between the *Origin of Species* and the Book of Genesis. He sought this in the writings of St. Augustine, and believed that the solution of the contradiction would be found in St. Augustine's idea of creation as a single act, containing in it *"potentialiter, causaliter, primordialiter,"* the necessary evolution of all organisms, as opposed to the common conception of successive special creations. According to this view, the act of creation by evolution proceeds continuously till the final summing up of all temporal things. [...] Fogazzaro in his zeal for the reconciliation of faith and knowledge was too sanguine in claiming Darwin on the side of orthodoxy; the problem of faith and knowledge cannot be solved by a harmony between Darwinism and the Church, so long as Darwinism looks upon mental and spiritual facts as subject to the laws of evolution, and the Church claims a direct and immutable message from God to man. Such a reconciliation may become part of the collective consciousness of humanity, and of Christendom as a part of humanity; and the collective consciousness of Christendom may create a creed, and (to use Fogazzaro's words) "express itself spontaneously in acts of authority": but the creed will not be the creed of Pius X. Authority must step down from its high estate before it can join hands with knowledge.

"Fogazzaro and Modernism," *Edinburgh Review* 214 (1911): 269–92, on 283.

Secondary Literature: Marina Marcolini, "Le conferenze scientifiche di Antonio Fogazzaro," in *Antonio Fogazzaro,* ed. M. Marcolini (Padua: Esedra, 1993).

EDWARD FORBES (1815–1854)
English Naturalist

Valentine
By a Palaeontologist

> Borne upon Pterodactyle's wing,
>   This heart, which once you deemed of stone,
> Model of maids, to thee I bring,
>   And offer it to thee alone!
>   Not Owen*, pondering o'er bone
> Of great Dinornis, fonder grew
>   Of mighty wingless birds unknown,
> Than I, sweet maid, of you.

The Glyptodon, which Darwin found
   Beside the South Atlantic main,
Was in no harder armour bound,
   Than that my spirit did enchain;
   Till, bade by thee, Love rent in twain
The fetters which my fancy tied
   To boulder, glacier, and moraine,
And bore me to thy side!

Like some fantastic Trilobite
   That perished in Silurian sea,
And long lay hid from mortal sight,
   So was the heart I yield to thee.
   Now from its stony matrix free,
Thy palaeontologic skill
   Once more hath call'd it forth to be
The servant of thy will.

[Forbes] refers to this song in a letter to Mr. Thomson.[†] "Look at last *Literary Gazette*," he says, "for a song of mine. Buckland[‡] amusingly enough quoted it in full at the geological anniversary dinner yesterday, not knowing the author. De la Beche[§] afterwards, in returning thanks for the Survey, wickedly proclaimed that it came from our staff, and urged it as a proof that any sort of article could be produced in the Museum of Economic Geology."

Poem Read at Geological Society of London dinner, 14 February 1845, quoted in *Memoir of Edward Forbes,* by George Wilson and Archibald Geikie (London: Macmillan, 1861), 387. [†]William Thomson, Lord Kelvin (1824–1907) argued against both uniformitarianism and Darwinian evolution, having concluded that the age of the earth was shorter than those theories had suggested. [‡]William Buckland (1784–1856), English vicar, paleontologist, and catastrophist geologist. [§]Henry de la Beche (1796–1855), English geologist.

Secondary Literature: Frank Egerton III, "Edward Forbes, Jr.," in *Dictionary of Scientific Biography,* ed. Charles Coulston Gillispie, 16 vols. (New York: Scribner's, 1970–80), 5:66–68.

FORD MADDOX FORD (1873–1939)
ENGLISH NOVELIST, CRITIC, AND EDITOR

I saw Samuel Butler* bearing down on us. I disliked Samuel Butler more than anyone I knew. He was intolerant and extraordinarily rude in conversation—particularly to old ladies and young persons. He had perhaps cause to be, for he was conscious of great gifts and of being altogether neglected. *Erewhon* was nearly forgotten, *The Way of All Flesh* unpublished. [. . .] Otherwise he was unfavorably known as having pirated the ideas of Darwin and as having behaved with extraordinary rudeness to that great and aged scientist. [. . .] I wanted to defend anyone from Darwin to a pirate who was attacked by that fellow who himself looked and behaved, in conversation, like a pirate. [. . .] Samuel Butler wrote an immense number of wasted words in the attempt to avenge himself from some fancied slight at the hands of Darwin. But, in spite of these follies *The Way of All Flesh* is vastly more important to us today than is *The Origin of Species*. Darwin as a scientist is as superseded as the poor alchemist in the Spessart Inn: so is Butler in the same department of human futility. But *The Way of All Flesh* cannot be superseded because it is a record of humanity. Science changes its aspect as every new investigator gains sufficient publicity to discredit his predecessors. The stuff of humanity is unchangeable.

Ford Maddox Ford, *Return to Yesterday* (1931; reprint, New York: Norton, 1972), 170–78.

Secondary Literature: Arthur Mizener, *The Saddest Story: A Biography of Ford Maddox Ford* (New York: World, 1971).

E. M. FORSTER (1879–1970)
ENGLISH NOVELIST

"I thought that Colonel Robert [G.] Ingersoll* says you must be strenuous."

At the sound of this name he whisked open a little cupboard, and declaring, "I haven't a moment to spare," took out of it a pile of *Clarion* and other reprints, adorned as to their covers with bald or bearded apostles of humanity. Selecting a bald one, he began at once to read, occasionally exclaimed, "That's got them," "That's knocked Genesis," with similar ejaculations of an aspiring mind. She glanced at the pile. [Ernest] Renan* minus the style. Darwin, minus the modesty. A comic edition of the book of Job, by "Excelsior," Pittsburgh, Pa. "The Beginning of Life," with diagrams. "Ape or Angel?" by Mrs. Julia P. Chunk. She was amused, and wondered idly what was passing within his narrow but not uninteresting brain.

E. M. Forster, *The Longest Journey* (New York: Knopf, 1922), 105.

Secondary Literature: John Keith Johnstone, *The Bloomsbury Group: A Study of E. M. Forster, Lytton Strachey, Virginia Woolf, and Their Circle* (London: Secker & Warburg, 1954).

HARRY EMERSON FOSDICK (1878–1969)
American Clergyman

If God exists at all, he must care for his creation, and if he cares at all, he must care for the crown of creation, personality. Charles Darwin tells us that at times he had a warm sense of a friendly God, but that at other times this feeling vanished. Yet even with so fugitive a faith in a universe that cared for its creatures, he wrote, "it is an intolerable thought that man and all other sentient beings are doomed to complete annihilation, after such long-continued slow progress." To one who is deeply convinced that Darwin's occasional and evanescent sense of a friendly God may be a man's reasonable and constant faith, such a conception of the world is not only intolerable; it is impossible. To talk about the fatherhood of a God, who begets children, only to annihilate them, is absurd. The goodness of God is plainly at stake when one discusses immortality, for if death ends all, the Creator is building men like sand houses on the shore, caring not a whit that the fateful waves will quite obliterate them all. If death ends all, the struggle and aspiration of humanity have meant no more to him than the mist that rests in the morning on the Alps and at noon is gone. If death ends all, there is no God of whom goodness, in any connotation imaginable to man, can be predicated.

Harry Emerson Fosdick, *The Assurance of Immortality* (New York: Macmillan, 1913), 121.

Secondary Literature: Robert Moats Miller, *Harry Emerson Fosdick: Preacher, Pastor, Prophet* (New York: Oxford University Press, 1985).

ANATOLE FRANCE (1844–1924)
French Novelist

Madame des Aubels, who had her reasons for doubting this, expressed at least one:

"You have no wings."

"Why should I, Madame? Am I bound to resemble the angels on your holy water stoups? Those feathery oars that beat the waves of the air in rhythmic cadences are not always worn by the heavenly messengers on their shoulders. Cherubim may be apterous. That all too beautiful angelic pair who spent an anxious night in the house of Lot compassed about by an oriental horde—they had no wings! No, they appeared just like men and the dust of the road covered their feet, which the patriarch washed with pious hand. I would beg you to observe, Madame, that according to the Science of Organic Metamorphosis created by Lamarck and Darwin, the wings of birds have been successively transformed into forefeet in the case of quadrupeds and into arms in the case of the Linnaean primates."

Anatole France, *The Revolt of the Angels,* trans. Mrs. Wilfred Jackson (London: John Lane, 1914), 89.

I have already said in these pages that eighteen years ago [ca. 1904] we were enthusiastic determinists. Among us were one or two neo-Catholics. But they felt very uneasy. The fatalists, on the other hand, displayed a serene confidence, which, alas, they have not retained. To-day we know well that this romance of the universe is as full of disappointments as the rest, but in those days Darwin's books were our Bible; the magnificent praise with which Lucretius extolled the divine Epicurus appeared to us inadequate to glorify the English naturalist. With burning faith we used to say: "A man has come who has freed mankind from vain terrors." I cannot resist once more recalling the fertile visits which we used to pay, with

Darwin under our arms, to the old Jardin des Plantes, where M. Paul Bourget[†] complaisantly parades the hero of his new novel, *Adrien Sixte.* Personally I used to enter the halls of the Museum as a sanctuary: halls filled with every species of organic form, from the fossil crinoids and the long jaws of the great primitive saurians to the wrinkled skin of the elephant and the gorilla's hand.

Anatole France, *Life and Letters,* 3rd ser. (London: John Lane, 1922), 54–55. [†]Paul Bourget (1852–1935), French novelist and critic.

Secondary Literature: Dushan Bresky, *The Art of Anatole France* (The Hague: Mouton, 1969).

EDWARD A. FREEMAN (1823–1892)
English Historian

I have taken a large sheet, because I am going to write on a great subject, at any rate a long subject, even Longus Burgo[†] your Dean. He praught this morning at New College, and altogether out-comicked the old "comic sermon" that used to be at St. Mary's. First of all, there are some changes in the ceremony. Surely in times past there was a pulpit wheeled in—then I rather fancy men praught from the flying eagle. But now Longus Burgo praught from a stall in the middle of *Decani,* and piled himself up a heap of pillows wherewithal to preach, one of which he stole from me, as I was unwittingly within two stalls of him. Text, creation of man, male and female, &c. First, refutation of evolution, and special woe against them that [. . .] stretched out one hand to Moses and one to Darwin. They that believed nothing were to thole [Old English, "suffer"] all revealed punishments, but they that tried to believe two things were to thole something further—particulars. It would seem known only to [John William Burgon]. Methought how, in the sixteenth century, haply some men talked in just the same way about holding one hand to Joshua and another to Copernicus. But the great point was male and female, we were to notice, not men and women: forwhy, woman was not yet, she was to come out of man, and so man, which potentially contained her, might be said to be created male and female. (N.B.—I can believe the Christian religion, and yet see that here and in the Eve-story we have two distinct versions; but then that entitles me to the unrevealed tholings.)

Freeman to W. R. W. Stephens, Trinity Sunday 1884, in *The Life and Letters of Edward A. Freeman,* ed. W. R. W. Stephens, 2 vols. (London: Macmillan, 1895), 2:321–22. †John William Burgon (1813–88), Anglican prelate, dean of Chichester Cathedral, and strenuous defender of scriptural inerrancy.

Secondary Literature: R. Bérard, "Edward Augustus Freeman and University Reform in Victorian Oxford," *History of Education* 9 (1980): 287–301.

SIGMUND FREUD (1856–1939)
Austrian Psychoanalyst

At the same time [in Gymnasium] the theories of Darwin, which were then of topical interest, strongly attracted me, for they held out hopes of an extraordinary advance in our understanding of the world; and it was hearing Goethe's beautiful essay on Nature read aloud at a popular lecture by Professor Carl Brühl† just before I left school that decided me to become a medical student.

When I further took into account Darwin's conjecture that men originally lived in hordes, each under the domination of a single powerful, violent and jealous male, there rose before me out of all these components the following hypothesis, or, I would rather say, vision. The father of the primal horde, since he was an unlimited despot, had seized all the women for himself; his sons, being dangerous to him as rivals, had been killed or driven away. One day, however, the sons came together and united to overwhelm, kill, and devour their father, who had been their enemy but also their ideal. After the deed they were unable to take over their heritage since they stood in each other's way. Under the influence of failure and regret they learned to come to an agreement among themselves, they banded themselves into a clan of brothers by the help of the ordinances of totemism, which aimed at preventing a repetition of such a deed, and they jointly undertook to forego the possession of the women on whose account they had killed their father. They were then driven to finding strange women, and this was the origin of exogamy which is closely bound up with totemism. The totem-feast was the commemoration of the fearful dead from which sprang men's sense of guilt (or "original sin") and which was the beginning at once of social organization, of religion and of ethical restrictions.

Sigmund Freud, *An Autobiographical Study* (New York: Norton, 1952), 7, 76. †Brühl (1820–99) was an Austrian zoologist.

Secondary Literature: Lucille B. Ritvo, *Darwin's Influence on Freud: A Tale of Two Sciences* (New Haven, CT: Yale University Press, 1990).

ROBERT FROST (1874–1963)
American Poet

Accidentally On Purpose

The Universe is but the Thing of things
The things but balls all going round in rings.
Some of them mighty huge, some mighty tiny,
All of them radiant and mighty shiny.

They mean to tell us all was rolling blind
Till accidentally it hit on mind
In an albino monkey in a jungle,
And even then it had to grope and bungle.

Till Darwin came to earth upon a year
To show the evolution how to steer.
They mean to tell us, though, the Omnibus
Had no real purpose till it got to us.

Never believe it, at the very worst
It must have had the purpose from the first
To produce purpose as the fitter bred:
We were just purpose coming to a head.

Whose purpose was it? His or Hers or Its?
Let's leave that to the scientific wits.
Grant me intention, purpose, and design—
That's near enough for me in the Divine.

And yet for all this help of head and brain
How happily instinctive we remain,
Our best guide upward further to the light,
Passionate preference such as love at sight.

The Robert Frost Reader, ed. Edward Connery Lathem and Lawrence Thompson
(New York: Macmillan, 2002), 196–97.

Secondary Literature: Robert Faggen, *Robert Frost and the Challenge of Darwin*
(Ann Arbor: University of Michigan Press, 2001).

FRANCIS GALTON (1822–1911)

English Psychometrician and Eugenicist

The publication in 1859 of the *Origin of Species* by Charles Darwin made a marked epoch in my own mental development, as it did in human thought generally. Its effect was to demolish a multitude of dogmatic barriers by a single stroke, and to arouse a spirit of rebellion against all ancient authorities whose positive and unauthenticated statements were contradicted by modern science.

I doubt, however, whether any instance has occurred in which the perversity of the educated classes in misunderstanding what they attempted to discuss was more painfully conspicuous. The meaning of the simple phrase "Natural Selection" was distorted in curiously ingenious ways, and Darwinism was attacked, both in the press and pulpit, by persons who were manifestly ignorant of what they talked about. This is a striking instance of the obstructions through which new ideas have to force their way. Plain facts are apprehended in a moment, but the introduction of a new Idea is quite another matter, for it requires an alteration in the attitude and balance of the mind which may be a very repugnant and even painful process. On my part, however, I felt little difficulty in connecting with the *Origin of Species,* but devoured its contents and assimilated them as fast as they were devoured, a fact which perhaps may be ascribed to an hereditary bent of mind that both its illustrious author and myself have inherited from our common grandfather, Dr. Erasmus Darwin.

I made occasional excursions to visit Charles Darwin at Down, usually at luncheon-time, always with a sense of the utmost veneration as well as of the warmest affection, which his invariably hearty greeting greatly encouraged. I think his intellectual characteristic that struck me most forcibly was the aptness of his questionings; he got thereby very quickly to the bottom of what was in the mind of the person he conversed with, and to the value of it.

Francis Galton, *Memories of My Life* (New York: Dutton, 1909), 287–88, 169.

You have made a convert of an opponent in one sense, for I have always maintained that, excepting fools, men did not differ much in intellect, only in zeal and hard work; and I still think this is an eminently important difference.

Darwin to Galton, 3 December 1869, quoted in Francis Galton, *Memories of My Life* (New York: Dutton, 1909), 290.

Secondary Literature: Nicholas Wright Gillham, *A Life of Sir Francis Galton: From African Exploration to the Birth of Eugenics* (New York: Oxford University Press, 2001).

MAHATMA GANDHI (1869–1948)
INDIAN STATESMAN

Darwin was a great Englishman of the last century who made great scientific discoveries. His memory and his power of observation were amazing. He has written some books which deserve to be read and pondered, with a mass of evidence and arguments, he has shown how man came into being after a large number of experiments and much sifting of evidence, he realized that there was not much difference between the anatomy of man and that of the ape. Whether this conclusion is correct or not has not much to do with ethics. Besides this, Darwin has shown how ideas of morality affect mankind. [. . .]

Though Darwin did not write as a moral philosopher, he has shown how close the connection is between morality and environment. Those who think that morality is unimportant and that physical strength and mental capacity are the only things that matter should read Darwin. [. . .]

Darwin shows further that moral strength is even superior to physical and intellectual strength; and we can see in various ways that a man who has moral qualities lasts longer than one who is devoid of then. Some hold that Darwin taught that strength is enough; that is, those who are physically strong ultimately survive. Superficial thinkers may believe that morality is of no use. But this is not Darwin's view at all. We find from the evidence of the early history of man that races without morality have completely disappeared. The people of Sodom and Gomorrah were extremely immoral and are now therefore completely extinct.

Gandhi, *Collected Works,* 100 vols. (Delhi: Ministry of Information, 1969–94), 6:316–17.

February 5, 1907
Now about what you have been reading. The belief which you refer to was held once. It is no longer held now. People have not understood some of Malthus's theories, and some others are faulty. The law which applies to lower animals does not apply to man. Lower animals live on one another. Man strives to grow out of that condition. It is the spirit of non-violence which prompts him to do so. [. . .] Man's excellence lies in his readiness to let others live and lay down his own life. As he progresses, his food also changes for the better. He has the capacity to grow still further. There have been many more discoveries after Darwin's. The book which you have been reading seems to be an old one. Whether it is old or new, the "Principle of the greatest good of the greatest number," or "survival of the fittest" is false.

Gandhi to Premabehn Kantak, in *Collected Works,* 100 vols. (Delhi: Ministry of Information, 1969–94), 50:309–10.

Secondary Literature: David Arnold, *Gandhi* (New York: Longman, 2001).

JAMES A. GARFIELD (1831–1881)
American President

This liberal [religious] tendency which he [Garfield] anticipated as the result of the wider culture was fully realized. He was emancipated from mere sectarian belief, and with eager interest pushed his investigations in the direction of modern progressive thought. He followed with quickening step into the paths of exploration and speculation so fearlessly trodden by Darwin, by Huxley*, by [John] Tyndall*, and by other living scientists of the radical and advanced type.

James Gillespie Blaine, *Eulogy on the Late President Garfield, delivered in the Hall of Representatives, Washington, D.C., February 27th, 1882* (Philadelphia: Hubbard Brothers, 1882), 34–35.

Secondary Literature: Ira M. Rutkow, *James A. Garfield* (New York: Times Books, 2006).

HAMLIN GARLAND (1860–1940)
American Author

I was led to take up the study of expression in connection with my work in literature, by the idea that all writers are sooner or later dependent upon an artist of expression, whether they get complete hearing or not. That is: *voice* and *action* cannot be written, they are only indicated, and upon the degree of their expressiveness are the authors ranked. The matter of expression has, however, been for the most part in the hands of "professors of Elocution" till Darwin in 1872 put it on a universal basis: and Spencer*, [Alexander] Bain*, Mantegazza[†] and many others have since pushed it into the region of science and into the splendid domain of causes. What I began as an aid to teaching literature I now recognize as a branch of psychology (scientifically) and a great art in the realm of esthetics.

Garland to Edwin Booth, 22 January 1886, in *Selected Letters of Hamlin Garland,* ed. Keith Newlin and Joseph B. McCullough (Lincoln: University of Nebraska Press, 1998), 12. [†]Paolo Mantegazza (1831–1910), Italian physiologist, anthropologist, and evolutionist.

He hung over the rail and looked at the minnows swimming there.

"I wonder if they're the same identical chaps that used to boil and glitter there when I was a boy—looks so. Men change from one generation to another, but the fish remain the same. The same eternal procession of types. I suppose Darwin'ud say their environment remains the same."

Hamlin Garland, *Main-Travelled Roads* (New York: Harper, 1899), 37–38.

Secondary Literature: Keith Newlin, *Hamlin Garland: A Life* (Lincoln: University of Nebraska Press, 2008).

WENDELL PHILLIPS GARRISON (1840–1907)
AMERICAN EDITOR AND AUTHOR

Brocklebank reached New York by Thanksgiving day, which he celebrated with his family in no perfunctory spirit, while they were not a little alarmed for his strange recital of what had befallen him. He paid an early visit to Professor [Othniel C.] Marsh* at New Haven, and was dumfounded by the sight of *Eohippus* already set up in its place. Fearing that scientist's incredulity, he kept silent about his experience with the "living fossil," and contented himself with volunteering to join the next Yale expedition to the Rockies. Meanwhile, he pondered much on the theological problem which had survived the Houyhnhnm cataclysm,[†] and set to work upon a treatise having for epigraph that query of [Thomas] Carlyle's*, in the "Latter-Day Pamphlets,"

Am I not a horse and *half*-brother?

this couplet of Baudelaire's,

Nous sommes des animaux,
Voilà mon système

And Darwin's

but man can do his duty.

Wendell Phillips Garrison, *The New Gulliver* (Jamaica, NY: Marion, 1898), 48–49.
[†]The Houyhnhnms are a race of intelligent horses in Swift's *Gulliver's Travels*.

My little tract is first of all an evolutionary spike, aimed at the abominable Calvinist theology.

Garrison to William R. Thayer, n.d., in *Letters and Memorials of Wendell Phillips Garrison*, ed. Joseph Hetherington McDaniels (Boston: Houghton Mifflin, 1909), 275.

MARCUS GARVEY (1887–1940)
AFRICAN AMERICAN LEADER

President [Warren G.] Harding should not have made that speech. That is not his business. It is for the Negro to speak for himself. We have been expecting others to do too much for us, and that is why everybody passes us by. The time has come for the Negro to speak for himself. [...] Speak for yourselves. That is, if you believe yourselves to be men. If you still think with Darwin, then you can allow someone else to speak for you. If you believe you are the missing link, then let someone else speak for you, but if you believe you are men, speak for yourselves as men.

Marcus Garvey, speech given in Washington, DC, on 20 November 1921, in *Marcus Garvey and the Universal Negro Improvement Association Papers,* ed. Robert A. Hill, 10 vols. (Berkeley and Los Angeles: University of California Press, 1983–2006), 4:205–6.

Secondary Literature: Colin Grant, *Negro with a Hat: The Rise and Fall of Marcus Garvey* (Oxford: Oxford University Press, 2008).

ELIZABETH GASKELL (1810–1865)
English Novelist

He was a tall powerfully-made young man, giving the impression of strength more than elegance. His face was rather square, ruddy-coloured (as his father had said), hair and eyes brown—the latter rather deep-set beneath his thick eyebrows; and he had a trick of wrinkling rather up his eyelids when he wanted particularly to observe anything, which made his eyes look even smaller still at such times. He had a large mouth, with excessively mobile lips; and another trick of his was, that when he was amused at anything, he resisted the impulse to laugh, by a droll manner of twitching and puckering up his mouth, till at length the sense of humour had its way, and his features relaxed, and he broke into a broad sunny smile; his beautiful teeth—his only beautiful feature—breaking out with a white gleam upon the red-brown countenance. These two tricks of his— of crumpling up the eyelids, so as to concentrate the power of sight, which made him look stern and thoughtful; and the odd twitching of the lips that was preliminary to a smile, which made him look intensely merry— gave the varying expressions of his face a greater range "from grave to gay, from lively to severe," than is common with most men.

Elizabeth Gaskell, *Wives and Daughters* (London: Dent, 1966), 94–95, description of Roger Hamley patterned after the young Charles Darwin. See Winifred Gérin, *Elizabeth Gaskell: A Biography* (Oxford: Clarendon, 1976), 98.

Secondary Literature: Winifred Gérin, *Elizabeth Gaskell: A Biography* (Oxford: Clarendon, 1976); Philip J. Waller, *Writers, Readers, and Reputations: Literary Life in Britain, 1870–1918* (Oxford: Oxford University Press, 2006).

PATRICK GEDDES (1854–1932)
English Biologist and Social Scientist

Here two little anecdotes of Darwin may make this normal naturalist's progress more vivid. One day, many years ago, the writer of this section— then an assistant to Burdon Sanderson and Schafer[†] at University College—was amusing a spare hour by searching a pond-sample with his microscope, and had drawn a comparative blank with his microscope, with only two or three common green Euglenas swimming amid a few motile bacilli. He was about to put this slide away, when he was gently pushed aside. A big bear came over his shoulder—here was Darwin! who had come in unnoticed. He said nothing, but looked closely into this—to me—barren microscopic-field: then suddenly broke out, positively shouting for joy; "I say! They're moving, they're *moving!* Sanderson! Sanderson! Come and see; they're MOVING! Look at that!"

Was not here a vivid and memorable lesson in biology—this literally Panic intoxication of ecstasy. In our oldest of veterans, greatest of masters, before this simplest spectacle of life!

The other story, from a year or so later, was told me by a close friend, George Murray, then keeper of Cryptograms in the Natural History Museum, as he returned from a week-end at Down, to which Darwin had invited a few other younger science-men from the Museum and from Cambridge. Murray told how Darwin had spent the previous evening questioning each, and drawing him out on his subject; for no man was more open and eager to learn. Then, leaning back in his chair, he said: "I am always feeling my ignorance, but never have I had it more strongly brought home to me than tonight. You have surprised me!—and again and again! What you (pointing in turn to each) "know about cryptogams, and you tell me about phanerogams, and you about bacteriology, and you about embryology, and you about fishes, and so on, is most interesting! It's something astonishing! You do indeed make me feel my ignorance, and what I have missed!" Pause: then jumping up from his chair, and with a thump on the table: "But—damn you!—there's not a Naturalist in the whole lot of you!"

Arthur J. Thompson and Patrick Geddes, *Life: Outlines of General Biology*, 2 vols. (New York: Harper & Brothers, 1931), 2:1454–55. †John Burdon Sanderson (1828–1906), English physiologist, and Edward Albert Sharpey-Schafer (1850–1935), English physiologist.

Secondary Literature: Lewis Mumford and Patrick Geddes: The Correspondence, ed. Frank Novak Jr. (New York: Routledge, 1995).

CARL GEGENBAUR (1826–1903)
German Anatomist

[Gegenbaur and Haeckel*] were very good friends, these two. They were drawn together by the strong magnetism of two true natures that understood each other to the golden core, though in other respects they were as different as possible. Gegenbaur was no enthusiast. His ideal was "to keep cool to the very heart." But he was at one with Haeckel in a feeling for a broad outlook in scientific research. He never shrank from large connections or vast deductions, as long as they were led up to by a sober and patient logic. This logical character he afterwards recognized in Darwin's idea of evolution, and so the friends once more found themselves in agreement, and for a long time they were a pair of real Darwinian Dioscuri. This feeling for moderation and at the same time for far-reaching logic was combined in Gegenbaur with a certain steady and unerring independence of character. He made little noise, but he never swerved from his aim. What he accomplished with all these qualities, in many other provinces besides Darwinism, cannot be told here. It may be read in the history of zoology. He had, as far as such a thing was possible, a restful influence of the most useful character on Haeckel. If we imagine what Darwinism would have become in the nineteenth century in the hands of such men as Gegenbaur, without Haeckel, we can appreciate the difference in temperament between the two men. With Gegenbaur evolution was always a splendid new technical instrument that no layman must touch for fear of spoiling it. With Haeckel it became a devouring wave, that will one day, perhaps, give its name to the century. In other natures these differences might have led to open conflict. But Haeckel and Gegenbaur show us that, like so many of our supposed "differences," they can at least live together in perfect accord in the freshest years of life, each bearing fruit in its kind.

Wilhelm Bölsche, *Haeckel, His Life and Work,* trans. Joseph McCabe (London: T. Fisher Unwin, 1906), 63–64.

Secondary Literature: Sander Gliboff, *H. G. Bronn, Ernst Haeckel, and the Origins of German Darwinism* (Cambridge, MA: MIT Press, 2008).

ARCHIBALD GEIKIE (1835–1924)
Scottish Geologist

The fundamental principles of Stratigraphy having been well established before the middle of last century, this branch of geological science has during the last fifty years undergone a remarkable expansion from four influences. Firstly, it has been profoundly modified by the writings of Darwin; secondly, it has been greatly affected by the introduction of zonal classification among the fossiliferous formations; thirdly, it has been augmented by the rise and extraordinary development of Glacial Geology; and lastly, it has enormously gained by the multiplication of detailed geological maps.

Charles Darwin (1809–1882) contributed several valuable works to the literature of geology. But it is not for these that I now cite his name. The two geological chapters in his *Origin of Species* produced the greatest revolution in geological thought which has occurred in my time. Young students, who are familiar with the ideas there promulgated, can hardly realize the effect of them on an older generation. They seem now so obvious and so well-established, that it may be difficult to conceive a philosophical science without them.

To most of the geologists of his day, Darwin's contention for the imperfection of the geological record, and his demonstration of it, came as a kind of surprise and awakening. They had never realized that the history revealed by the long succession of fossiliferous formations, which they had imagined to be so full, was in reality so fragmentary. And yet when Darwin pointed out this fact to them, they were compelled, sometimes rather reluctantly, to admit that he was right. Some of them at once adopted the idea, as [Andrew Crombie] Ramsay* did, and carried it further into detail. Until Darwin took up the question, the necessity for vast periods of time, in order to explain the characters of the geological record, was very inadequately comprehended.

Archibald Geikie, *The Founders of Geology,* 2nd ed. (London: Macmillan, 1905), 438–39.

Secondary Literature: David Roger Oldroyd, *The Highlands Controversy: Constructing Geological Knowledge in Fieldwork in Nineteenth-Century Britain* (Chicago: University of Chicago Press, 1990); John Challinor, "Archibald Geikie," in *Dictionary of Scientific Biography,* ed. Charles Coulston Gillispie, 16 vols. (New York: Scribner's, 1970–80), 5:333–38; Sandra Herbert, *Charles Darwin, Geologist* (Ithaca, NY: Cornell University Press, 2005).

JAMES GEIKIE (1839–1915)
Scottish Geologist

I heard all about your Shetland work. It did my heart good, and right glad I am that it has been done by a Survey man. [...] You would hear about [Sydney] Skertchly's† find. I was down there again ten days ago at [Andrew Crombie] Ramsay's* request, to see the evidence again. [...] In my new edition, which is out (and selling well!), I go much more fully into the English Drifts. I got to-day a long letter from Darwin, along with a copy of his new edition of *Geological Observations.* His letter is very complimentary, and of course that is gratifying to me, for I look upon Darwin as a real genius.

Geikie to John Home, 18 November 1877, in *James Geikie, the Man and the Geologist,* by Marion I. Newbigin and J. S. Flett (London: Gurney & Jackson, 1917), 73.
†Sidney Skertchly (1850–1926), English-Australian geologist.

Secondary Literature: John Challinor, "James Geikie," in *Dictionary of Scientific Biography,* ed. Charles Coulston Gillispie, 16 vols. (New York: Scribner's, 1970–80), 5:338–39.

ISIDORE GEOFFROY SAINT-HILAIRE (1805–1861)
French Zoologist

Is natural selection, as Darwin believes and he has ingeniously undertaken to demonstrate, the means generally employed by nature of creating new types? There is *at least* reason to doubt it; but even granting this, one could still, despite generally incontestable analogy, refuse to accept the parallel between *natural selection* and the selection practiced by your agriculturalists, which Darwin uses to explain the multiplication of species. The work of the agriculturalists is so little natural, that not only has nature never brought new races to birth in this way. But also tends ceaselessly to cause such variations to disappear.

Isidore Geoffroy Saint-Hilaire, *Histoire naturelle général,* 3 vols. (Paris: V. Masson, 1854–62), 3:522–23, quoted in Robert Stebbins, "France," in *The Comparative Reception of Darwinism,* ed. Thomas F. Glick, 2nd ed. (Chicago: University of Chicago Press, 1988), 124.

Secondary Literature: Toby Appel, *The Cuvier-Geoffroy Debate: French Biology in the Decades before Darwin* (New York: Oxford University Press, 1987).

ALFRED GIARD (1846–1908)
French Marine Biologist

In 1867 I entered the École Normal [Superièure]. Darwin, undervalued by our official scientists, was made known to me only by an article published in *Le Magasin Pittoresque.* The translation of the *Origin of Species* had just appeared, and when I bought it near the Odéon it was a revelation for me. It had an enormous personal influence on the orientation of my ideas and on all my scientific life. So, without ever having met her personally, I contracted an enormous debt of recognition towards Madame Clémence Royer*.

Giard to organizers of a banquet honoring Clémence Royer, 22 February 1897, quoted in Joy Harvey, *"Almost a Man of Genius": Clémence Royer, Feminism, and Nineteenth-Century Science* (New Brunswick, NJ: Rutgers University Press, 1997), 171.

Certainly the conceptions of Darwin were in many respects justified by experiment, even in the strictest sense of the word, and Darwin has proved it himself by his beautiful investigations concerning self- and cross-fertilization and concerning climbing plants and carnivorous plants, etc. But it is necessary to recognize how many experimental verifications relative to natural selection, to heredity, demand conditions rarely realized, a length of time which renders them easy of accomplishment only by a group of persons (societies of scholars), or necessitate large resources which most investigators cannot command.

Apart from some brilliant exceptions [. . .] the disciples of Darwin who have followed most closely the tendencies of their master have understood experimentation in the very large sense that we give to this word as applied to a great number of investigations relative to secondary factors.

The importance of the study of primary factors in evolution did not escape Darwin, but, excellent observer though he was, he was undoubtedly dismayed by the complexity of the role of these factors and did not attempt to disentangle the mechanisms which give rise to the numberless variations of living beings.

Alfred Giard, "The Present Tendencies of Morphology and its Relations to the Other Sciences," in *Congress of Arts and Sciences: Universal Exposition, St. Louis, 1904,* ed. Howard J. Rogers, vol. 5 (Boston: Houghton Mifflin, 1906), 258–82, on 272–73.

Secondary Literature: François Bouyssi, *Alfred Giard, 1846–1908, et ses élèves: Un cenacle de philosophes biologistes. Aux origines du scientisme?* (Paris: École Pratique des Hautes Études, 1998), microfiche.

JAMES CARDINAL GIBBONS (1834–1921)
AMERICAN PRELATE

Some belittle Mr. [Thomas A.] Edison as a mere mechanic. I have no patience with such views. He is the representative of American inventive genius and has brought glory upon our country in the whole world; and he has paid the penalty just as so many of our great men do, just as Darwin did.

Darwin bemoaned at the end of his life that his intense devotion to scientific study had atrophied his sense of poetry, of music, and I know not what; I would add, his sense of religion, for the religious spirit, if not cultivated, will die. So has it been with Mr. Edison; he has maimed his own mind, just as Darwin did, by a too one-sided exercise of its powers. He talks with great freedom, and, I may say, with not a little contempt, of theology; but one suspects that he has been too occupied, and perhaps too contemptuous, of theology, to devote much time to its study. One suspects that his acquaintance with it is almost limited to fragmentary reminiscences of sermons heard in boyhood days.

Even the Pope does not dogmatize until the question has been discussed for centuries and settled by the voice of experts. But here is a scientist who proclaims dogmas to the public; and he seems to ask them to believe them—because he believes them. [. . .] I do not know for whom he speaks.

"Gibbons answers Edison's Soul Talk; Inventor has Maimed his Mind, as Darwin did, Declares the Cardinal," *New York Times,* 19 February 1911.

ANDRÉ GIDE (1869–1951)
FRENCH AUTHOR

Our two letters crossed. That makes a false pass, and all is to be done again. There was no use trying to surround my illness with mystery—my mother could not ignore it; her anxiety was even too great, I am already much better but just escaped being gravely ill. Nevertheless, I am going to remain in Biskra all winter and I must take good care of myself. I have never been less bored. I am reading a lot of German; I amuse myself with Darwin.

Gide to Paul Valéry, December 1893, in Paul Valéry, *Moi,* trans. Marthiel Matthews and Jackson Mathews (Princeton, NJ: Princeton University Press, 1975), 148.

W. S. GILBERT (1836–1911)
English Poet and Lyricist

The Ape And The Lady

A LADY fair, of lineage high,
Was loved by an Ape, in the days gone by —
The Maid was radiant as the sun,
The Ape was a most unsightly one —
So it would not do —
His scheme fell through;
For the Maid, when his love took formal shape,
Expressed such terror
At his monstrous error,
That he stammered an apology and made his 'scape,
The picture of a disconcerted Ape.

With a view to rise in the social scale,
He shaved his bristles, and he docked his tail,
He grew moustachios, and he took his tub,
And he paid a guinea to a toilet club.
But it would not do,
The scheme fell through —
For the Maid was Beauty's fairest Queen,
With golden tresses,
Like a real princess's,
While the Ape, despite his razor keen,
Was the apiest Ape that ever was seen!

He bought white ties, and he bought dress suits,
He crammed his feet into bright tight boots,
And to start his life on a brand-new plan,
He christened himself Darwinian Man!
But it would not do,
The scheme fell through —
For the Maiden fair, whom the monkey craved,
Was a radiant Being,
With a brain far-seeing —

While a Man, however well-behaved,
At best is only a monkey shaved!

W. S. Gilbert, *Songs of a Savoyard* (London: Routledge, 1891), 51–52.

Secondary Literature: Philip J. Waller, *Writers, Readers, and Reputations: Literary Life in Britain, 1870–1918* (Oxford: Oxford University Press, 2006).

CHARLOTTE PERKINS GILMAN (1860–1935)
American Novelist and Feminist

Not woman, but the condition of woman, has always been a doorway of evil. The sexuo-economic relation has debarred her from the social activities in which, and in which alone, are developed the social virtues. She was not allowed to acquire the qualities needed in our racial advance; and, in her position of arrested development, she has maintained the virtues and the vices of the period of human evolution at which she was imprisoned. At a period of isolated economic activity—mere animal individualism—at a period when social ties ceased with the ties of blood, woman was cut off from personal activity in social economics, and confined to the functional activities of her sex.

Charlotte Perkins Gilman, *Women and Economics: A Study of the Economic Relation Between Men and Women as a Factor in Social Evolution* (Boston: Small Maynard, 1898), 329–30.

If our business here is social progress and improvement, they are the most virtuous; the most noble, who lead most people farther onward. Jesus, with his vision of human unity and human love; Lamarck, Wallace* and Darwin, with their vision of Life in Motion—the great world-hope of evolution; Lester Ward*, with his vision of the true relation of the sexes—such as these stand highest for our psychic advancement. The mechanical world helpers we are more familiar with, but do not properly rate their social value. Such men as the Wright Brothers, [Luther] Burbank*, Edison and Marconi—in all their splendid numbers, are world benefactors of the highest degree.

Charlotte Perkins Gilman, *Social Ethics: Sociology and the Future of Society* (1914; reprint, Westport, CT: Praeger, 2004), 102.

Secondary Literature: Minna Doskow, "Charlotte Perkins Gilman: The Female Face of Social Darwinism," *Weber: The Contemporary West* 14, no. 3 (1997).

GEORGE GISSING (1857–1903)
English Novelist

Now you speak of Father [Thomas Gissing]. During his lifetime I was of course too young to understand his thoughts, but in looking back I see clearly that he knew & accepted all these truths. Look at the books he collected. When he was quite a young man he read *The Leader*, a radical and free-thinking paper edited by George Henry Lewes*. Later on he bought Darwin, he bought Lewes' *Hist. of Philosophy*, & many other books tending to throw light on deep questions. I remember once he told me distinctly as we were sitting together that the old theories of the creation of the world etc. were quite untenable.

Gissing to Algernon Gissing, 9 May 1880, in *The Collected Letters of George Gissing*, ed. Paul F. Mattheisen, Arthur C. Young, and Pierre Coustillas, vol. 1 (Athens: Ohio University Press, 1990), 270.

[Auguste] Comte is for me the supplement to Darwin; the theories of both point to the same result, and must be true!

George Gissing, *Workers in the Dawn: A Novel*, 3 vols. (London: Remington, 1880), 1:216.

Spent the evening in a troubled frame of mind, occasionally glancing at Darwin's *Origin of Species.*—a queer jumble of thoughts.

Gissing's Diary, 7 November 1889, quoted in *George Gissing: A Life*, by Paul Delaney (London: Weidenfeld & Nicolson, 2008), 168.

I have just been reading in the *Spectator* a notice of [Benjamin] Kidd's* new book [*Principles of Western Civilization*], where I find that Huxley* is spoken of as a retrograde force, as one who misinterpreted, & spoilt, the message of Darwin. This, I fancy, is rubbish. It is all very well for the sun to be shining in heaven; but what if a lot of pestilent chimneys vomit so much smoke that you walk about gasping in gloom? Someone must *have at* the folk who so befoul the atmosphere, & make them consume their evil vapors. Huxley was a great "cloud-compeller." He did, to a wonderful extent, let in the light.

Gissing to Edward Clodd, 1 March 1902, in *The Collected Letters of George Gissing*, ed. Paul F. Mattheisen, Arthur C. Young, and Pierre Coustillas, vol. 8 (Athens: Ohio University Press, 1996), 350.

"Why don't people write about the really important things of life?" Gissing makes one of his characters explain, and at the unexpected cry the horrid burden of fiction begins to slip from the shoulders. [...] Here in Gissing is a gleam of recognition that Darwin had lived, that science was developing, that people read books, that once upon a time there was such a place as Greece. It is the consciousness of these things that makes his books such painful reading.

Virginia Woolf, *The Second Common Reader*, ed. and intro. Andrew McNeillie (New York: Harcourt, 1986), 223.

Secondary Literature: London and the Life of Literature in Late Victorian England: The Diary of George Gissing, ed. Pierre Coustillas (Hassocks, UK: Harvester, 1978).

WILLIAM GLADSTONE (1809–1898)
English Prime Minister

On the Sunday afternoon [in 1877] Sir John Lubbock*, our host, took us all up to the hilltop whence in his quiet Kentish village Darwin was shaking the world. The illustrious pair, born in the same year, had never met before. Mr. Gladstone as soon as seated took Darwin's interest in lessons of massacre for granted and launched forth his thunderbolts with unexhausted zest. His great, wise, simple, and truth-loving listener, then, I think, busy with the digestive powers of the *Drosera*† in his green-house, was intensely delighted. When we broke up, watching Mr. Gladstone's erect alert figure as he walked away, Darwin, shading his eyes with his hand against the evening rays, said to me in unaffected satisfaction, "What an honor that such a great man should come to visit me!" Too absorbed in his one overwhelming conflict with the powers of evil, Mr. Gladstone makes no mention of his afternoon call, and only says of the two days that "he found a notable party, and made interesting conversation," and that he could not help liking one of the company, then a stranger to him.

John Morely, *The Life of William Ewart Gladstone,* new ed., 2 vols. (New York: Macmillan, 1911), 2:562. †*Drosera rotundifolia,* an insectivorous plant, the subject of experiments by Darwin.

Open Letter to Robert G. Ingersoll* [1888]

[T]he system of Mr. Darwin is hurled against Christianity as a dart which cannot but be fatal: "His discoveries, carried to their legitimate conclusion, destroys the creeds and sacred Scriptures of mankind." [...] On what ground [...] and for what reason, is the system of Darwin fatal to Scriptures and creeds? [...] It is not possible to discover from [Ingersoll's] random language [...] whether the scheme of Darwin is to sweep way all theism, or is to be content with extinguishing revealed religion. If the latter is meant, I should reply that the moral history of man [...] has been distinctly an evolution from the first until now; and that the succinct though grand account of the Creation in Genesis is singularly accordant with the same idea, but is wider than Darwinism, since it concludes in the gradual progression of the inanimate world as well as the history of organisms. [...]

If, however, the meaning be that thesis is swept away by Darwinism, I observe that, as before, we have only an unreasoned dogma or dictum to deal with, and, dealing perforce with the unknown, we are in danger of striking at a will of the wisp. Still, I venture on remarking that the doctrine of Evolution has acquired both praise and dispraise which it does not deserve. It is lauded in the skeptical camp because it is supposed to get rid of the shocking idea of what are termed sudden acts of creation; and it is as unjustly dispraised, because it is thought to bridge over the gap between man and the inferior animals, and to give emphasis to the relationship between them. [...]

One striking effect of the Darwinian theory of descent is [...] to reduce the breadth of all intermediate distinctions in the scale of animal life. It does not bring all creatures into a single lineage, but all diversities are traced back, at some point in the scale and by stage indefinitely minute, to a common ancestry. All is done by steps, nothing by strides, leaps, or bounds; all from protoplasm up to Shakespeare.

The Works of Robert G. Ingersoll, Dresden Memorial Edition, 12 vols. (New York: C. P. Farell, 1900), 6:233–35.

Secondary Literature: Owen Chadwick, *The Victorian Church,* 2 vols. (New York: Oxford University Press, 1966–70).

ELLEN GLASGOW (1873–1945)
AMERICAN NOVELIST

It is true, nevertheless, that even in my efforts and my failure, in my belief and my skepticism, I had arrived at the basis of what I may call a determining point of view, if not a philosophy. In my endless curiosity about life, I had fallen by a happy accident (how else could this occur in my special environment?) upon a strayed copy of *The Origin of Species;* and this single book had led me back, through biology, to the older theory of evolution. The Darwinian hypothesis did not especially concern me, nor was I greatly interested in the scientific question of its survival. In any case, it was well able, I felt, to take care of itself and fight its own battles. What did interest me, supremely, was the broader synthesis of implications and inferences. On this foundation of probability, if not of certainty, I have found—or so it still seems to me—a permanent resting place: and in the many years that have come and gone, I have seen no reason, by and large, to reject this cornerstone of my creed.

Clifton Fadiman, ed., *I Believe: The Personal Philosophies of Certain Eminent Men and Women of Our Time* (New York: Simon & Schuster, 1939), 101–2.

Secondary Literature: J. R. Raper, *Without Shelter: The Early Career of Ellen Glasgow* (Baton Rouge: Louisiana State University Press, 1971).

EDMUND GOSSE (1849–1928)
Englᴉsн Poᴇᴛ

This was the great moment in the history of thought when the theory of the mutability of species was preparing to throw a flood of light upon all departments of human speculation and action. It was becoming necessary to stand emphatically in one army or the other. Lyell* was surrounding himself with disciples who were making strides in the direction of discovery. Darwin had long been collecting facts with regard to the variation of animals and plants. Hooker* and Wallace*, Asa Gray* and even Agassiz*, each in his own sphere, were coming closer and closer to a perception of that secret which was first to reveal itself clearly to the patient and humble genius of Darwin. In the year before, in 1856, Darwin, under pressure from Lyell*, had begun that modest statement of the new revelation, that "abstract of an Essay," which developed so mightily into *The Origin of Species*. [Thomas Vernon] Wollaston's* *Variation of Species* had just appeared, and had been a nine days' wonder in the wilderness.

On the other side, the reactionaries, although never dreaming of the fate which hung over them, had not been idle. In 1857 the astounding question had for the first time been propounded with contumely, "What, then, did we come from an orang-outang?" The famous *Vestiges of Creation* had been supplying a sugar-and-water panacea for those who could not escape from the trend of evidence, and who yet clung to revelation. [Richard] Owen* was encouraging reaction by resisting, with all the strength of his prestige, the theory of the mutability of species.

In this period of intellectual ferment, as when a great political revolution is being planned, many possible adherents were confidentially tested with hints and encouraged to reveal their bias in a whisper. It was the notion of Lyell, himself a great mover of men, that, before the doctrine of natural selection was given to a world which would be sure to lift up at it a howl of execration, a certain bodyguard of sound and experienced naturalists, expert in the description of species, should be privately made aware of its tenor. Among those who were thus initiated, or approached with a view towards possible illumination, was my Father [Philip Henry Gosse]. He was spoken to by Hooker, and later on by Darwin, after meetings of the Royal Society in the summer of 1857.

My Father's attitude towards the theory of natural selection was critical in his career, and oddly enough, it exercised an immense influence on my own experience as a child. Let it be admitted at once, mournful as the admission is, that every instinct in his intelligence went out at first to greet the new light. It had hardly done so, when a recollection of the opening chapter of "Genesis" checked it at the outset. He consulted with [William Benjamin] Carpenter*, a great investigator, but one who was fully as incapable as himself of remodelling his ideas with regard to the old, accepted hypotheses. They both determined, on various grounds, to have nothing to do with the terrible theory, but to hold steadily to the law of the fixity of species. It was exactly at this juncture that we left London, and the slight and occasional but always extremely salutary personal intercourse with men of scientific learning which my Father had enjoyed at the British Museum and at the Royal Society came to an end.

Edmund Gosse, *Father and Son* (London: Heinemann, 1907), 117–19.

Secondary Literature: Philip J. Waller, *Writers, Readers, and Reputations: Literary Life in Britain, 1870–1918* (Oxford: Oxford University Press, 2006).

STEPHEN JAY GOULD (1941–2002)
American Paleontologist

Darwin's central theory of natural selection is about advantages [. . .] that accrue to individuals, not to species. In fact, this counterintuitive proposal—that individual bodies, not "higher" groups like species, act as units and targets of natural selection—lies at the heart of Darwin's radicalism, and explains a large part of our difficulty in grasping and owning his powerful idea. Natural selection may lead to benefits for the species, but these "higher" advantages can only arise as *sequelae,* or, side consequences, of natural selection's causal mechanism: differential reproductive success *of individuals.*

Warm and fuzzy ideas about direct action for the good of species represent a classical form of spin doctoring that has precluded proper understanding of natural selection for more than a century. If evolution worked explicitly for species, then we could soften the blow of Darwin's radicalism. The transition from God's overt beneficence *toward* species to evolution's direct operation *on* species permits a soft landing in transferring allegiance from creationism to evolution—for the central focus on "higher" good as *raison d'etre* remains unchanged.

But Darwin's real theory of natural selection is a theory of ultimate individualism. Darwin's mechanism works through the differential reproductive success of individuals who, by fortuitous possession of features rendering them more successful in changing local environments, leave more surviving offspring. Benefits accrue thereby to species in the same paradoxical and indirect sense that Adam Smith's economic theory of laissez faire may lead to an ordered economy by freeing individuals to struggle for personal profit alone—no accident in overlap, because Darwin partly derived his theory of natural selection as a creative intellectual transfer from Smith's ideas.

Stephen Jay Gould, "Spin Doctoring Darwin," *Natural History,* July 1995, 6–9, 70–71, on 8.

Secondary Literature: Warren D. Allmonn, Patricia H. Kelley, and Robert M. Ross, eds., *Stephen Jay Gould: Reflections on His View of Life* (New York: Oxford University Press, 2008).

MOUNTSTUART E. GRANT DUFF (1829–1906)
Scottish Politician and Author

15 January 1871: Went over from High Elms with [John] Lubbock*, Huxley* and the chancellor of the Exchequer, to call on Darwin, whom Lowe† had never seen since they met as quite young men, on two neighboring reading [*sic*] parties forty years ago. We stayed as long as it was safe, for a very little too much talking brings on an attack of the violent sickness which has been the bane of the great philosopher's life. As we returned, Huxley expressed the opinion, which was probably correct that no man now living had done so much to give a new direction to the human mind. "Ah," said Lowe, "you think him the top-sawyer of these times." "Yes," said the other.

Mountstuart E. Grant Duff, *Notes from a Diary, 1851–1872,* 2 vols. (London: John Murray, 1897), 2:188. †Robert Lowe, Lord Sherbrooke (1811–92), English-Australian statesman.

4 October 1874: I am spending some days at High Elms, my wife being at the sea, and to-day we have a large and interesting party—[John] Tyndall*, fresh from his address at Belfast, which has been making such a sensation, the Millers,[†] Herbert Spencer*, and Mr. Sorby,[‡] who has been working lately at the subject of color, but is also known as a geologist and mineralogist, for his service to which sciences he has received the Hugo [*sic*] Boerhaave medal—a mass of gold of the value of 1000 francs, but not, I thought, very beautiful. In the afternoon we went over to see Darwin, who was extremely lively and talked a good deal about the insect-eating plants on which he has been lately at work.

15 December 1880: Drove with my hostess to Liverpool. She told me that she had lately explained to Darwin the state of her sight, which is very peculiar. "Ah! Lady Derby,"[§] said the great philosopher, "how I should like to dissect you!"

Lowe had asked him, some little time ago, on what he was engaged? "Chiefly at present upon radicles," was the reply. "H'm," said the other, "I know something about *them*."

16 January 1881: In the afternoon we walked up [from John Lubbock's*] to see Darwin. He has of late been studying earthworms, and said to Lubbock, "You antiquarians ought to have great respect for them; they have done more to preserve tessellated pavements than any other agency. I have ascertained, by careful examination, that the worms on a single acre of land bring up ten tons of dry earth to the surface in a year."

26 March 1881: Bywater[††] quoted a passage from a sermon of Burgon's[‡‡] against Darwin: "If they leave me my ancestors in Paradise, I am content to leave them theirs in the zoological Gardens!"

14 August 1881: Walked from High Elms with Lubbock to see Darwin. Spottiswoode, Rothery the Wreck Commissioner, Lockyer,[§§] and many others were there.

Mountstuart E. Grant Duff, *Notes from a Diary, 1873–1881*, 2 vols. (London: John Murray, 1898), 1:85, 2:283, 289, 300, 347. [†]Probably William Hallows Miller (1801–80), English crystallographer and mineralogist, and his wife. [‡]Henry Clifton

Sorby (1826–1908), English geologist and microscopist. [§]See Thomas Carlyle entry. [††]Ingram Bywater (1840–1914), English classical scholar. [‡‡]See Edward A. Freeman entry. [§§]Henry Cadogan Rothery (1817–88), commissioner of wrecks (government official charged with investigating shipwrecks); and Joseph Norman Lockyer (1836–1920), English astronomer, editor of *Nature*. For William Spottiswoode, *see* Thomas Carlyle entry.

Secondary Literature: Max Jones, "Measuring the World: Exploration, Empire and the Reform of the Royal Geographical Society, c. 1874–93," in *The Organisation of Knowledge in Victorian Britain,* ed. Martin J. Daunton (Oxford: Oxford University Press, 2005), 313–36.

ASA GRAY (1810–1888)
American Botanist

This morning [22 January 1839] we breakfasted with Richard Taylor[†] in the city and went afterwards to the College of Surgeons, by appointment Sir Wm. Hooker[‡] had made to see Prof. [Richard] Owen* and the fine museum of the College under his charge (John Hunter's[§] originally); a magnificent collection it is, in the finest possible order, and the arrangement and plan of the rooms is far, far, prettier and better than any I have seen. We there met Mr. Darwin, the naturalist who accompanied Captain Fitz-Roy in the Beagle. I was glad to form the acquaintance of such a profound scientific scholar as Professor Owen—the best comparative anatomist living, still young, and one of the most mild, gentle, childlike men I ever saw. He gave us a great deal of interesting information, and showed us personally throughout the whole museum.

From Gray's journal, in *Letters of Asa Gray,* ed. Jane Loring Gray, 2 vols. (Boston: Houghton Mifflin, 1893–94), 1:117. [†]Richard Taylor (1781–1858), English naturalist and editor of *Scientific Memoirs* from 1837 to 1852. [‡]William Jackson Hooker (1785–1865), English botanist, father of Joseph Dalton Hooker. [§]John Hunter (1728–93), surgeon, built a famous museum to house his anatomical collection.

Dr. Gray made his fifth journey to Europe in the fall of 1868. He landed in September, and went at once to Kew, where he remained most of the time at work in the herbarium until November. He made a short round of visits, first to Mr. Church, who was then rector of Whatley, a village of Somersetshire, where, with Mrs. Gray, he enjoyed to the full his stay in one of the loveliest parts of rural England. They went also to Down to pay a visit to Darwin, and with them went Dr. and Mrs. Hooker*, with their two eldest children, and Professor [John] Tyndall*. Those were days never to be forgotten. [. . .] In England he again worked at Kew, and repeated the visits at Whatley and Down, sailing for America, November 9, 1869.

Letters of Asa Gray, ed. Jane Loring Gray, 2 vols. (Boston: Houghton Mifflin, 1893–94,) 2:565.

Secondary Literature: A. Hunter Dupree, *Asa Gray* (New York: Atheneum, 1968).

JOHN RICHARD GREEN (1837–1883)
ENGLISH HISTORIAN

[Oxford, July 1860]

I saw your "friend" Dr. Falconer† the other day. A good-humored, jocular Irishman, whom Lyell* styled second as a paleontologist to [Richard] Owen* only! So you measure swords with a creditable antagonist. He has not as yet read a paper, but he rose to speak on one of the most notable which I have as yet heard at the B. A. [British Association for the Advancement of Science] A Mr. C. Moore who lives at Bath, found in a quarry in his neighborhood a small drift-deposit of the Triassic Epoch. He carted two tons of it home, a distance of twenty miles, and spent two years in washing, sorting, and microscopically examining it. He was thus enabled to exhibit about three hatfuls of fish teeth; a similar quantity of scales, etc.,—but what was of real importance some twenty small jaw-bones, etc., of *mammals*—unmistakable mammals, *judicibus professore Eugeaco et Doctore Auceps* (is not that the Latin for Falconer?). This brings them far lower, you see, than even the Stonesfield slate. Some two such remains have been found in the Muschelkalk in Germany—with which this may be about contemporaneous—but they have been fought over and disputed. These twenty put an extinguisher on all question. Lyell made a beautiful speech on the matter. Paucity of remains, he argued, do not argue paucity of animal life. Were we left to infer the animal creation of the present day from the deposits of the Ganges or the Nile, should we be content merely with the few species we might light on? Rather (and here he brought beautifully in the principle of the correlation of life) should we not be bound to infer from these few a large quantity of species as yet unfound? Important too—he said—was the fact that up to that time all the animals thus discovered were very minute, while in this last deposit were found remains which must have been of an animal as large as a pole-cat, a size which at once sweeps away all hypotheses founded on this fact of minuteness, and gives us an ordinary link in the common series of animal life. Strongly Darwinian, he? and strongly common-sense too. I have (after finishing Lyell) been reading Hugh Miller's[‡] posthumous book, his sketches (originally intended, had he lived, to form the basis of a geologic History of Scotland), and I have been much struck with the utter weakness of the theory to which he clings so very fondly, of the "definiteness" of organic life, its "dead stops," etc. Such a theory required and justified that dioramic view of geology which Miller adopts—picture succeeding picture in strong contrast—but which seems to me utterly unwarrantable and unscientific. Read after Lyell he strikes me as a man who gathered up the researches of others and gave them a dash of the picturesque. I am afraid I am boring you (I always bore my friends with the subject I have on hand), but now I am on Hugh Miller it reminds me that I have made extracts from him—one of which (of the period of the Tertiaries down to the post-Pliocene and human epochs) seems truthful and good. It will be useful for your sketch of the bone-cave period, as it is drawn out in great detail. His oolite reminded me funnily of yours (a great compliment by the way), but has an Iguanodon in its menagerie which I don't think you possessed. It may be fun to read, so I will bring my notebook down. I was introduced to Robert Chambers (the supposed author of the *Vestiges*) the other day, and heard him chuckle over the Episcopal defeat. I haven't told you that story, have I? On Saturday morning I met Jenkin[§] going to the Museum. We joined company, and he proposed going to Section D, the Zoology, etc., "to hear the Bishop of Oxford [Samuel Wilberforce*] smash Darwin." "Smash Darwin! Smash the Pyramids," said I, in great wrath, and muttering something about "impertinence," which caused Jenkin to explain that "the Bishop was a first-class in mathematics, you know, and so has a right to treat on scientific matters," which of course silenced my cavils. Well, when Professor Draper[††] had ceased, his hour and a half of nasal Yankeeism, up rose "Sammivel," and proceeded to act the smasher; the white chokers, who were abundant, cheered lustily, a sort of "Pitch it into him" cheer, and the smasher got so uproarious as to pitch into Darwin's friends—Darwin being smashed—and especially Professor Huxley[*]. Still the white chokers cheered, and the smasher rattled on. "He had been told that Professor Huxley had said that he didn't see that it mattered much to a man whether his grandfather was an ape or not. Let the learned Professor speak for himself" and the like. Which being ended—and let me say that such rot never fell from episcopal lips before—arose Huxley, young, cool, quiet, sarcastic, scientific in fact and in treatment, he gave his lordship such a smashing as he may meditate on with profit over his port at Cuddesdon. This was the exordium, "I asserted, and I repeat—that a man has no reason to be ashamed of having an ape for his grandfather. If there were an ancestor whom I should feel shame in recalling, it would rather be a man, a man of restless and versatile intellect, who, not content with an equivocal success in his own sphere of activity, plunges into scientific questions with which he has no real acquaintance, only to obscure them by an aimless rhetoric, and distract the attention of his hearers from the real point at issue by eloquent digressions and skilled appeals to religious prejudice."

Green to W. Boyd Dawkins, 3 July 1860, in *Letters of John Richard Green*, ed. Leslie Stephen (London: Macmillan, 1902), 43–45. [†]Hugh Falconer (1808–65), Scottish geologist and botanist. [‡]Hugh Miller (1802–56), Scottish geologist and evangelical, anti-evolutionist. [§]Fleeming Jenkin (1833–85), Scottish engineering professor, critic of Darwin's theory. [††]John William Draper (1811–82), American chemist, author of *History of the Conflict between Religion and Science* (1874), which was repeatedly cited in the evolution polemic.

Secondary Literature: Anthony Brundage, *The People's Historian: John Richard Green and the Writing of History in Victorian England* (Westport, CT: Greenwood, 1994).

JOHN THOMAS GULICK (1832–1923)
AMERICAN MISSIONARY AND NATURALIST

We find it difficult to point to any of those active causes of accumulated variation, classed by Darwin as illustrations of "Natural Selection." The conditions under which they live are so completely similar, that it does not appear what ground there can be for difference in the characters best fitting the possessors for survival in the different valleys in which they are found. [. . .] There is no reason to doubt that some varieties less fitted to survive have disappeared; but it does not follow that the "Survival of the Fittest" (those best fitted when compared with those dying prematurely, but equally fitted when compared with each other) is the determining cause which has led to these three species being separated from each other in adjoining valleys. The "Survival of the Fittest" still leaves a problem concerning the distribution of those equally fitted. It cannot be shown that the "Survival of the Fittest" is at variance with the survival, under one set of external circumstances, of varieties differing more and more widely from each other in each successive generation. The case of the three species under consideration does not seem to be one in which difference of "Environment" has been the occasion of different forms preserved in the different localities.

John T. Gulick, "On the Variation of Species as Related to their Geographical Distribution, as Illustrated by the *Achatinellinae*," *Nature* 6 (1872): 222–24, on 223–24.

In March 1868, Moritz Wagner[†] read a paper before the Royal Academy of Sciences at Munich on "The Law of the Migration of Organisms," and in 1873 an English translation of a fuller paper by him entitled "The Darwinian Theory and the Law of the Migration of Organisms" was published by Edward Stanford,[‡] of London. It was through this pamphlet that I became acquainted with his theory concerning the impossibility of the production of new species except when and where migration establishes a colony geographically isolated from the original stock. [. . .] The following are some of the facts with which his general theory as well as his special statements were in conflict:

That, through change of climate and of other conditions due to geological changes, the whole fauna and flora of an island like Iceland might be subjected to new forms of selection producing a complete change of many species into new species, without any chance for migration. After the publication of my article in *Nature*, July 18, 1872 [see excerpt, above], in which I emphasized the importance of isolation, *I met Darwin at his home*, and he called my attention to Wagner's theory, and suggested that it did not correspond with the facts of nature, especially on this point.

John T. Gulick, "Isolation and Selection in the Evolution of Species: The Need of Clear Definition," *American Naturalist* 42 (1908): 48–57, on 55, italics mine. [†]Moritz Wagner (1813–87), German naturalist and geographer who defined the process of speciation through geographical isolation. [‡]Edward Stanford (1827–1904), publisher of maps.

Secondary Literature: Ron Amundson, "John T. Gulick and the Active Organism: Adaptation, Isolation, and the Politics of Evolution," in *Darwin's Laboratory: Evolutionary Theory and Natural History in the Pacific,* ed. Roy MacLeod and Philip F. Rehbock (Honolulu: University of Hawaii Press, 1994), 110–39.

LUDWIG GUMPLOWICZ (1838–1909)
Polish-Austrian Sociologist

Organic types seem to arise in two distinct ways and the solution of the whole anthropological problem depends upon setting aside the one or harmonizing the two. Is the principle of perpetual growth in organic bodies heredity, or adaptation, or what?

The wisdom of the ages which must not be despised answers heredity; radical modern materialism answers adaptation: "The man is what he eats"; Darwinism to reconcile the difference says: Both. Let us see which answer is nearest the truth. A superficial glance at organic structures is enough to show that heredity is the mightiest principle of their growth. It is clearly the rule that such structures are as their progenitors were. However there are some exceptions, for which the cleverest and at present the most widely accepted explanation is Darwin's theory of adaptation; what cannot be explained by heredity must be referred to the property of adaptation to external conditions, which organisms possess and to which the struggle for existence forces them to resort.

Some human types originated in adaptation and evolution, but not necessarily all. If the geographical character of the habitat is sufficient still to modify a type of organism how much greater must its influence have been upon the origin of varieties, for once it produced, so to speak, genetic differences, but though still active the original genetic effect proves to be more permanent. This might seem to justify one argument used in support of evolution, viz., that if the period be indefinitely extended the supposition of an original method becomes superfluous. But this is only arithmetically correct, it is insufficient to refute the supposition of an original genetic origin when so many other considerations support it.

The struggle between social groups, the component parts of the state, is as inexorable as that between hordes or states. The only motive is self-interest. In *Der Rassenkampf* [1875] we described the conflict as a "race-war" for such is its inexorable animosity that each group that is able tends to become exclusive like a caste, to form a consanguineous circle, in short to become a race.

Ludwig Gumplowicz, *The Outlines of Sociology,* trans. Frederick Wightman Moore (Philadelphia: American Academy of Political and Social Science, 1899), 134, 135, 145.

I wish to protest in the strongest possible terms against the application of the term Darwinism to the race struggle. I know of no ethnologist, historian, or sociologist among those who see the real effect of the struggle of races, who has accepted this designation for that law. The general character of that struggle has always been known, and therefore it no more belongs to Darwin's teachings than does the law of parasitism. But the great discovery of precisely how the race struggle operates in the process of civilization, though clearly formulated by Gumplowicz in 1875, a pamphlet of whose existence Darwin could have known nothing, was not fully worked out until 1883, one year after Darwin's death. That principle is to be ranked with the principle of natural selection, and may be appropriately called its sociological homologue, because, although an entirely different principle, it agrees with the latter in constituting a strictly scientific explanation of a great natural process, never before understood. I call it the principle of social synergy. It certainly is not social Darwinism nor Darwinism in any form. It would be difficult to find even an adumbration of it in any of Darwin's works, or, for that matter, in the works of any author prior to 1875 or even to 1883. But Ratzenhofer[†] in 1893 and especially in 1898 took it up and greatly expanded it. But he acknowledges that it was Gumplowicz who succeeded in first establishing sociology as the science which forms the foundation of all political teachings.

Lester Ward, "Social and Biological Struggles," *American Journal of Sociology* 13 (1907): 289–99, on 293–94. [†]Gustav Ratzenhofer (1842–1904), Austrian field marshal, sociologist, and Social Darwinist.

Secondary Literature: Brian Porter, *When Nationalism Began to Hate: Imagining Modern Politics in Nineteenth-Century Poland* (New York: Oxford University Press, 2002).

AHAD HA-ʿAM
(ASHER ZVI HIRSCH GINSBERG)
(1856–1927)

RUSSIAN JEWISH ESSAYIST

Thus it is with profane books; but with sacred books it is otherwise. Here the content sanctifies the book, and subsequently the book becomes the essential, and the content the accident. The book remains unchanged forever; the content changes ceaselessly with the progress of life and culture. What is there that men have not found in our sacred books, from Philo's day to this? In Alexandria they found Plato in them; in Spain, Aristotle; the Cabbalists found their own teaching, and the followers of other religions theirs; nay, some pious scholars have even found in them Copernicus and Darwin. All these men sought in Scripture only the truth—each one his own truth—and all found that which they sought. They found it because they had to find it: because if they had not found it, then truth would not have been truth, or the Scriptures would not have been holy.

Scientific development has shaken the foundations of every faith, and the Jewish faith has not escaped: so much so that even the editor of *La Gerbe* confesses, with a sigh, that "the scientific heresy which bears the name of Darwin" is gaining ground, and it is only from a feeling of *noblesse oblige* that he still continues to combat it. What, then, are those Jews to do who have nothing left but this theoretical religion, which is itself losing its hold on them?

I at least can speak my mind concerning the beliefs and the opinions which I have inherited from my ancestors, without fearing to snap the bond that unites me to my people. I can even adopt that "scientific heresy which bears the name of Darwin," without any danger to my Judaism. In a word, I am my own, and my opinions and feelings are my own. I have no reason for concealing or denying them, for deceiving others or myself.

On more general grounds, too, these writers of ours should have studied the laws of historical evolution a little more deeply before trying their hands at pulling down and building up. It is true that [Friedrich] Nietzsche* himself hated historians, and stigmatized Darwin and Spencer*, the authors of the evolutionary theory, as mediocrities. But this did not prevent even him from inventing historical hypotheses in order to explain the progress of morality, or from taking the cornerstone of his new system from Darwin.

Ahad Ha-ʿAm, *Selected Essays,* trans. Leon Simon (Philadelphia: Jewish Publication Society of America, 1912), 43–44, 183, 194, 237.

Secondary Literature: Steven J. Zipperstein, *Elusive Prophet: Ahad Haʿam and the Origins of Zionism* (Berkeley and Los Angeles: University of California Press, 1993).

ERNST HAECKEL (1834–1919)
German Zoologist

I cannot let this opportunity pass without giving expression to the considerable astonishment I felt over Darwin's exciting theory about the origin of species. I am moved to do this even more because the German professionals have found this epoch making work to be an unhappy presumption; they make this charge partly because they seem to misunderstand his theory completely. Darwin himself wished that his theory might be tested from every possible side and he looked "with confidence toward the young and striving naturalists who will be able to judge both sides of the question without partiality. Whoever is inclined to view species as changeable will, through the conscientious admission of his conviction, do a good service to science; only thereby can the mountain of prejudice under which this subject is buried be generally avoided." I share this view completely and believe for this reason that I must express my conviction that species are changeable and that organisms are really related genealogically.

Though I have some reservations about extending Darwin's insight and hypothesis in every direction and about all his attempts to demonstrate his theory, yet I must admire in his work the first, earnest and scientific effort to explain all appearances of organic nature from one excellent, unitary view point and his attempt to bring all sorts of inconceivable wonders under a conceivable law of nature. Perhaps there is in Darwin's theory, as the first effort of this sort, more error than truth. [...] The greatest confusion of the Darwinian theory lies probably herein, that it does not rest upon the origin of the urorganism—most probably a simple cell—whence all others have been developed. When Darwin assumes for this first species a special creative act, it seems of little consequence, and it seems to me not seriously meant. Aside from this and other confusions, Darwin's theory already has performed the immortal service of having brought the entire doctrine of relationships of organisms to sense and understanding. When one considers how every great reform, every strong advance has found a mighty opposition, the more he will oppose without caution the rooted prejudice and battle against the ruling dogma; so one will, indeed, not wonder that Darwin's ingenious theory has, instead of well deserved recognition and test, found only attack and rebuff.

Ernst Haeckel, *Die Radiolarien*, 2 vols. (Berlin: Reimer, 1862), 1:231–32, trans. Robert J. Richards, in *The Tragic Sense of Life: Ernst Haeckel and the Struggle of Evolutionary Theory* (Chicago: University of Chicago Press, 2008), 71–72.

Haeckel's Visit to Down, October 1866

As the coach pulled up to Darwin's ivy-covered house, shaded by elms, out of the shadows of the vine-covered entrance came the great scientist to meet me. He had a tall, worthy form with the broad shoulders of Atlas, who carries a world of thought. He had a Jupiter-like forehead, high and broadly domed, similar to Goethe's, and with deep furrows from the habit of mental work. His eyes were the friendliest and kindest, beshadowed by the roof of a protruding brow. His sensitive mouth was surrounded by a great silver-white full beard. The welcoming, warm expression of his whole face, the quiet and soft voice, the slow and thoughtful speech, the natural and open flow of ideas in conversation—all of this captured my whole heart during the first hours of our discussion. It was similar to the way his great book on first reading conquered my understanding by storm.

I believed I had before me the kind of noble worldly wisdom of the Greek ancients, that of a Socrates or an Aristotle.

Wilhelm Bölsche, *Ernst Haeckel: Ein Lebensbild* (Berlin: Georg Bondi, 1909), 179, quoted in *The Tragic Sense of Life: Ernst Haeckel and the Struggle over Evolutionary Thought*, by Robert J. Richards (Chicago: University of Chicago Press, 2008), 174.

On Sunday we had a gt visitation. One of Papa's most thoroughgoing disciples, a Jena professor, came to England on his way to Madeira & asked to come down & see Papa. We didn't know whether he cd speak English & our spirits was [*sic*] naturally rather low. He came quite early on Sunday & when first he entered he was so agitated he forgot all the little English he knew & he & Papa shook hands repeatedly, Papa reiteratedly remarking that he was very glad to see him & Haeckel receiving it in dead silence. However afterwards it turned out that he could stumble on very decently—some of his sentences were very fine. Talking of dining in London—"I like a good bit of flesh at a restoration." Of the war in Germany he remarked as an advantage the Russians had from their good education, "Zat ven ze officers are deeded ze commons take ze cheap," i.e. when the officers are killed the private takes the chief command. He told us that there are more than 200 medaillons [*sic*] of Papa made by a man from Wm's photo in circulation amongst the students in Jena. Papa has just begun his gt Pangenesis chapter.

Henrietta Darwin to George Darwin, 21 October 1866, Darwin Correspondence, DAR 251–326, Manuscript Room, Cambridge University Library, quoted in *The Tragic Sense of Life: Ernst Haeckel and the Struggle over Evolutionary Thought*, by Robert J. Richards (Chicago: University of Chicago Press, 2008), 174–75.

The monumental greatness of Charles Darwin, who surpasses every other student of science in the nineteenth century by the loftiness of his monistic conception of nature and the progressive influence of his ideas, is perhaps best seen in the fact that not one of his many successors has succeeded in modifying his theory of descent in any essential point or in discovering an entirely new standpoint in the interpretation of the organic world. Neither Nägeli[†] nor [August] Weismann[*], neither De Vries [Hugo DeVries][*] nor Roux,[‡] has done this. Nägeli, in his *Mechanisch-Physiologische Theorie der Abstammungslehre* (Munich, 1884), which is to a great extent in agreement with Weismann, constructed a theory of the idioplasm, that represents it (like the germ-plasm) as developing continuously in a definite direction from internal causes. But his internal "principle of progress" is at the bottom just as teleological as the vital force of the Vitalists, and the micellar structure of the idioplasm is just as hypothetical as the "dominant" structure of the germ-plasm. In 1889 Moritz Wagner[§] sought to explain the origin of species by migration and isolation, and on that basis constructed a special "migration-theory." This, however, is not out of harmony with the theory of selection. It merely elevates one single factor in the theory to a predominant position. Isolation is only a special case of selection, as I had pointed out in the fifteenth chapter of my *Natural History of Creation*. The "mutation-theory" of De Vries (*Die Mutationstheorie*, Leipzig, 1903), that would explain the origin of species by sudden and saltatory variations rather than by gradual modification, is regarded by many botanists as a great step in advance, but it is generally rejected by zoologists. It affords no explanation of the facts of adaptation, and has no causal value.

Ernst Haeckel, "Charles Darwin as an Anthropologist," in *Essays in Commemoration of the Centenary of the Birth of Charles Darwin and of the Fiftieth Anniversary of the Publication of the Origin of Species*, ed. A. C. Seward (Cambridge: Cambridge University Press, 1909), 137–51. [†]Karl Wilhelm von Nägeli (1817–91), Swiss botanist. [‡]Wilhelm Roux, German zoologist and embryologist. [§]See John Thomas Gulick entry.

Secondary Literature: Mario A. Di Gregorio, *From Here to Eternity: Ernst Haeckel and Scientific Faith* (Göttingen: Vandenhoeck & Ruprecht, 2005).

JAMES DUNCAN HAGUE (1836–1906)
American Mining Engineer

Early in 1871, while passing a few months in London, it being my good fortune to know Sir Charles Lyell*, I, through his introduction, made the personal acquaintance of Mr. Darwin. [. . .] I received a message from Sir Charles saying that Mr. Darwin, accompanied by some of his family, was spending a few days in town at his brother's house in Queen Anne Street, where he would be pleased to see me at luncheon on the following Sunday. This pleasant invitation was accompanied with a thoughtful warning in a note from Lady Lyell to the effect that although I might find Mr. Darwin looking well and strong, I should remember his really delicate health, and not stay too long.

On my arrival at the door, at the appointed hour, the servant evidently recognized the name of an expected visitor, and took me upstairs at once to a small library room, where I found Darwin alone. He came forward very cordially, putting out his hand in a way to make a stranger feel welcome, and entered directly into conversation in a manner lively enough to relieve any apprehension of his suffering as a great invalid. At that time he was about sixty-two, and he already wore the full gray beard visible in the portraits taken in his later years. We had a little to say about his health, and he spoke of the opinion of some of his friends that its present rather feeble state might be attributed to long-continued seasickness on his voyages years ago; and then said "So you've been a voyager too?" and asked me in what part of the Pacific Ocean I had been, and what islands I had seen. I thereupon related to him some of my experiences and observations, which led to a pleasant talk on his part about the coral islands. He spoke with great vivacity and interest of his voyage in the Beagle, and especially of his work on coral reefs and atolls, of the wonderful impression that those islands make upon the mind of an observer, and of the charm and poetry they possess in their singular beauty and their peculiar origin and structure. [. . .]

In the course of conversation I expressed to Mr. Darwin the regret I had ever since felt that when I went out to the South-sea Islands I was so poorly qualified to observe many of the most interesting features of the places visited, as I had studied very little of natural history, and knew nothing of birds, fishes, corals, shells, or anything of that sort. "Well," said he, "you need not think yourself unique in that respect. I never knew a man who had a rare opportunity for observation who did not regret his imperfect qualifications. It was my own experience. If I could only go now, with my head sixty years old and my body twenty-five, I could do something." Then he said that his visit to the Pacific, or rather his voyage in the Beagle, was the beginning of his scientific career; that he had not before given much serious attention to science, or studied with a definite purpose; that when the Beagle was fitting out he was a young man, fond of sport, shooting and fishing, and with a strong liking for natural history; and it seemed to him a pleasant thing to go as volunteer on the party with his friend Captain Fitzroy. Even after he came back, he said, it was only after many talks with Lyell, who always heard with interest whatever he had to say, and who urged him to work up and publish his results, that he determined to devote himself to scientific studies.

After a while he spoke of some of his American friends, saying that Mr. Charles Eliot Norton* and family had been his neighbors in the country for a time, and that he had enjoyed their society very much. He also mentioned visits which he had received from the younger Agassiz† and from Dr. [Asa] Gray*, saying of the latter, "Gray often takes me to task for making hasty generalizations; but the last time he was here talking that way I said to him: 'Now, Gray, I have one more generalization to make, which is not hasty, and that is, the Americans are the most delightful people I know.'"

The conversation having turned upon his last book, *The Descent of Man,* which had made its first appearance only shortly before, he spoke of the reviews which he had so far seen, saying that most of them were, of course, somewhat superficial, and doubtless others might come later of a different tone, but he had been much impressed by the general assent with which his views had been received. "Twenty years ago," said he, "such ideas would have been thought subversive of everything good, and their author might have been hooted at, but now not only the press, but, from what my friends who go into society say, everybody, is talking about it without being shocked."

"That is true," I said, "but, according to Mr. Punch, while the men seem to accept it without dissent, the women are inclined to protest."

"Ah, has Punch taken me up?" said Mr. Darwin, inquiring further as to the point of the joke, which, when I had told him, seemed to amuse him very much. "I shall get it to-morrow," he said: "I keep all those things. Have you seen me in the Hornet?" As I had not seen the number referred to, he asked one of his sons to fetch the paper from upstairs. It contained a grotesque caricature representing a great gorilla having Darwin's head and face, standing by the trunk of a tree with a club in his hand. Darwin showed it off very pleasantly, saying, slowly and with characteristic criticism, "The head is cleverly done, but the gorilla is bad: too much chest; it couldn't be like that."

The humorists have done much to make Mr. Darwin's features familiar to the public, in pictures not so likely to inspire respect for the author of *The Descent of Man* as they are to imply his very close relation to some slightly esteemed branches of the ancestry he claims; but probably no one has enjoyed their fun more than he. [...]

Being again in London in 1878, I was very glad to receive one day about the middle of May a message from Mrs. Darwin inviting me to go down into the country to dine and pass the night at their house in company with one other guest, Colonel T. W. [Thomas Wentworth] Higginson*, of Boston. [...]

The dwelling of the Darwin family, as I recall it is a spacious and substantial old-fashioned house, square in form and plain in style, but pleasing in its comfortable and home-like appearance. The approach seems now to my memory to have been by a long lane, as though the house stood remote from any much-traveled highway, and without near neighbors, surrounded by trees and shrubbery, and commanding a far-reaching view of green fields and gently undulating country. A portion of the house, the front, has, I believe, been built long enough to be spoken of as old even in England, to which in the rear some modern additions have been made. Entering a broad hall at the front, we passed, on the right, the door of the room the interior of which has since been made known in pictures as "Mr. Darwin's Study"; and a little further on were welcomed immediately by Mr. and Mrs. Darwin to a spacious and cheerful parlor or family room, whose broad windows and outer door opened upon a wide and partly sheltered piazza at the rear of the house, evidently a favorite sitting-place, judging from the comfortable look of easy-chairs assembled there, beyond which was a pleasing vista of fresh green lawn, bright flower beds, and blossoming shrubbery, gravel-paths, and a glass greenhouse, or perhaps botanical laboratory, and, further yet, a garden wall, with a gate leading to pleasant walks in fields beyond. All this one could see standing before the hearth, from which, although past the middle of May, a slowly smoldering fire gave out a pleasant warmth, by no means unwelcome to one whose clothing had been dampened by the brisk shower encountered on the way from the station. [...]

Mr. Darwin sat by the fireside, where he seemed to have been reading. Dressed in an easy-fitting, short, round coat, he looked as though he might have spent the day at work in the garden or laboratory. Though seven years had passed since I had seen him in London, he appeared as well as then and hardly older. We had some pleasant personal chat for a little while, when Colonel Higginson, whose arrival had been delayed, came in, and, soon after, we went to our rooms to dress for dinner, Mr. Darwin himself going on before to show his guests the way, and to see that their needs had been duly provided for.

Re-assembled a half-hour later, we were soon seated at the dinner table, six in all, George and Francis being the only younger members of the family then at home. The dining room was a handsome, spacious apartment, with windows opening upon the lawn in the rear of the house; its walls hung with pictures, among which I seem to recall some dim, dark portraits of the Darwin ancestry, though nothing, if I remember rightly, more remote than the distinguished grandfather, Dr. Erasmus. Our dinner, though noteworthy for its wholesome simplicity in these days of excessive luxury, was served with care and elegance by two men-servants in livery, the elder butler, a man of advanced years, being evidently of long service in the family. The dinner-table talk was for the greater part light, cheerful, personal; to some extent political, suggested by current events in England and the United States; and touching somewhat upon social reforms, as might indeed be readily imagined in the presence of my distinguished *vis-à-vis,* Colonel Higginson. [. . .]

When Mrs. Darwin rose to leave the gentlemen with their cigars, she reminded her husband that he too must go and take his customary nap. He accordingly withdrew, but about half an hour later rejoined us in the drawing-room, where he remained until half past ten or eleven. During this time he made inquiry for American friends, mostly of Cambridge, but also [Othniel C.] Marsh*, of New Haven. He spoke, too, with particular interest of Mark Twain*, from whose writings he had evidently derived much entertainment.

James Duncan Hague, "A Reminiscence of Mr. Darwin," *Harpers* 69 (1884): 759–63.

†Alexander Agassiz (1835–1910), marine biologist, son of Louis.

Secondary Literature: Julia Adams Stratton and Loretta H. Mannix, *Mind and Hand: The Birth of MIT* (Cambridge, MA: MIT Press, 2005).

J. B. S. HALDANE (1892–1964)
ENGLISH BIOLOGIST

Darwinism has been a subject of embittered controversy ever since its inception. The period up till Darwin's death saw a vast mass of criticism. This was mostly an attack on the doctrine of evolution and was almost entirely devoid of scientific value. The few really pertinent attacks were lost amid a jabber of ecclesiastical bombominations. The criticism was largely dictated by disgust or fear of this doctrine, and it was natural that the majority of scientific men rallied to Darwin's support. By the time of Darwin's death in 1882, Darwinism had become orthodox in biological circles. The next generation saw the beginnings of a more critical attitude among biologists. It was possible to criticize Darwin without being supposed to be supporting the literal authenticity of the Book of Genesis. The criticism came from all sides. Paleontologists, geneticists, embryologists, psychologists, and others, found flaws of a more or less serious character in Darwin's statements. But they almost universally accepted evolution as a fact.

The rising generation of biologists, to which I belong, may now perhaps claim to make its voice heard. We have this advantage over our predecessors, that we get no thrill from attacking either theological or biblical orthodoxy; for eminent theologians have accepted evolution and eminent biologists denied natural selection.

J. B. S. Haldane, *The Causes of Evolution* (1932; reprint, Princeton, NJ: Princeton University Press, 1990), 2.

Secondary Literature: Ronald Clark, *J.B.S.: The Life and Work of J.B.S. Haldane* (London: Hodder & Stoughton, 1968).

GRANVILLE STANLEY HALL (1844–1924)
AMERICAN PSYCHOLOGIST

A former test of a good bird singer, Darwin tells us, was to see if it will continue to sing while the cage is swung around the owner's head, and birds matched in rivalry will sometimes sing for hours till one drops exhausted or dead. A canary sang continuously for fourteen hours. The best singers are commonly not brilliant in hue, but charm with their voice. If song is not confined to the breeding season, the very voice often changes then. Drumming, rattling quills, the whirring of the birds made by feathers especially shaped to cut the air, as they plunge or turn in it, like most noises in the insect world which is so similar to that of birds, primarily serve the reproductive function. The larynx of some animals enlarges during rut, and others are mute save in the breeding season. The voice is often to strike terror before battle or in challenge. Some monkeys make the woods vocal in the spring. Darwin holds that music, instead of originating in speech cadences, as Spencer* thinks, sprung from and is reminiscent of the psychoses of old courtships of a long-past age. However this may be, sound in both the animal and human world is a potent agent of love. The song of crickets, birds, and the pleasure of the other sex in hearing it, suggests to [August] Weismann* that not only the voice but other kinds of musical organs have a sexual origin as mediations of selection. Whether we hold with Darwin that song was developed by sexual selection and language was evolved from it, or with Spencer, Schweibe, and others, that speech was primary, or even with Weismann that the musical sense has no necessary relation to sexual life, but was a complementary product of the organ of audition, we know that timbre alone has great power in arousing or arresting sexual feeling and that music and love are closely associated.

G. Stanley Hall, *Adolescence*, 2 vols. (New York: Appleton, 1905), 2:25–26.

Secondary Literature: Dorothy Ross, *G. Stanley Hall: The Psychologist as Prophet* (Chicago: University of Chicago Press, 1972); John R. Morss, *The Biologising of Childhood: Developmental Psychology and the Darwinian Myth* (London: Taylor & Francis, 1990).

THOMAS HARDY (1840–1928)
English Novelist and Poet

Next this strange message Darwin brings,
  (Though saying his say
  in a quiet way);
We all are one with creeping things;
  And apes and men
  Blood-brethren.

Thomas Hardy, "Drinking Song," in *The Complete Poems*, ed. James Gibson (London: Palgrave Macmillan, 2002), 906–7.

Epitaph for G. K. Chesterton

Here lies nipped in this narrow cyst
The literary contortionist
Who prove and never turn a hair
That Darwin's theories were a snare
He'd hold as true with tongue in jowl
That Nature's geocentric rule
... true and right
And if one with him could not see
He'd shout his choice word "blasphemy."

Thomas Hardy, *The Complete Poems*, ed. James Gibson (London: Palgrave Macmillan, 2002), 954.

Secondary Literature: Philip J. Waller, *Writers, Readers, and Reputations: Literary Life in Britain, 1870–1918* (Oxford: Oxford University Press, 2006).

WILLIAM HENRY HARVEY (1811–1866)
IRISH BOTANIST

And now for what you say of the famous theory. I saw in the first few lines, by your using the word "sweeping," what the *undercurrent* of your thoughts was, and I felt glad that there was *one*, at least, of my botanical friends whose feet were not quite whipt from under him. I am fully disposed to admit natural selection as a "vera causa" of much change, but not as the "vera causa" of species. I fully admit the impossibility of defining the limits of species, genera, or orders; but this does not shake my belief in the existence of *limits*—unseen by our eyes, undefinable by our philosophy. But how many "natural species" were *created*, and what they were like, I trow not. That they were vastly fewer than the forms we now call species, I think probable, but also that they were vastly more numerous than Darwin would have us believe.

I have read your Darwin papers with great pleasure and profit. Almost thou persuadest me to be a *Grayite*. I have no objection *per se* to a doctrine of derivative descent. Why should I? *One* mode of creation is as feasible to the Almighty as another, and, as *put by you,* is very consonant to sound doctrine. I have had a short friendly correspondence with Darwin on the subject, but without much result one way or the other. I confess, however, since I have read the *whole book,* to a somewhat changed view. His latter chapters are those which have most impressed me, and particularly that on geographical distribution, and the geological-geographical distribution successively through ages. Certainly there are many *broad facts* which can be read by a supposition of descent with variation. *How broad* those facts are, and how broad the limits of descent with variation may be, are questions which I do not think his theory affords answer to. It opens vistas vast, and so it evidently points whence, through time, light may come by which to see the objects in those vistas, but to my mind it does no more. When he passes this true deductive inference, and proceeds to build further inductions on it, and to force all things to converge on one point, then I draw back, thinking with Hamlet, that there may be things *in the scheme of creation* which are not explained, although (they may be "dreamt of") in our philosophy. A good deal of Darwin reads to me like an ingenious dream.

Harvey to Asa Gray, 20 May and 3 November 1860, *Memoir of W. H. Harvey* (London: Bell & Daldy, 1869), 333–34, 337–38.

Secondary Literature: James H. Price, "Goody Two-Shoes or a Monument to Industry? Aspects of the Phycologia Britannica of William Henry Harvey," *Bulletin of the British Museum (Natural History), Historical Series* 16 (1988): 87–216.

WILLET M. HAYS (1859–1916)
American Plant Breeder

Probably the most important work to be carried on in the grass garden is that of the improvement of the standard grasses and clovers which are of such immense importance in our agriculture. No one thinks strange that pansy fanciers and propagators go to the trouble to improve pansies that they may have rich and novel combinations of colors. But who has even carefully studied out the variation in the plants of timothy or red clover or other varieties upon which so many of our people depend for sustenance and means of obtaining the enjoyments of life. Who can tell us as to whether material improvements can easily be obtained by simply selecting the best plants of timothy for a number of generations. Who among us is really reaping for the farmers the fruits of the seed sown and the harvests of fame made possible by the great work of Charles Darwin. He showed us how by crossing plants carrying distinct blood lines yet close enough to unite that we could increase productiveness and by selecting desirable variations we can secure sorts which better suit our needs. We are not keeping apace with our horticultural brethren who are cross-fertilizing apples, grapes, strawberries, and many other classes of plants.

W. M. Hays, "Grass Gardens—Methods and Purposes," in U.S. Department of Agriculture, *Proceedings of the Fifth Annual Convention of the Association of American Agricultural Colleges and Experiment Stations* (Washington, DC: Government Printing Office, 1892), 130–32, on 131.

Scientists were long prone to confine their studies of heredity to wild forms of plant and animal life. It is most strange that those who took up the work of Charles Darwin failed to see that he secured his best evidences of organic evolution from domesticated species, and strange, too, that they have nearly all devoted their energies to studying the wild species. It is no less strange that for half a century so few scientists have comprehended the full significance in economic production of Darwin's generalizations. Not until about the first of the new century did the scientific world begin fully to comprehend the immense scientific opportunities in artificial evolution. And even now there is needed an effort to direct, scientific research more into the study of such important matters as the flowering habits of our important plants, into the hybridizing possibilities among animals, and into practical and business methods of breed and variety formation and breed and variety improvement. Biologists are being forced to see that by uniting the scientific interests more closely with the economic interests there will be a far greater development of science.

W. M. Hays, "American Methods of Breeding," in *Annual Report, Nebraska State Board of Agriculture for the Year 1905* (Fremont, NE: Hammond Printing Co., 1906), 111–15, on 112.

Secondary Literature: M. C. Judd, "Willet M. Hays: Exponent of the New Agriculture," *American Review of Reviews* 39 (1909): 689–95.

LAFCADIO HEARN (1850–1904)
American Journalist and Author

I felt glad for divers reasons on receiving your letter and the little parcel—firstly, because I felt that you were not very angry at my foolish fable; and secondly, because I always feel happy on having something nice to read. I had already read considerable of Darwin's *Voyages;* but just now I happened to desire a work of just that kind in order to educate myself in regard to certain ethnological points. I accept Darwin fully [...] I do not believe in God—neither god of Greece nor of Rome nor any other god.

"Letters to a Lady" (1875), in *Letters from the Raven, Being the Correspondence of Lafcadio Hearn and Henry Watkin,* ed. Milton Bronner (London: Archibald Constable, 1908), 141.

You are not quite correct in saying that Spencer* could not obtain a hearing before Darwin. Before Darwin, Spencer had already been recognized by [George Henry] Lewes* as the mightiest of all English thinkers, with the remarkable observation that he was too large and near to be justly estimated even in his lifetime. Darwin did much, of course, to illuminate one factor of evolution; but I hardly say that one factor, though the most commonly identified with evolution, is but one of myriads. Natural selection can explain but a very small part of the thing. The colossal brain which first detected the necessity of evolution as a cosmic law—governing the growth of a solar system as well as the growth of a gnat—the brain of Spencer, discerned that law by pure mathematical study of the laws of force. And the work of the Darwins and Huxleys* and [John] Tyndalls* is but detail—small detail—in that tremendous system which has abolished all preexisting philosophy and transformed all science and education.

Lafcadio Hearn to Basil Hall Chamberlain, April 1895, in *The Life and Letters of Lafcadio Hearn,* by Elizabeth Bisland, 2 vols. (Boston: Houghton Mifflin, 1906), 2:235.

Secondary Literature: Paul Murray, *A Fantastic Journey: The Life and Literature of Lafcadio Hearn* (Ann Arbor: University of Michigan Press, 1997).

WERNER HEISENBERG (1901–1976)
German Physicist

Just as in the case of chemistry, one learns from simple biological experience that the living organisms display a degree of stability which general complicated structures consisting of many different types of molecules could certainly not have on the basis of the physical and chemical laws alone. Therefore, something has to be added to the laws of physics and chemistry before the biological phenomena can be completely understood.

With regard to this question two distinctly different views have frequently been discussed in the biological literature. The one refers to Darwin's theory of evolution in its connection with modern genetics. According to this theory, the only concept which has to be added to those of physics and chemistry in order to understand life is the concept of history. The enormous time interval of roughly four thousand million years that has elapsed since the formation of the earth has given nature the possibility of trying an almost unlimited variety of structures of groups of molecules. [...] Accidental changes in the structures provided a still larger variety of the existing structures. Different structures had to compete for the material drawn from the surrounding matter and in this way, through the "survival of the fittest," the evolution of living organisms finally took place. [...]

On the other hand, it is just this argument that has lost much of its weight through quantum theory. Since the concepts of physics and chemistry form a closed and coherent set, namely, that of quantum theory, it is necessary that wherever these concepts can be used to describe phenomena the laws connected with the concepts must be valid too.

Werner Heisenberg, *Physics and Philosophy: The Revolution in Modern Science* (New York: Harper & Brothers, 1958), 102–4.

Secondary Literature: David C. Cassidy, *Uncertainty: The Life and Science of Werner Heisenberg* (New York: W. H. Freeman, 1991).

HERMANN VON HELMHOLTZ (1821–1894)
German Physicist

At the same time, we should not forget the clear interpretation Darwin's grand conception has supplied of the till then mysterious notions respecting natural affinity, natural systems, and homology of organs in various animals; how by its aid the remarkable recurrence of the structural peculiarities of lower animals in the embryos of others higher in the scale, the special kind of development appearing in the series of paleontological forms, and the peculiar conditions of affinity of the faunas and floras of limited areas have, one and all, received elucidation. Formerly natural affinity appeared to be a mere enigmatical, and altogether groundless similarity of forms; now it has become a matter for actual consanguinity. The natural system certainly forced itself as such upon the mind, although theory strictly disavowed any real significance to it; at present it denotes an actual genealogy of organisms. The facts of paleontological and embryological evolution and of geographical distribution were enigmatical wonders so long as each species was regarded as the result of an independent act of creation, and cast a scarcely favorable light on the strange tentative method which was ascribed to the Creator. Darwin has raised all these isolated questions from the condition of a heap of enigmatical wonders to a great consistent system of development, and established definite ideas in the place of such a fanciful hypothesis as, among the first, had occurred to Goethe, respecting the facts of the comparative anatomy and the morphology of plants.

Hermann von Helmholtz, *Popular Lectures on Scientific Subjects,* trans. E. Atkinson (New York: Appleton, 1883), 387–88.

Secondary Literature: Helmut Pulte, "Darwin's Relevance for Nineteenth-Century Physics: A Comparative Study," in *The Reception of Charles Darwin in Europe,* ed. Eve-Marie Engels and Thomas F. Glick, 2 vols. (London: Continuum, 2008), 1:116–34.

JOHN STEVENS HENSLOW (1796–1861)
English Botanist and Clergyman

It was not till 1859 [. . .] that Mr. Darwin's book appeared *On the Origin of Species.* It may be supposed with what interest he read this work, embracing the whole question of species in its most comprehensive aspect, and coming from a naturalist of such high reputation, and who had enjoyed rare opportunities of making observations in so many different parts of the world, independently of experiments closely carried on at home, connected with the breeding of animals and the cultivation of plants. [Henslow] has nowhere formally recorded any detailed opinion of this work, though what he thought of it may be gathered, in a general way, from a letter of his to the editor of *Macmillan's Magazine,* in reply to some who had placed him in the ranks of those who supported Darwin's views (*Macmillans Magazine,* vol. iii., 336). He says, "The manner in which my name is noticed in a review of Mr. Darwin's work in your number for December (1860), is liable to lead to a misapprehension of my view of Mr. Darwin's *Theory on the Origin of Species* [*sic*]. Though I have always expressed the greatest respect for my friend's opinions, I have told himself that I cannot assent to his speculations without seeing stronger proofs than he has yet produced." He then alludes to a letter of mine to himself, in which I had stated that I could imagine that many of the smaller groups, both of animals and plants, might at some remote period have had a common parentage, though I thought that there was no satisfactory evidence to show that such had been the case with the higher groups; I added that I did not, with some, say that the whole of Darwin's theory *could* not be true, but that it was very far from proved, and I doubted its ever being possible to prove it. To all this he said that it "very nearly expressed the views he at that time entertained in regard to Darwin's theory, or rather hypothesis, as he should prefer calling it." In this instance, as in so many other instances, he showed his philosophic caution. He declined going a step further than he could well see his way; at the same time that he would not take upon himself to say that such and such things could not be, where there were no sufficient data for establishing the truth one way or the other. In previous letters to myself, he had told me he thought there were in Darwin's book too many suppositions, too many things assumed, which might or might not he true. Moreover, the further the question was followed up towards its source, the more it was beset with difficulties which were never likely to be solved. In fact, he said, he considered an inquiry into the origin of species about as hopeless as an inquiry into the origin of evil.

With reference to the religious aspect of this question, he stated, during his last illness, that he thought it an objection to the hypothesis of all animal and vegetable forms having been evolved in succession, through countless ages, from one primitive germ, or from a few such germs, that it did not allow for the interposition of the Almighty. "God," he observed, "did not set the creation going like a clock, wound-up to go by itself, but from time to time interposes and directs things as he sees fit." Yet he had always defended Darwin from some of his opponents, whom he considered as having shown him ungenerous treatment in fastening upon him opinions of an infidel or irreligious tendency, which those who are acquainted with his real sentiments on matters of religion know to be utterly without foundation.

Leonard Jenyns, *Memoir of the Rev. John Stevens Henslow* (London: John van Voorst, 1862), 211–13.

Secondary Literature: Stuart Max Walters and Elizabeth Anne Stow, *Darwin's Mentor: John Stevens Henslow* (Cambridge: Cambridge University Press, 2001).

JOHN HERSCHEL (1792–1871)
English Astronomer

I have heard, by a roundabout channel, that Herschel says my book "is the law of higgledy-piggledy." What this exactly means I do not know, but it is evidently very contemptuous. If true this is a great blow and discouragement.

Darwin to Charles Lyell, 21 December 1859, in *Charles Darwin: His Life Told in an Autobiographical Chapter, and in a Selected Series of his Published Letters,* ed. Francis Darwin (London: John Murray, 1892), 220.

Secondary Literature: David S. Evans, "John Frederick William Herschel," *Dictionary of Scientific Biography,* ed. Charles Coulston Gillispie, 16 vols. (New York: Scribner's, 1970–80), 6:323–28.

THEODOR HERZL (1860–1904)
Hungarian Jewish Journalist and Zionist

Anti-Semitism, which is a strong and unconscious force among the masses, will not harm the Jews. I consider it to be a movement useful to the Jewish character. It represents the education of a group by the masses, and will perhaps lead to its being absorbed. Education is accomplished only though hard knocks. A Darwinian mimicry will set in. The Jews will adapt themselves. They are like the seals, which an act of nature cast into the water. These animals assume the appearance and habits of fish, which they certainly are not. Once they return to dry land again and are allowed to remain there for a few generations, the will turn their fins into feet again.

The Complete Diaries of Theodor Herzl, ed. Raphael Patai, 3 vols. (New York: Herzl Press, 1960), 1:20.

Secondary Literature: Julius H. Schoeps, *Theodor Herzl and the Zionist Dream* (London: Thames & Hudson, 1997).

THOMAS WENTWORTH HIGGINSON
(1823–1911)
American Clergyman and Abolitionist

I visited Darwin twice in his own house at an interval of six years, once passing the night there. On both occasions, I found him the same, but with health a little impaired after the interval,—always the simple, noble, absolutely truthful soul. Without the fascinating and boyish eagerness of Agassiz*, he was also utterly free from the vehement partisanship which this quality brings with it, and he showed a mind ever humble and open to new truth. Tall and flexible, with the overhanging brow and long features best seen in Mrs. Cameron's[†] photograph, he either lay half reclined on the sofa or on high cushions, obliged continually to guard against the cruel digestive trouble which haunted his whole life. I remember that at my first visit, in 1872, I was telling him of an address before the Philological Society by Dr. Alexander J. Ellis,[‡] in which he had quoted from *Through the Looking Glass* the description of what were called portmanteau words, into which various meanings were crammed. As I spoke, Mrs. Darwin glided quietly away, got the book, and looked up the passage. "Read it out, my dear," said her husband; and as she read the amusing page, he laid his head back and laughed heartily. It was altogether delightful to see the man who had revolutionized the science of the world giving himself wholly to the enjoyment of Alice and her pretty nonsense. Akin to this was his hearty enjoyment of Mark Twain*, who had then hardly begun to be regarded as above the Josh Billings grade of humorist; but Darwin was amazed that I had not read "The Jumping Frog," and said he always kept it by his bedside for midnight amusement. I recall with a different kind of pleasure the interest he took in my [Civil War] experience with the colored race, and the faith which he expressed in the negroes. This he afterward stated more fully in a letter to me, which may be found in his published memoirs. [...]

At my second visit [May 1878] Darwin was full of interest in the Peabody Museum at Yale College and quoted with approval what Huxley* had told him, that there was more to be learned from that one collection than from all the museums of Europe. But for his chronic seasickness, he said, he would visit America to see it. He went to bed early that night, I remember, and the next morning I saw him, apparently returning from a walk through the grounds—an odd figure, with white beard, and with a short cape wrapped round his shoulders, striding swiftly with his long legs. He said that he always went out before breakfast—beside breakfasting at the very un-English hour of half-past seven—and that he was also watching some little experiments. His son added reproachfully, "There it is: he pretends not to be at work, but he is always watching some of his little experiments, as he calls them, and gets up in the night to see them." Nothing could be more delightful than the home relations of the Darwin family; and the happy father once quoted to me a prediction made by some theological authority that his sons would show the terrible effects of such unrighteous training, and added proudly, looking round at them, "I do not think I have much reason to be ashamed."

Thomas Wentworth Higginson, *Cheerful Yesterdays* (Cambridge, MA: Riverside Press, 1900), 283–86. †Julia Cameron (1815–79), pioneer English photographer, known for her portraits of famous people. ‡Alexander John Ellis (1814–90), English mathematician and philologist.

It must be remembered that Anglomania is confined among us to a very limited class, and to certain very limited pursuits and interests of that class. It does not exist, for instance, among our men of science, inasmuch as they go to Germany in shoals for study, and rarely visit England since the death of Darwin.

Thomas Wentworth Higginson, *Women and Men* (New York: Harper, 1888), 23.

Secondary Literature: Brenda Wineapple, *White Heat: The Friendship of Emily Dickinson and Thomas Wentworth Higginson* (New York: Knopf, 2008).

CHARLES HODGE (1797–1878)
AMERICAN THEOLOGIAN

We have not forgotten Mr. Darwin. It seemed desirable, in order to understand his theory, to see its relation to other theories of the universe and its phenomena, with which it is more or less connected. His work on the *Origin of Species* does not purport to be philosophical. In this aspect it is very different from the cognate works of Mr. Spencer*. Darwin does not speculate on the origin of the universe, on the nature of matter, or of force. He is simply a naturalist, a careful and laborious observer; skillful in his descriptions, and singularly candid in dealing with the difficulties in the way of his peculiar doctrine. [. . .]

To account for the existence of matter and life, Mr. Darwin admits a Creator. This is done explicitly and repeatedly. Nothing, however, is said of the nature of the Creator and of his relation to the world, further than is implied in the meaning of the word.

Charles Hodge, *What is Darwinism?* (New York: Scribner, Armstrong, 1874), 26–27.

Secondary Literature: *Charles Hodge Revisited: A Critical Appraisal of His Life and Work,* ed. John W. Stewart and James H. Moorhead (Grand Rapids, MI: Eeerdmans, 2002).

THOMAS HODGKIN (1831–1913)
ENGLISH HISTORIAN AND BANKER

I think that to me the greatest help in dealing with these difficulties has come from the doctrine of Evolution, so much dreaded by most of us fifty years ago. Now that I see that the Almighty has been patiently evolving our material world through infinite ages out of Chaos, I can the better understand that in the spiritual and ethical world also He has been gradually evolving His creature Man. I see now that our Lord and Master knew this and intended to convey this or something like it by such words as "He suffered it because of the hardness of their hearts," an idea which His servant Paul expressed by the words "The times of this ignorance God winked at." So that I can truly say "Thank God for Charles Darwin," little as he himself may have known of the wide spiritual import of his discoveries.

Hodgkin to Richard Westlake, 31 July 1912, in *Life and Letters of Thomas Hodgkin*, ed. Louise Creighton, 2nd ed. (London: Longmans, Green, 1918), 340–41. Hodgkin had read *Origin of Species* in the mid-1860s (ibid., 71).

Secondary Literature: Edward Pickard and Edwin Tregelles, *George Fox and his Latest Biographer* (Falmouth, UK, 1896), pamphlet.

HARALD HOFFDING (1843–1931)
DANISH PHILOSOPHER

In accentuating the struggle for life Darwin stands as a characteristically English thinker: he continues a train of ideas which Hobbes and Malthus had already begun. Moreover in his critical views as to the conception of species he had English forerunners; in the middle ages Occam and Duns Scotus, in the eighteenth century Berkeley and Hume. In his moral philosophy, as we shall see later, he is an adherent of the school which is represented by Hutcheson, Hume and Adam Smith. Because he is no philosopher in the stricter sense of the term, it is of great interest to see that his attitude of mind is that of the great thinkers of his nation.

Darwin declared himself an agnostic, not only because he could not harmonize the large amount of suffering in the world with the idea of a God as its first cause, but also because he "was aware that if we admit a first cause, the mind still craves to know whence it came and how it arose." (*Life and Letters*, Vol. I. page 306.) He saw, as Kant had seen before him and expressed in his *Kritik der Urtheilskraft*, that we cannot accept either of the only two possibilities which we are able to conceive: chance (or brute force) and design. Neither mechanism nor teleology can give an absolute answer to ultimate questions. The universe, and especially the organic life in it, can neither be explained as a mere combination of absolute elements nor as the effect of a constructing thought. Darwin concluded, as Kant, and before him Spinoza, that the oppositions and distinctions which our experience presents, cannot safely be regarded as valid for existence in itself. And, with Kant and Fichte, he found his stronghold in the conviction that man has something to do, even if he cannot solve all enigmas. "The safest conclusion seems to be that the whole subject is beyond the scope of man's intellect; but man can do his duty." (Ibid. page 307.)

Harald Hoffding, "The Influence of the Conception of Evolution on Modern Philosophy," in *Essays in Commemoration of the Centenary of the Birth of Charles Darwin and of the Fiftieth Anniversary of the Publication of the Origin of Species*, ed. A. C. Seward (Cambridge: Cambridge University Press, 1909), 448, 464.

Secondary Literature: Frithiof Brandt, "Harald Hoffding," in *Encyclopedia of Philosophy*, ed. Paul Edwards, 8 vols. (New York: Macmillan, 1967), 4:48–49.

OLIVER WENDELL HOLMES SR. (1809–1894)
American Physician and Author

There was a tribunal once in France, as you may remember, called the Chambre Ardente, the Burning Chamber. It was hung all round with lamps, and hence its name. The burning chamber for the trial of young maidens is the blazing ball-room. What have they full-dressed you, or rather half-dressed you for, do you think? To make you look pretty, of course! Why have they hung a chandelier above you, flickering all over with flames, so that it searches you like the noonday sun, and your deepest dimple cannot hold a shadow? To give brilliancy to the gay scene, no doubt!—No, my dear! Society is inspecting you, and it finds undisguised surfaces and strong lights a convenience in the process. [. . .] For, look you, there are dozens, scores, hundreds, with whom you must be weighed in the balance; and you have got to learn that the "struggle for life" Mr. Charles Darwin talks about reaches to vertebrates clad in crinoline, as well as to mollusks in shells, or articulates in jointed scales, or anything that fights for breathing-room and food and love in any coat of fur or feather! Happy they who can flash defiance from bright eyes and snowy shoulders back into the pendants of the insolent lustres!

Oliver Wendell Holmes Sr., *Elsie Venner: A Romance of Destiny* (1861; reprint, [Whitefish, MT?]: Kessinger, 2005), 122–23.

Oliver Wendell Holmes [Sr.], born in the same year [as Darwin], delighted to speak of the good company in which he came into the world. On January 27th, 1894, I had the great pleasure of sitting next to him at a dinner of the Saturday Club in Boston, and he then spoke of the subject with the same enthusiasm with which he deals with it in his writings. [. . .] Dr. Holmes further said that he remembered with much satisfaction an occasion on which he was able to correct Darwin on a matter of scientific fact. He could not remember the details, but we may hope for their ultimate recovery, for he said that Darwin had written a courteous reply accepting the correction.

Edward B. Poulton, *Charles Darwin and the Theory of Natural Selection* (New York: Macmillan, 1896), 9.

Secondary Literature: Edwin Palmer Hoyt, *The Improper Bostonian: Dr. Oliver Wendell Holmes* (New York: Morrow, 1979).

OLIVER WENDELL HOLMES JR. (1841–1935)
American Jurisprudent

It has always seem to us a singular anomaly that believers in the theory of evolution and in the natural development of institutions by successive adaptation to the environment should be found laying down a theory of government intended to establish its limits once for all by a logical deduction from axioms. But the objection which we wish to express at the present time is that this presupposes an identity of interest between the different parts of a community which does not exist in fact. Consistently with his views, however, Mr. Spencer* is forever putting cases to show that the reaction of legislation is equal to its action. By changing the law, he argues, you do not get rid of any burden, but only change the mode of bearing it; and if the change does not make it easier to bear for society, considered as a whole, legislation is inexpedient. This tacit assumption of the solidarity of the interests of society is very common, but seems to us to be false. The struggle for life, undoubtedly, is constantly putting the interest of men at variance with those of the lower animals. And the struggle does not stop in the ascending scale with the monkeys, but is equally the law of human existence.

Oliver Wendell Holmes Jr., "The Gas-Stokers Strike," *American Law Review* 7 (1873): 582, quoted in *The Essential Holmes*, ed. Richard A. Posner. (Chicago: University of Chicago Press, 1997), 121–22. Holmes, although an evolutionist, did not read Darwin until 1907. Mark DeWolfe Howe, *Justice Oliver Wendell Holmes: The Shaping Years, 1841–1870* (Cambridge, MA: Harvard University Press, 1957), 156.

With regard to philosophy, I didn't know that anything was happening but if it is, I should suppose the fundamental explanation to be the increasing realization that Darwin had not exhausted the question of variations and that, as was pointed out long ago, the physicists before they know it are talking metaphysics.

Holmes to Felix Frankfurter, 25 July 1921, in *Holmes and Frankfurter: Their Correspondence, 1912–1934,* ed. Robert M. Mennel and Christine L. Compston (Hanover, NH: University Press of New England, 1996), 118.

Secondary Literature: G. Edward White, *Oliver Wendell Holmes, Jr.* (New York: Oxford University Press, 2006); Louis Menand, *The Metaphysical Club* (New York: Farrar, Straus, & Giroux, 2001).

JOSEPH DALTON HOOKER (1817–1911)
English Botanist

The next act in the drama of our lives opens with personal intercourse. This began with an invitation to breakfast with him at his brother's (Erasmus Darwin's) house in Park Street; which was shortly afterwards followed by an invitation to Down to meet a few brother Naturalists. In the short intervals of good health that followed the long illnesses which oftentimes rendered life a burthen to him, between 1844 and 1847, I had many such invitations, and delightful they were. A more hospitable and

more attractive home under every point of view could not be imagined—
of Society there were most often Dr. Falconer,[†] Edward Forbes[*], Professor
Bell,[‡] and Mr. Waterhouse[§]—there were long walks, romps with the children on hands and knees, music that haunts me still. Darwin's own hearty
manner, hollow laugh, and thorough enjoyment of home life with friends;
strolls with him all together, and interviews with us one by one in his study,
to discuss questions in any branch of biological or physical knowledge that
we had followed; and which I at any rate always left with the feeling that I
had imparted nothing and carried away more than I could stagger under.
Latterly, as his health became more seriously affected, I was for days and
weeks the only visitor, bringing my work with me and enjoying his society
as opportunity offered. It was an established rule that he every day pumped
me, as he called it, for half an hour or so after breakfast in his study, when
he first brought out a heap of slips with questions botanical, geographical,
etc., for me to answer, and concluded by telling me of the progress he had
made in his own work, asking my opinion on various points. I saw no more
of him till about noon, when I heard his mellow ringing voice calling my
name under my window—this was to join him in his daily forenoon walk
round the sand-walk. On joining him I found him in a rough grey shooting-
coat in summer, and thick cape over his shoulders in winter, and a stout
staff in his hand; away we trudged through the garden, where there was
always some experiment to visit, and on to the sand-walk, round which a
fixed number of turns were taken, during which our conversation usually
ran on foreign lands and seas, old friends, old books, and things far off to
both mind and eye. In the afternoon there was another such walk, after
which he again retired till dinner if well enough to join the family; if not,
he generally managed to appear in the drawing-room, where seated in his
high chair, with his feet in enormous carpet shoes, supported on a high
stool—he enjoyed the music or conversation of his family.

The Life and Letters of Charles Darwin, ed. Francis Darwin, 2 vols. (London: John
Murray, 1887), 1:387–88. [†]See John Richard Green entry. [‡]Thomas Bell (1792–1880),
English zoologist. [§]George Waterhouse (1810–88), English zoologist and entomologist, curator of the museum of the Zoological Society of London when Darwin entrusted him with the study of the mammals and beetles from the voyage of
the *Beagle,* later "keeper" (curator) of geology at the British Museum, author of
volume on mammals of *The Zoology of the Voyage of the Beagle.*

[31 December 1859]

We have risen from the perusal of Mr. Darwin's book much impressed with its importance, and have moreover found it to be so dependent on the phenomena of horticultural operations, for its facts and results, and so full of experiments that may be repeated and discussed by intelligent gardeners, and of ideas that may sooner fructify in their minds than in those of any other class of naturalists, that we shall be doing them (and we hope also science) a service by dwelling in some detail upon its contents. [. . .] Mr. Darwin's work [. . .] is devoted to the inquiry of how it is that the world has come to be peopled by so many and such various kinds of plants and animals as now inhabit it; and why it is that these do not all present a continuous instead of an interrupted series, capable of being divided into varieties, species, genera, orders &c. To all careless and many careful naturalists the assertion is sufficient, that they were so created, and have been endowed with power to remain unchanged. [. . .] In Mr. Darwin's opinion the objections to [the creation theory] are numerous and some of them are more conclusive than the arguments in its favor. In the first there is the fact, that gardeners and cattle and bird breeders have made races which are not only more dissimilar than many species in a state of nature are, but if found in a state of nature would unquestionably be ranked as species and even genera.

Joseph Dalton Hooker, review of *Origin of Species,* in *Gardeners' Chronicle,* 31 December 1859, 1051–52, reprinted in *Darwin and his Critics,* by David Hull (Cambridge, MA: Harvard University Press, 1973), 82–83.

Secondary Literature: Jim Endersby, *Imperial Nature: Joseph Hooker and the Practices of Victorian Science* (Chicago: University of Chicago Press, 2008).

GERARD MANLEY HOPKINS (1844–1899)
English Poet

I see the *Academy* no more but among the last things I read in it was [John] Tyndall's* address. I thought it interesting and eloquent, though it made me "most mad." It is not only that he looks back to an obscure origin, he looks forward with the same content to an obscure future—to be lost "in the infinite azure of the past" (fine phrase by the by). I do not think, do you know, that Darwinism implies necessarily that man is descended from any ape or ascidian or maggot or what not but only from the common ancestor maggots, and so on: these common ancestors, if lower animals, need not have been repulsive animals. What Darwin himself says about this I do not know. You should read St. George [Jackson] Mivart's* *Genesis of Species:* he is an Evolutionist though he combats downright Darwinism and is very orthodox.

Hopkins to his mother, 20 September 1874, in *Gerard Manley Hopkins: Selected Letters,* ed. Catherine Phillips (Oxford: Clarendon, 1990), 79.

Secondary Literature: Tom Zaniello, *Hopkins in the Age of Darwin* (Ames: Iowa State University Press, 1988).

FENTON JOHN ANTHONY HORT (1828–1892)
Irish Theologian

[March 1860]
Have you read Darwin? How I should like a talk with you about it! In spite of difficulties, I am inclined to think it unanswerable. In any case it is a treat to read such a book.

Hort to Rev. B. F. Westcott, 10 March 1860, in *Life and Letters of Fenton John Anthony Hort,* ed. Arthur Fenton Hort, 2 vols. (London: Macmillan, 1896), 1:414.

[April 1860]
[T]he book which has most engaged me is Darwin. Whatever may be thought of it, it is a book that one is proud to be contemporary with. I must work out and examine the argument more in detail, but at present my feeling is strong that the theory is unanswerable. If so, it opens up a new period in I know not what not.

Hort to Rev. John Ellerton Hardwick, 3 April 1860, in *Life and Letters of Fenton John Anthony Hort,* ed. Arthur Fenton Hort, 2 vols. (London: Macmillan, 1896), 1:416.

The *Life of Darwin,* which I dare say you have seen on our drawing-room table, is a very interesting book; but I am afraid you will find many things in it hard to understand, and it is a big book. Some day you may like to have a talk about it.

Hort to his youngest son, 24 November 1888, in *Life and Letters of Fenton John Anthony Hort*, ed. Arthur Fenton Hort, 2 vols. (London: Macmillan, 1896), 2:398.

Secondary Literature: Owen Chadwick, *The Victorian Church*, 2 vols. (New York: Oxford University Press, 1966–70).

WILLIAM DEAN HOWELLS (1837–1920)
AMERICAN AUTHOR AND CRITIC

Outside of his scientific circle in Cambridge he [Agassiz*] was more friends with [Henry Wadsworth] Longfellow* than with any one else, I believe, and Longfellow told me how, after the doctors had condemned Agassiz to inaction, on account of his failing health he had broken down in his friend's study, and wept like an "Europaer," and lamented, "I shall never finish my work!" Some papers which he had begun to write for the Magazine, in contravention of the Darwinian theory, or part of it, which it is known Agassiz did not accept, remained part of the work which he never finished. After his death, I wished Professor Jeffries Wyman* to write of him in the Atlantic, but he excused himself on account of his many labors, and then he voluntarily spoke of Agassiz's methods, which he agreed with rather than his theories, being himself thoroughly Darwinian. I think he said Agassiz was the first to imagine establishing a fact not from a single example, but from examples indefinitely repeated. If it was a question of something about robins for instance, he would have a hundred robins examined before he would receive an appearance as a fact.

William Dean Howells, *Literary Friends and Acquaintances: A Personal Retrospect of American Authorship* (1911; reprint, Charleston, SC: BiblioLife, 2008), 201.

Secondary Literature: Susan Goodman and Carl Dawson, *William Dean Howells: A Writer's Life* (Berkeley and Los Angeles: University of California Press, 2005).

WILLIAM HENRY HUDSON (1841–1922)
ARGENTINE-ENGLISH AUTHOR

What effect had Darwin produced on me? I had to confess that I had not read a line of his work, that with the exception of Draper's *History of Ci-*

vilisation, which had come by chance in my way, I had during all these five years read nothing but the old books which had always been on our shelves. [My brother] said he knew Draper's *History,* and it was not the sort of book for me to read at present. He had a store of books with him, and would lend me the *Origin of Species* to begin with [in 1860].

When I had read and returned the book, and he was eager to hear my opinion, I said it had not hurt me in the least, since Darwin had to my mind only succeeded in disproving his own theory with his argument from artificial selection. He himself confessed that no new species had ever been produced in that way. That, he said in reply, was the easy criticism that anyone who came to the reading in a hostile spirit would make. They would fasten on that apparently weak point and not pay much attention to the fact that it is fairly met and answered in the book. When he first read the book it convinced him; but he had come to it with an open mind and I with a prejudiced mind on account of my religious ideas. He advised me to read it again, to read and consider it carefully with the sole purpose of getting at the truth. "Take it," he said, "and read it again in the right way for you to read it—as a naturalist." [...] I read it again in the way he had counseled, and then refused to think any more on the subject. I was sick of thinking.

William Henry Hudson, *Far Away and Long Ago: A History of My Early Life* (1923; reprint, New York: AMS, 1968), 344.

However close an observer that naturalist [Darwin] may be, it was not possible for him to know much of a species from seeing perhaps one or two individuals, in the course of a rapid ride across the pampas. Certainly if he had truly known the habits of the bird, he would not have attempted to adduce from it an argument in favor of this theory of the *Origin of Species,* as so great a deviation from the truth in this instance might give the opponents of his book a reason for considering other statements in it erroneous or exaggerated.

Hudson to Spencer Fullerton Baird, 1870, quoted in *W. H. Hudson, a Biography,* by Ruth Tomalin (London: Faber & Faber, 1982), 89.

Secondary Literature: Felipe Arocena, *William Henry Hudson: Life, Literature, and Science* (London: McFarland, 2003).

ALDOUS HUXLEY (1894–1963)
English Author

I am debating what to get out of the library next. Darwinism will last till next Wednesday, when I can go round in person.

Aldous Huxley to Leonard Huxley (his father), March 1912, in *Letters of Aldous Huxley*, ed. Grover Smith (London: Chatto & Windus, 1969), 41.

Laforgue[†] was also a hearty Darwinian and liked the thought of being a developed beast [...] hence you see him all for instinct and the like, and down on consciousness and intellect as rather supererogatory excrescences of the Unconsh.

Huxley to Frances Petersen, August 1915, in *Letters of Aldous Huxley*, ed. Grover Smith (London: Chatto & Windus, 1969), 76. [†]Jules Laforgue (1860–87), French symbolist poet who used Darwinian themes in a famous 1883 essay on impressionism.

I wish you'd tell me what's a good book on the effects of modern genetic discoveries on the Darwinian theory. How do organisms ever get out of that hereditary predestination exhibited by Mendelism?

Huxley to Julian Huxley (brother), 21 June 1925, in *Letters of Aldous Huxley*, ed. Grover Smith (London: Chatto & Windus, 1969), 250.

I've just bought a little house down here in the Mediterranean [...] I hate living in big towns [...] because French literary society is really a bit too literary and, owing to stringent economic conditions in the intellectual world, terribly Darwinian. If you want to see nature red in tooth and claw go to a French literary salon. The only more Darwinian spectacle is a collection of French women.

Huxley to Flora Strousse, 14 June 1930, in *Letters of Aldous Huxley*, ed. Grover Smith (London: Chatto & Windus, 1969), 338.

I was interested to hear of the Darwin film. Will it be in the nature of a biography? Or of a documentary about Natural Selection? Or a mixture of both? Darwin himself is a richly comic character, with his mixture of enormous daring and timidity, his calculated use of psycho-somatic illness in the service of his vocation. I hope you will be able to bring him in as a person as well as a man of science and revolutionary thinker.

Aldous Huxley to Matthew (son) and Ellen Huxley, 27 March 1957, in *Letters of Aldous Huxley*, ed. Grover Smith (London: Chatto & Windus, 1969), 819.

Secondary Literature: Nicholas Murray, *Aldous Huxley: A Biography* (New York: St. Martin's, 2003).

JULIAN HUXLEY (1887–1975)
English Biologist

The reaction against Darwinism set in during the nineties of the [nineteenth] century. The younger zoologists of that time were disconnected with the trends of their science. The major school still seemed to think that the sole aim of zoology was to elucidate the relationship of larger groups. Had not Kovalevsky [Alexander O. Kovalevski]* demonstrated the vertebrate affinities of the sea-squirts, and did not comparative embryology prove the common ancestry of groups so unlike as worms and mollusks? Intoxicated with such earlier successes of evolutionary phylogeny, they proceeded (like some Forestry Commission of science) to plant wildernesses of family trees over the beauty-spots of biology. [...] Late nineteenth century Darwinism came to resemble the early nineteenth-century school of Natural Theology *redivivus,* one might say, but philosophically upside-down, with Natural Selection instead of a Divine Artificer as the *Deus ex machina.* There was little contact of evolutionary speculation with the concrete facts of cytology and heredity, or with actual experimentation.

Julian Huxley, *Evolution: The Modern Synthesis* (New York: Harper, 1943), 22–23.

Secondary Literature: C. Kenneth Waters and Albert van Helden, eds., *Julian Huxley: Biologist and Statesman of Science* (Houston: Rice University Press, 1992).

THOMAS H. HUXLEY (1825–1895)
English Zoologist

[January 1860]
Darwin is the great subject just at present, and everybody is talking about it. The thoroughly orthodox hold up their hands and lift up their eyes, but

know not how to crush the enemy, barred as they are by finding all of us (whatever our views as to the truth of Darwin's opinions) fully agreed that the book shall not be hunted down but shall have fair play, on fair scientific grounds.

Huxley to William Barton Rogers, 17 January 1860, in *Life and Letters of William Barton Rogers*, edited by his wife, with the assistance of William T. Sedgwick, 2 vols. (Boston: Houghton, Mifflin, 1896), 2:25.

[September 1860]
Samuel [Wilberforce]* thought it was a fine opportunity for chaffing a savan [*sic*]—However he performed the operation vulgarly and I determined to punish him—partly on that account and partly because he talked pretentious nonsense. So when I got up I spoke pretty much to the effect—that I had listened with great attention to the Lord Bishop's speech but had been unable to discover either a new fact or a new argument in it—except indeed the question raised as to my personal predilections in the matter of ancestry—That it would not have occurred to me to bring forward such a topic as that for discussion myself, but that I was quite ready to [meet] the Right Revd. prelate even on that ground—If then, said I, the question is put to me would I rather have a miserable ape for a grandfather or a man highly endowed by nature and possessed of great means of influence and yet who employs those faculties and that influence for the mere purpose of introducing ridicule into a grave scientific discussion—I unhesitatingly affirm my preference for the ape. Whereupon there was inextinguishable laughter among the people—and they listened to the rest of my argument with the greatest attention. [. . .] I believe I was the most popular man in Oxford for full four and twenty hours afterwards.

Huxley to Frederick Dyster, 9 September 1860, http://alepho.clarku.edu/huxley/letters/60.html#9sep1860 (accessed 14 January 2009), partial transcription in J. R. Lucas, "Wilberforce and Huxley: A Legendary Encounter," *Historical Journal* 22 (1979): 313–30, on 326.

His [Samuel Wilberforce's*] words are quite misquoted by you (which your father [Thomas H. Huxley*] refuted). They did not appear vulgar, nor insolent nor personal, but flippant. He had been talking of the perpetuity of species in birds: and then denying *a fortiori* the derivation of the species Man from Ape, he rhetorically invoked the help of *feeling:* and said (I swear to the *sense* and the *form* of the sentence, if not to the words) "If anyone were to be willing to trace his descent through an ape as his *grandfather,* would he be willing to trace his descent similarly on the side of his *grandmother.*" It was (you see) the [*sic*] arousing the antipathy about degrading *women* to the Quadrumana. It was not to the point, but it was the purpose. It did not sound insolent, but unscientific and unworthy of the zoological argument which he had been sustaining. It was a *(bathos).*

Your father's reply, (Remember, he did not use the word "*prostituting* his abilities,*" but (I believe) "degrading." But I will swear to the absence of the former low word. (Also *equivocal* was not used).) He showed that there was a vulgarity as well as a folly in the Bishop's words; and the impression distinctly was, that the Bishop's party as they left the room, felt abashed; and recognized that the Bishop had forgotten to behave like a gentleman. The victory of your father, was not the ironical dexterity shown by him, but the fact that he had got a victory in respect of *manners* and *good breeding.* You must remember that the whole audience was made up of gentlefolk, who were not prepared to endorse anything vulgar. The speech which really left its mark *scientifically* on the meeting, was the short one of *Hooker**, wherein he said "he considered that Darwin's views were true in the field of Botany; and that he must claim that students should 'provisionally accept them as a *working hypothesis* in the field of the Animal Kingdom.'" I am confident, in the above statements, not only that I have given the true impression, but I can corroborate my quotations of the words used by the exact memory of the late Canon T. S. Evans of Durham, who about twelve years ago, talked over with me, the details of the meeting. [...]

The blank look of Sir B. Brodie[†] to your father's remark, corroborates my view that the insolence and personality of Bishop Wilberforce's remark was not caught by the meeting, until your father remarked it. [...]

The spiteful narrative which you quote from J. R. [John Richard] Green* (the *historical* writer) is hardly worthy of him! I should say that to fair minds, the *intellectual* impression left by the discussion was that the Bishop had stated some facts about the perpetuity of Species, but that no one had really contributed any valuable point to the opposite side except Hooker; but that your father had scored a victory over Bishop Wilberforce in the question of good *manners.*

Frederic William Farrar to Leonard Huxley, 12 July 1899, Huxley Papers, Imperial College, London, quoted in J. R. Lucas, "Wilberforce and Huxley: A Legendary Encounter," *Historical Journal* 22 (1979): 313–30, on 326–27. †Benjamin Collins Brodie (1783–1862), English physiologist and surgeon.

Mr. Darwin's long-standing and well-earned scientific eminence probably renders him indifferent to that social notoriety which passes by the name of success; but of the calm spirit of the philosopher have not yet wholly superseded the ambition and vanity of the carnal man within him, he must be well satisfied with the results of his venture in publishing the *Origin of Species*. Overflowing the narrow bounds of purely scientific circles, the "species question" divides with Italy and the Volunteers the attention of general society. Everybody has read Mr. Darwin's book, or, at least, has given an opinion upon its merits or demerits; pietists, whether lay or ecclesiastic, decry it with the mild railing which sounds so charitable; bigots denounce it with ignorant invective; old ladies of both sexes consider it a decidedly dangerous book, and even savants, who have no better mud to throw, quote antiquated writers to show that its author is no better than an ape himself; while every philosophical thinker hails it as a veritable Whitworth gun in the armory of liberalism; and all competent naturalists and physiologists, whatever their opinions as to the ultimate fate of the doctrines put forth, acknowledge that the work in which they are embodied is a solid contribution to knowledge and inaugurates a new epoch in natural history.

T. H. Huxley, *Darwiniana* (London: Macmillan, 1893), 22–23.

The gradual lapse of time has now separated us by more than a decade from the date of the publication of the *Origin of Species*—and whatever may be thought or said about Mr. Darwin's doctrines, or the manner in which he has propounded them, this much is certain, that, in a dozen years, the *Origin of Species* has worked as complete a revolution in biological sciences as the *Principia* did in astronomy—and it has done so, because, in the words of [Hermann von] Helmholtz*, it contains "an essentially new creative thought."

T. H. Huxley, *Darwiniana* (London: Macmillan, 1893), 121.

I am not likely to take a low view of Darwin's position in the history of science, but I am disposed to think that Buffon[†] and Lamarck would run him hard in both genius and fertility. In breadth of view and in extent of knowledge these two men were giants, though we are apt to forget their services. Von Baer[‡] was another man of the same stamp; Cuvier, in a somewhat lower rank, another; and J. Müller[§] another.

Huxley to George John Romanes, 9 May 1882, in *Life and Letters of Thomas Henry Huxley,* ed. Leonard Huxley, 2 vols. (London: Macmillan, 1900), 2:253. [†]Georges-Louis Leclerc, comte de Buffon (1707–88), French naturalist, author of a 44-volume *Histoire naturelle,* the most influential natural-history work of his age. [‡]Karl Ernst von Baer (1792–1876), German zoologist and embryologist who discovered the human *ovum.* [§]Johannes Müller (1801–58), German physiologist and comparative anatomist.

Secondary Literature: Paul White, *Thomas Huxley: Making the "Man of Science"* (Cambridge: Cambridge University Press, 2003).

I

INDEX OF PROHIBITED BOOKS
The Vatican

[1878]

This work merits serious and special attention. In it Darwinism is expounded and partly approved, [stating] that it has many points of contact with religious doctrine, especially with Genesis and other books of the Bible. Until now the Holy See has rendered no decision on the system mentioned. Therefore, if Caverni's work[†] is condemned, as it should be, Darwinism would be indirectly condemned. Surely there would be cries against this decision: the example of Galileo would be held up; it will be said that this Holy Congregation is not competent to emit judgments on physiological and ontological doctrines or theories of change. But we should not focus on the probable clamor. With his system, Darwin destroys the bases of revelation and openly reaches pantheism and an abject materialism. Thus, an indirect condemnation of Darwin is not only useful, but even necessary, together with that of Caverni, his defender and propagator among Italian youth.

Report of the General Congregation of the Index, meeting of 1 July 1878 in Rome to consider banning Raffaello Caverni's book, *De' nuovi studi della filosofia: Discorsi di Raffaello Caverni a un giovane studente,* translated and quoted in *Negotiating Darwin: The Vatican Confronts Evolution, 1877–1902,* by Mariano Artigas, Thomas F. Glick, and Rafael Martínez (Baltimore: Johns Hopkins University Press, 2006), 47. [†]Raffaelo Caverni (1837–1900), Catholic priest, author of *History of the Experimental Method in Italy,* 6 vols. (1891–1900).

ROBERT G. INGERSOLL (1833–1899)
American Freethinker and Orator

[1884]

Mr. Ingersoll discussed Charles Darwin, "one of the greatest men who ever touched this globe." He has explained more, he said, of the phenomena of life than all the religious teachers. His doctrine of evolution, of the survival of the fittest and of the origin of species has removed from every thinking mind the last vestige of Orthodox Christianity. This light has broken in on some of the clergy, and the greatest man who to-day occupies a pulpit in an orthodox church, Henry Ward Beecher*, admits that Darwin can be relied on and that Moses made a mistake.

"Ingersoll on Orthodoxy," *New York Times,* 28 April 1884.

[1888]

Open Letter to William Gladstone*

You have discovered a way by which [. . .] the neck of orthodoxy can escape the noose of Darwin. [. . .] I insist, that the discoveries of Darwin do away absolutely with the inspiration of scriptures—with the account of creation in Genesis, and demonstrate not simply the falsity, not simply the wickedness, but the foolishness of the "sacred volume." [. . .]

The Church demonstrated the falsity and folly of Darwin's theories by showing that they contradicted the Mosaic account of creation, and now the theories of Darwin having been fairly established, the church says that the Mosaic account is true, because it is in harmony with Darwin. Now, if it should turn out that Darwin was mistaken, what then?

The Works of Robert G. Ingersoll, Dresden Memorial Edition, 12 vols. (New York: C. P. Farell, 1900), 6:274–75.

Secondary Literature: Orvin Prentiss, *American Infidel, Robert G. Ingersoll: A Biography* (New York: Citadel, 1962).

J

JOSEPH JACOBS (1854–1916)
English Jewish Folklorist and Anthropologist

The earlier period at which "the custom of woman" (Gen. xxxi, 35) appears among Jewesses [. . .] is another trait which, if substantiated by wider induction, must be regarded as distinctly racial. If Darwin's explanation of its origin (*Descent*, 1st edit., I, 212) be correct, it must have preserved its periodicity for an incalculable time, and it may be surmised that any other temporal relation, such as the age of its appearance would be equally persistent. If it appears among Jewesses of St. Petersburg at the same early age as among Southern Asiatics, the Eastern origin of the former may be considered well established.

Joseph Jacobs, "On the Racial Characteristics of Modern Jews," *Journal of the Anthropological Institute of Great Britain and Ireland* 15 (1886): 23–62, on 51.

Most persons will be equally surprised to observe equality [between "Europeans" and "Jews"] in science, both in what Mr. [Francis] Galton* calls natural science and in science pure and simple, chiefly mathematical. As regards the former, of course Jews have no Darwin. It took England 180 years after Newton before she could produce a Darwin, and as Britishers are five times the number of Jews, even including those of Russia, it would take on the same showing 900 years before they produce another Spinoza, or, even supposing the double superiority were true, 450 years would be needed. But in the lower ranks even of biology Jews have done and are doing good work.

Joseph Jacobs, "The Comparative Distribution of Jewish Ability," *Journal of the Anthropological Institute of Great Britain and Ireland* 15 (1886): 351–79, on 363–64.

Secondary Literature: Simon Rabinovitch, "Jews, Englishmen, and Folklorists: The Scholarship of Joseph Jacobs and Moses Gaster," in *The "Jew" in Late-*

Victorian and Edwardian Culture: Between the East End and East Africa, ed. Eitan Bar-Yosef and Nadia Valman (Basingstoke, Hampshire, UK: Palgrave Macmillan, 2009), 113–30.

HENRY JAMES (1843–1916)
AMERICAN NOVELIST

I found the Charles [Eliot] Nortons* settled for the time in London, with social contacts and penetrations, a give and take of hospitality, that I felt as wondrous and of some elements of which they offered me, in their great kindness, the benefit; so that I was long to value having owed them in the springtime of '69 five separate impressions of distinguished persons, then in the full flush of activity and authority, that affected my young provincialism as a positive fairytale of privilege. I had a Sunday afternoon hour with Mrs. Lewes at North Bank [. . .] and then the opportunity of dining with Mr. [John] Ruskin* at Denmark Hill, an impression of uneffaced intensity and followed by a like—and yet so unlike—evening of hospitality from William Morris† in the medieval mise-en-scene of Queen Square. This had been preceded by a luncheon with Charles Darwin, beautifully benignant, sublimely simple, at Down; a memory to which I find attached our incidental wondrous walk—Mrs. Charles Norton, the too near term of her earthly span then smoothly out of sight, being my guide for the happy excursion—across a private park of great oaks, which I conceive to have been the admirable Holwood and where I knew my first sense of a matter afterwards, through fortunate years, to be more fully disclosed: the springtime in such places, the adored footpath, the first prim roses, the stir and scent of renascence in the watered sunshine and under spreading boughs that were somehow before aught else the still reach of remembered lines of [Alfred Lord] Tennyson*, ached over in nostalgic years. The rarest hour of all perhaps, or at least the strangest, strange verily to the pitch of the sinister, was a vision, provided by the same care, of D. G. Rossetti‡ in the vernal dusk of Queen's House Chelsea—among his pictures, amid his poetry itself, his whole haunting "esthetic," and yet above all bristling with his personality, with his perversity, with anything, as it rather awfully seemed to me, but his sympathy, though it at the same time left one oddly desirous of more of him.

Henry James, *Autobiography: A Small Boy and Others, Notes of a Son and Brother, the Middle Years,* ed. Frederick W. Dupee (New York: Criterion Books, 1956), 515. †William Morris (1834–96), English architect, graphic designer, and historical preservationist. ‡Dante Gabriel Rosetti (1828–82), English poet and translator.

Secondary Literature: Paul Fisher, *House of Wits: An Intimate Portrait of the James Family* (New York: Henry Holt, 2008).

WILLIAM JAMES (1842–1910)
AMERICAN PSYCHOLOGIST

Darwin's book [*Variation of Plants and Animals under Domestication*] has just come to me. As it is of course too late for the next N. A. R.,† I will send review of it at my leisure.

William James to Henry James, 4 March 1868, in *The Correspondence of William James,* vol. 1 (Charlottesville: University Press of Virginia, 1992), 36. †*North American Review,* the key journal of the Boston intelligentsia.

Here is the Darwin for Chas. [Charles Eliot] Norton* who I spoke of in my letter of 3 days ago. [. . .] The more I think of Darwin's idea the more weighty do they appear to me—tho' of course my opinion is worth little— still I *believe* that that scoundrel Agassiz* is unworthy either intellectually or morally for him to wipe his shoes on, & I feel a certain pleasure in yielding to the feeling.

William James to Henry James, 9 March 1868, in *The Correspondence of William James,* vol. 1 (Charlottesville: University Press of Virginia, 1992), 38–39.

Darwin's new book [*Descent of Man*] appears to make a good deal of noise in the papers. I suppose it makes hardly less in Germany notwithstanding the war. I haven't yet seen it. I had the pleasure of hearing Agassiz* call it "rubbish" the other day.

William James to Henry James, 8 April 1871, in *The Correspondence of William James,* vol. 4 (Charlottesville: University Press of Virginia, 1995), 38–39.

Old fashioned theism was bad enough, with its notion of God as an exalted monarch, made up of a lot of unintelligible or preposterous "attributes"; but, so long as it held strongly by the argument from design, it kept some touch with concrete realities. Since, however, Darwinism has once for all displaced design from the minds of the "scientific," theism has lost that foothold; and some kind of an immanent or pantheistic deity working in things rather than above them is, if any, the kind recommended to our contemporary imagination.

William James, *Pragmatism, a New Name for Some Old Ways of Thinking* (New York: Longmans, Green, 1907), 70.

Secondary Literature: Robert D. Richardson, *Henry James: In the Maelstrom of American Modernism* (New York: Houghton Mifflin Harcourt, 2006); Louis Menand, *The Metaphysical Club* (New York: Farrar, Straus, & Giroux, 2001).

RICHARD JEFFERIES (1847–1887)
English Nature Writer

There might be said to be two principal theories as regards the present condition of the universe: one referred it mainly to design, the other mainly to evolution. Jefferies unhesitatingly rejected both.

John Lubbock, *Essays and Addresses, 1900–1903* (Leipzig: Bernhard Tauchnitz, 1904), 71.

The belief that the human mind was evolved, in the process of unnumbered years, from a fragment of palpitating slime through a thousand gradations, is a modern superstition, and proceeds upon assumption alone. Nothing is evolved, no evolution takes place, there is no record of such an event; it is pure assertion. The theory fascinates many, because they find, upon study of physiology, that the gradations between animal and vegetable are so fine and so close together, as if a common web bound them together. But although they stand so near they never change places. They are like the figures on the face of a clock; there are minute dots between, apparently connecting each with the other, and the hands move round over all. Yet ten never becomes twelve, and each second even is parted from the next, as you may hear by listening to the beat. So the gradations of life, past and present, though standing close together never change places. Nothing is evolved. There is no evolution any more than there is any design in nature. By standing face to face with nature, and not from books, I have convinced myself that there is no design and no evolution. What there is, what was the cause, how and why, is not yet known; certainly it was neither of these.

Richard Jefferies, *The Story of My Heart* (London: Longmans, Green, 1883), 125–26.

Two books occur to me at this moment which would be greatly appreciated in every country home, from that of the peasant who has just begun to read to the houses of well-educated and well-to-do people, if they only knew of their existence and their contents—of course provided they were cheap enough, for country people have to be careful of their money nowadays. I allude to Darwin's *Climbing Plants* and to his *Earthworms;* these are astonishing works of singular patience and careful observation. The first gives most fascinating facts about such a common plant, for example, as the hedge bryony and the circular motion of its tendrils. Any farmer, for instance, will tell you that the hop-bine will insist upon going round the pole in one direction, and you cannot persuade it to go the other. These circular movements seem almost to resemble those of the planets about their centre, all things down to the ether seem to have a rotatory motion; and some foreign plants which he grew send their far-extended tendrils round and round with so patent a movement that you can see it hour by hour like the hand of a clock. Perhaps the little book on earthworms is a yet more wonderful achievement of this great genius, who had not only untiring patience to observe and verify, but also possessed imagination, and could thereby see the motive idea at work behind the facts. At first it has a repellent sound, but we quickly learn how clumsy and prejudiced have been our views of the despised worm thrown up by every ploughshare.

Richard Jefferies, *Field and Hedgerow: Being the Last Essays of Richard Jefferies* (New York: Longmans, Green, 1897), 149–50.

[Jefferies'] worst fault would appear to have been a false and distorted sentimentality. It is difficult to realize he was a contemporary of Darwin.

He was, indeed, hopelessly out of touch with current scientific thought and speculation. While [Alfred Lord] Tennyson* was writing of Nature "red in tooth and claw with ravine," Jefferies was indicting panegyrics on the friendliness of the flowers.

T. Michael Pope, "Richard Jefferies," *Academy* 74 (1908): 617.

Secondary Literature: Nick Freeman, "Edward Thomas, Swinburne, and Richard Jefferies," *English Literature in Transition* 51 (2008): 164–83.

WILLIAM STANLEY JEVONS (1835–1882)
ENGLISH ECONOMIST

The theories of Darwin and Spencer* are doubtless not demonstrated; they are to some extent hypothetical, just as all the theories of physical science are to some extent hypothetical, and open to doubt. Judging from the immense numbers of diverse facts which they harmonize and explain, I venture to look upon the theories of evolution and natural selection in their main features as two of the most probable hypotheses ever proposed. I question whether any scientific works which have appeared since the *Principia* of Newton are comparable in importance with those of Darwin and Spencer, revolutionizing as they do all our views of the origin of bodily, mental, moral, and social phenomena.

Granting all this, I cannot for a moment admit that the theory of evolution will destroy theology. That theory embraces several laws or uniformities which are observed to be true in the production of living forms; but these laws do not determine the size and figure of living creatures, any more than the law of gravitation determines the magnitudes and distances of the planets. Suppose that Darwin is correct in saying that man is descended from the Ascidians: yet the precise form of the human body must have been influenced by an infinite train of circumstances affecting the reproduction, growth, and health of the whole chain of intermediate beings. No doubt, the circumstances being what they were, man could not be otherwise than he is, and if in any other part of the universe an exactly similar earth, furnished with exactly similar germs of life, existed, a race must have grown up there exactly similar to the human race. [. . .]

Theologians have dreaded the establishment of the theories of Darwin and Huxley* and Spencer, as if they thought that those theories could explain everything upon the purest mechanical and material principles, and exclude all notions of design. They do not see that those theories have opened up more questions than they have closed. The doctrine of evolution gives a complete explanation of no single living form. While showing the general principles which prevail in the variation of living creatures, it only points out the infinite complexity of the causes and circumstances which have led to the present state of things. Any one of Mr. Darwin's books, admirable though they all are, consists but in the setting forth of a multitude of indeterminate problems. He proves in the most beautiful manner that each flower of an orchid is adapted to some insect which frequents and fertilizes it, and these adaptations are but a few cases of those immensely numerous ones which have occurred in the lives of plants and animals. But why orchids should have been formed so differently from other plants, why anything, indeed, should be as it is, rather than in some of the other infinitely numerous possible modes of existence, he can never show. The origin of everything that exists is wrapped up in the past history of the universe. At some one or more points in past time there must have been arbitrary determinations which led to the production of things as they are.

W. Stanley Jevons, *The Principles of Science* (London: Macmillan, 1887), 762–65.

Secondary Literature: Bert Mosselmans, *William Stanley Jevons and the Cutting Edge of Economics* (London: Routledge, 2007); Joseph Schumpeter, *History of Economic Analysis* (New York: Oxford University Press, 1954).

SARAH ORNE JEWETT (1849–1909)
American Novelist

I wondered, as I looked at him, if he had sprung from a line of ministers; he had the refinement of look and air of command which were the heritage of the old ecclesiastical families of New England. But as Darwin says in his autobiography, "there is no such king as a sea-captain; he is greater even than a king or a schoolmaster."

Sarah Orne Jewett, "Captain Littlepage," in Jewett, *Novels and Stories,* ed. Michael Davitt Bell (New York: Library of America, 1994), 387–92, on 38.

Secondary Literature: Paula Blanchard, *Sarah Orne Jewett: Her World and Her Work* (Reading, MA: Addison-Wesley, 1994).

ERNEST JONES (1879–1958)
Welsh Neurologist and Psychoanalyst

Unlike both Mr. W. and Mr. A., however, the figure in the dream was quite bald and there was a wart by the side of the nose. These characteristics, and the upper half of the head in general, at once reminded the subject of Charles Darwin. The singular appropriateness will be admitted of the problem of personal origin being expounded by the author of *The Descent of Man* and *The Origin of Species*. As a student of biology the subject had greatly revered Darwin, who had, so to speak, answered the question he had propounded in vain to his own father; it was evident that unconsciously he had identified the two men, Darwin being to him what he had wished his father to be—an expounder of the problem of origin. Strangely enough, Darwin had on the dream-day been the topic of conversation between the subject and his wife. Being concerned at his overworking she had urged him to give up some of his routine teaching work so that he might devote himself more peacefully to his favorite pursuit of scientific research, and had considerately volunteered to do with less money. The talk drifted on to the endowment of research, and the subject remarked what a fortunate thing it was for mankind that Darwin had inherited enough money from his father to enable him to pursue his investigations undeterred by material considerations. He here was evidently identifying himself in his unconscious—that realm of unlimited egoism with Darwin, i.e., once more with his father—and was at the same time mutely reproaching his father for not having bestowed him with more worldly goods.

Ernest Jones, "A Forgotten Dream," in *Papers on Psycho-Analysis,* rev. ed. (New York: William Wood, 1918), 238, originally published in *Journal of Abnormal Psychology* 7 (1912–13): 5–16.

Secondary Literature: Brenda Maddox, *Freud's Wizard: Ernest Jones and the Transformation of Psychoanalysis* (New York: Da Capo, 2007).

JOHN JONES (1818–1881)
ENGLISH GEOLOGIST

Although, oysters have been found in much older formations, as exemplified by the unique specimen of *Osirca nobilis,* from the carboniferous limestone of Belgium, which may be seen in the British Museum, with others from the Triassic *Saliferian* of St. Cassian, they amount in number of species, in the opinion of Mr. S. P. Woodward,[†] to three only, and it is first in Jurassic strata that they make their appearance in any remarkable number or variety. Taking into consideration this fact, with that of the universally admitted variety of forms attributable to one species: to those who have interested themselves in the theory of transmutation, as originally propounded by Lamarck, subsequently by the author of the *Vestiges,* and since, more practically by Darwin, in his treatise *On the Origin of Species,* the elaboration of figures, and the minute details here presented, although apparently uselessly repeated, may yet assume an aspect of interest which they could not otherwise possess.

The Geologist, January 1863, 150–51. [†]Samuel P. Woodward (1821–65), English geologist, informant of Darwin's.

DAVID STARR JORDAN (1851–1931)
AMERICAN ZOOLOGIST

Though I accepted [Agassiz's[*]] philosophy regarding the origin and permanence of species when I began serious study in Zoology, as my work went on their impermanence impressed me more and more strongly. Gradually I found it impossible to believe that the different kinds of animals and plants had been separately created in their present forms. Nevertheless, while I paid tribute to Darwin's marvelous insight, I was finally converted to the theory of divergence through Natural Selection and other factors not by his arguments, but rather by the special facts unrolling themselves before my own eyes, the rational meaning of which [Darwin] had plainly indicated. I sometimes said that I went over to the evolutionists with the grace of a cat the boy "leads" by its tail across the carpet!

David Starr Jordan, *The Days of a Man*, 2 vols. (Yonkers-on-Hudson, NY: World Book, 1922), 1:114 (referring to 1873).

Secondary Literature: Edward McNall Burns, *David Starr Jordan: Prophet of Freedom* (Stanford, CA: Stanford University Press, 1953); Elizabeth Noble Shor, "David Starr Jordan," in *Dictionary of Scientific Biography*, ed. Charles Coulston Gillispie, 16 vols. (New York: Scribner's, 1970–80), 7:169–70.

JAMES PRESCOTT JOULE (1818–1889)
English Physicist

I am glad you feel disposed to expose some of the rubbish which has been thrust on the public lately. Not that Darwin is so much to blame because I believe he had no intention of publishing any finished theory but rather to indicate the difficulties to be solved. [. . .] It appears that nowadays the public care for nothing unless it be of a startling nature. Nothing pleases them more than parsons who preach against the efficacy of prayer and philosophers who find a link between mankind and monkey or gorilla— certainly a most pleasing example of what *muscular* Christianity may lead to.

Joule to William Thomson, 13 May 1861, quoted in *Energy and Empire: A Biographical Study of Lord Kelvin*, by Crosbie Smith and M. Norton Wise (Cambridge: Cambridge University Press, 1989), 525.

Secondary Literature: L. Rosenfeld, "James Prescott Joule," in *Dictionary of Scientific Biography*, ed. Charles Coulston Gillispie, 16 vols. (New York: Scribner's, 1970–80), 7:180–82.

BENJAMIN JOWETT (1817–1893)
English Classicist and Theologian

I have read some of [David Friedrich] Strauss's* *Alte und Neue Glaube* [The Old Faith and the New, 1872]. I am surprised to see that he pins his faith upon "Darwinism," which seems to me not so much as untrue, as an utterly inadequate account of the world. I have for a long time past thought that miracles had no sufficient evidence. But what I regret in these German critics is that they seem never to consider the proportion which their discoveries bear to the whole truth. Are we to be sunk in materialism and sensualism, feebly rising into a sort of sentimentalism, because Strauss and others have shown that the Gospels partake of the character of other ancient writings, or because Darwin has imagined a theory by which one species may pass into another?

Jowett to R. B. D. Morier, 16 January 1873, in *The Life and Letters of Benjamin Jowett,* ed. Evelyn Abbott and Lewis Campbell, 3rd ed., 2 vols. (London: John Murray, 1897), 2:89.

Secondary Literature: Peter Bingham Hinchliff, *Benjamin Jowett and the Christian Religion* (London: Oxford University Press, 1987).

JAMES JOYCE (1882–1941)
Irish Author

To revert to Mr Bloom who, after his first entry, had been conscious of some impudent mocks which he however had bourne with as being the fruits of that age upon which it is commonly charged that it knows not pity. The young sparks, it is true, were as full of extravagancies as overgrown children: the words of their tumultuary discussions were difficultly understood and not often nice: their testiness and outrageous MOTS were such that his intellect resiled from: nor were they scrupulous sensible of the proprieties though their fund of strong animal spirits spoke in their behalf. But the word of Mr Costello was an unwelcome language for him for he nauseated the wretch that seemed to him a cropeared creature of a misshapen gibbosity, born out of wedlock and thrust like a crookback toothed and feet first into the world, which the dint of the surgeon's pliers in his skull lent indeed a colour to, so as to put him in thought of that missing link of creation's chain desiderated by the late ingenious Mr Darwin.

To Master Percy Apjohn at High School in 1880 he had divulged his disbelief in the tenets of the Irish (protestant) church (to which his father Rudolf Virag (later Rudolph Bloom) had been converted from the Israelitic faith and communion in 1865 by the society for promoting Christian-

ity among the jews) subsequently abjured by him in favour of Roman ca-
tholicism at the epoch of and with a view to his matrimony in 1888. To
Daniel Magrane and Francis Wade in 1882 during a juvenile friendship [. . .] he had advocated during nocturnal perambulations the political
theory of colonial (e.g. Canadian) expansion and the evolutionary theory
of Charles Darwin, expounded in *The Descent of Man* and *The Origin of
Species.*

James Joyce, *Ulysses* (New York: Penguin Classics, 2000), 532.

Secondary Literature: Robert E. Spoo, *James Joyce and the Language of History* (New
York: Oxford University Press, 1994).

JOHN WESLEY JUDD (1840–1916)
English Geologist

I remember once expressing surprise to Darwin that, after the views which
he had originated concerning the existence of areas of elevation and others of subsidence in the Pacific Ocean, and in face of the admitted difficulty of accounting for the distribution of certain terrestrial animals and
plants, if the land and sea areas had been permanent in position, he still
maintained that theory. Looking at me with a whimsical smile, he said: "I
have seen many of my old friends make fools of themselves, by putting
forward new theoretical views or revising old ones, *after they were sixty
years of age;* so, long ago, I determined that on reaching that age I would
write nothing more of a speculative character." [. . .]

Both Lyell* and Darwin were cautious, but perhaps Lyell carried his
caution to the verge of timidity. I think Darwin possessed and Lyell lacked,
what I can only describe by the theological term "faith—the substance of
things hoped for, the evidence of things not seen." Both had been constrained to feel that the immutability of species could not be maintained.
Both, too, recognized the fact that it would be useless to proclaim this
conviction, unless prepared with a satisfactory alternative to what Huxley*
called "the Miltonic hypothesis." But Darwin's conviction was so far vital
and operative that it sustained him while working unceasingly for twenty-
two years in collecting evidence bearing on the subject. Till at last he was
in the position of being able to justify that conviction to others.

J. W. Judd, "Darwin and Geology," in *Darwin and Modern Science: Essays in Commemoration of the Birth of Charles Darwin and of the Fiftieth Anniversary of the Publication of the Origin of Species* (Cambridge: Cambridge University Press, 1909), 337–84, on 377–82.

Secondary Literature: David Roger Oldroyd, *The Highlands Controversy: Constructing Geological Knowledge in Fieldwork in Nineteenth-Century Britain* (Chicago: University of Chicago Press, 1990).

CARL JUNG (1875–1951)
Swiss Psychologist

The great news of the day [1880s] was the work of Charles Darwin. Shortly before this I had been living in a still medieval world with my parents, where the world and man were still presided over by divine omnipotence and providence. My Christian faith had been relativized by my encounter with Eastern religions and Greek philosophy. [. . .] My then historical interests had developed from my original preoccupation with comparative anatomy and paleontology when I worked as an assistant at the Anatomical Institute. I was fascinated by the bones of fossil man, particularly by the much-discussed Neathderthalensis and the still more controversial skull of [Eugène] Dubois'* Pithecanthropus.

Carl Jung, *The Symbolic Life* (London: Routledge, 1977), 213–14.

[T]he idea of the Superman who was to be created by man was very much helped by the ideas of Darwin, which were modern in those days [roughly from 1859 to Nietzsche's death in 1900]. Of course Darwin doesn't suggest that a Superman could be produced at will; he simply shows the possibility of the transformation of a species, say from ape to man. But then at once the question is raised, if the ape has developed to man, what can man develop into? Man can go on and produce a being superior to the actual man. And then the Protestant ideal leaps in and says: That is what you *ought* to do. You see, if there was a Protestant ape, he might once on a Sunday morning say, "Now I really ought to produce man"—which is exactly was Nietzsche is proposing to do here, of course not in one generation: he gives it at least three generations.

Carl Gustav Jung, *Nietzsche's Zarathustra* (Princeton, NJ: Princeton University Press, 1988), 924.

The theory of development cannot do without the final point of view. Even Darwin, as [Wilhelm] Wundt* points out, worked with final concepts, such as adaptation. The palpable fact of differentiation and development can never be explained exhaustively by causality; it requires also the final point of view, which man produced in the course of his psychic evolution, as he also produced the causal.

C. G. Jung, *The Structure and Dynamics of the Psyche* (Princeton, NJ: Princeton University Press, 1960), 23.

Secondary Literature: Thomas T. Lawson, *Carl Jung: Darwin of the Mind* (London: Karnac, 2008).

PAUL KAMMERER (1880–1926)
AUSTRIAN BIOLOGIST

The race theory boasts of the rigid unchangeability of national character, but nevertheless professes to be convinced of the theory of evolution. Abusing the name of Darwin, the race theorists claim that the pitiless struggle for existence is their only guide, but at the same time they coquette with an international religion of love. The race theorists claim to shun the mixing of politics and science, but they creep around the throne and altar, and even sometimes "color" the truth for partisan advantage.

Paul Kammerer, *The Inheritance of Acquired Characteristics,* trans. A. Paul Maerker-Branden (New York: Boni & Liveright, 1914), 267.

Secondary Literature: Sandor Gliboff, "'Protoplasm [. . .] is soft wax in our hands': Paul Kammerer and the Art of Biological Transformation," *Endeavour* 29 (2005): 162–67.

KARL KAUTSKY (1854–1938)
GERMAN MARXIST THEORIST

I had applied myself to history at the University [ca. 1899 or 1900], but was also enthusiastic over Darwinism. My ideal was the introduction of Darwinism into history. As a student I formed a plan, which was never carried out, to write a Universal History, in which the leading idea should be the struggle for existence of races and classes, my idea resembled [Ludwig] Gumplowicz's* race-war theory. But as a Socialist I could not be content to limit myself to the race-struggle as a factor of progress. I could not ignore the factor of economic development, which forms the classes and the class struggle. The more I occupied myself with the economic history the more had, in my view, the purely Darwinian factor of struggle for existence of races given way to the Marxian of the struggle of classes.

"Karl Kautsky [autobiography]," *Social Democrat* 6, no. 12 (15 December 1902), 355–60, on 357.

The name [Darwin] itself is already a program.

Karl Kautsky, quoted in "Darwinism and the Working Class in Wilhelmian Germany," by Alfred H. Kelly, in *Political Symbolism in Modern Europe: Essays in Honor of George Mosse,* ed. Seymour Drescher, David Sabean, and Allan Sharlin (New York: Transaction, 1982), 156.

Secondary Literature: Gray P. Steenson, "Karl Kautsky: Early Assumptions, Preconceptions, and Prejudices," *International Journal of Comparative Sociology* 30 (1989): 33–43.

MAY KENDALL (1861–1943)
English Poet and Social Reformer

"Do stop that!" said the poet. "How can a fellow write a sonnet with you two for ever sparring away at your musty scholasticisms? Haven't we heard enough about Paley and Darwin? You have frightened away the fairy between you, and that is plenty of mischief for one day."

Andrew Lang and May Kendall, *That Very Mab* (London: Longmans, Green, 1885), 48.

How all your faiths are ghosts and dreams,
    How in the silent sea
Your ancestors were monotremes
    Whatever these may be;
How you evolved your shining lights
    Of wisdom and perfection
From Jelly-Fish and trilobites
    By Natural Selection.

May Kendall, "The Lay of the Trilobite," in *Dreams to Sell* (London: Longmans, Green, 1887), 7.

Sometimes it explodes at high pressure

Of some overwhelming demand,
But plied in unmerciful measure

Tis wonderful what it will stand!
In college, in cottage, in mansion,

Bear witness, the girls and the boys,
How great are its powers of expansion,

How very peculiar its joys!

Oh Brain that is bulgy with learning,

Oh wisdom of women and men,
Oh Maids for a First that are yearning,

Oh youths that are lectured by Wren!
You're acquainted with Pisces and Taurus,

And all sorts of beasts in the sky,
But the brain of the Ichthyosaurus

Was never so good as his eye!
Reconstructed by Darwin or Owen*,

We dwell in sweet Bloomsbury's halls,
But we couldn't have passed Little go in

The Schools, we'd have floundered in Smalls!
Though so cleverly people restore us,

We are bound to confess with a sigh
That the brain of the Ichthyosaurus

Was never so good as his eye!

May Kendall, "Ballad of the Ichthyosaurus," in *Dreams to Sell* (London: Longmans, Green, 1887), 14.

Secondary Literature: Diana Maltz, "Sympathy, Humor, and the Abject Poor in the Work of May Kendall," *English Literature in Transition, 1880–1920* 50 (2007): 313–31.

JOHN MAYNARD KEYNES (1883–1946)
English Economist

By the time that the influence of Paley and his like was waning, the innovations of Darwin were shaking the foundations of belief. Nothing could seem more opposed than the old doctrine and the new—the doctrine which looked on the world as the work of the divine watchmaker and the doctrine which seemed to draw all things out of Chance, Chaos, and Old Time. But at this one point the new ideas bolstered up the old. The economists were teaching that wealth, commerce, and machinery were the children of free competition—that free competition built London. But the Darwinians could go one better than that—free competition had built man. The human eye was no longer the demonstration of design, miraculously contriving all things for the best; it was the supreme achievement of chance, operating under conditions of free competition and *laissez-faire*. The principle of the survival of the fittest could be regarded as a vast generalization of the Ricardian economics. Socialist interferences became, in the light of this grander synthesis, not merely inexpedient, but impious, as calculated to retard the onward movement of the mighty process by which we ourselves had risen like Aphrodite out of the primeval slime of ocean.

Therefore I trace the peculiar unity of the everyday political philosophy of the nineteenth century to the success with which it harmonized and diversified warring schools and united all good things to a single end. Hume and Paley, Burke and Rousseau, Godwin and Malthus,† Cobbett and Huskisson, Bentham and Coleridge, Darwin and the Bishop of Oxford [Samuel Wilberforce*], were all, it was discovered, preaching practically the same thing—individualism and *laissez-faire*. This was the Church of England and those her apostles, whilst the company of the economists were there to prove that the least deviation into impiety involved financial ruin. [...]

The parallelism between economic *laissez-faire* and Darwinianism, already briefly noted, is now seen, as Herbert Spencer* was foremost to recognize, to be very close indeed. Darwin invoked sexual love, acting through sexual selection, as an adjutant to natural selection by competition, to direct evolution along lines which should be desirable as well as effective, so the individualist invokes the love of money, acting through the pursuit of profit, as an adjutant to natural selection, to bring about the production on the greatest possible scale of what is most strongly desired as measured by exchange value.

John Maynard Keynes, *The End of Laissez-Faire: The Economic Consequences of the Peace* (Amherst, NY: Prometheus, 2004), 31. [†] Thomas Malthus (1766–1834), political economist whose *Essay on the Principle of Population* (1798) gave Darwin the key to the survival of the fittest.

Secondary Literature: Robert Skidelsky, *John Maynard Keynes, 1883–1946: Economist, Philosopher, Statesman* (New York: Penguin, 2005); Joseph Schumpeter, *History of Economic Analysis* (New York: Oxford University Press, 1954).

BENJAMIN KIDD (1858–1916)
ENGLISH SOCIOLOGIST

As they [a German professor and students] passed me [in the British Museum], my interest was excited by overhearing the remark in English: "Now we will see where the English keep their national copy of the greatest book of the century" [. . .] my curiosity got the better of me, and I followed them to see what in the opinion of the German was the great book of the age. He was taking out the end volume in the fifth row from the top. I saw them look at it thoughtfully, and turn over the leaves without reading: then they put it respectfully back in its place. When they had gone, I drew the little volume from its resting-place. [. . .] It was Darwin's *Origin of Species.* I took the book back to my seat, for the remark of the German had given a new interest to its familiar pages. [. . .] I was thinking that if I were a poet, I might indeed choose many a meaning theme for inspiration from that same small item of the great national collection.

Benjamin Kidd, "Glimpses of the Reading Room at the British Museum," *Chamber's Journal* 62 (1885): 363–65.

Any one who has kept in touch with the work which down to the present time has been done, in England, Germany, and America, in slowly organizing the evidence upon which the evolutionary view rests, will be conscious of a peculiar extension which has been taking place in the conception of the law of Natural Selection. Like nearly all important departures, the change has been effected gradually and under a number of phases, so that many of those who are well acquainted with the details of knowledge, which under one or more heads have contributed to it, have remained unconscious of the character and significance of the process of movement as a whole.

At the present day any close student of the *Origin of Species* can hardly rise from the study of that book without having left on his mind at least one clear and definite impression. He will in all probability feel, over and above everything else, how steadily and consistently Darwin kept before him the vision of the keen, long-drawn-out, and never-relaxed struggle in which every form of life is of necessity engaged; and the conception of the dominating importance of every feature and quality contributing to success and survival in this supreme rivalry.

Benjamin Kidd, *Principles of Western Civilisation* (New York: Macmillan, 1902), 40.

Secondary Literature: David Paul Crook, *Benjamin Kidd: Portrait of a Social Darwinist* (Cambridge: Cambridge University Press, 1984).

MARTIN LUTHER KING JR. (1929–1968)
African American Clergyman

[Metz's][†] statement that Darwin was no Darwinian is essentially true in the sense that Darwin never set out to establish any metaphysical or philosophical conclusions. He wrote as a biologist and not as a metaphysician. The one exception of a deviation from his biological interest was his attempt to delve into ethical theory. But certainly Darwin never set forth many of the philosophical theories that later became attached to his system. A case in point is Herbert Spencer*. After Darwin published his *Origin of Species,* Herbert Spencer welcomed it and proceeded to apply its underlying theories to the whole of society. We find Haeckel* attempting to define everything in terms of the Darwinian theory of evolution along with the law of substance. Many other examples could be cited. But these are adequate enough to show the many philosophical tenets developed from Darwin's system that he never realized. So Metz is essentially right: "Darwin was no Darwinian."

There are mainly four reasons why Darwin's evolutionary hypothesis raised such a furor:

(1) It seems to contradict the traditional view of the immutability of species.
(2) It contradicted those who accepted a literal account of the Bible.
(3) It seemed to take teleology from the universe. A first cause was also cast aside.
(4) It seemed to lessen man's status.

So we can see that Darwin's theory raised a deal of furor because it upset certain habits of mind. Of course most of the above accusations did not necessarily follow from the Darwinian hypothesis.

Examination answers for Richard Marion Millard's course on the history of recent philosophy, Boston University, 5 July 1952, in *The Papers of Martin Luther King, Jr.*, ed. Clayborne Carson, Ralph Luker, and Penny A. Russell (Berkeley and Los Angeles: University of California Press, 1992), 154. †Rudolph Metz (b. 1891), German historian of philosophy.

Secondary Literature: Troy Jackson, *Becoming King: Martin Luther King, Jr. and the Making of a National Leader* (Lexington: University of Kentucky Press, 2008).

CHARLES KINGSLEY (1819–1875)
English Clergyman and Novelist

[November 1859]
Dear Sir:
I have to thank you for the unexpected honor of your book [*Origin of Species*]. That the Naturalist whom, of all naturalists living, I most wish to know & learn from, should have sent a sciolist like me his book, encourages me at least to observe more carefully, & think more slowly.
I am so poorly (in brain) that I fear I cannot read your book just now as I ought. All I have seen of it *awes* me; both with the heap of facts, & the prestige of your name, & also with the clear intuition, that if you be right, I must give up much that I have believed & written. [...]
I have long since, from watching the crossing of domesticated animals

& plants, learnt to disbelieve the dogma of the permanence of species. [. . .] I have gradually learnt to see that it is just as noble a conception of Deity, to believe that he created primal forms capable of self-development into all forms needful pro tempore & pro loco, as to believe that He required a fresh act of intervention to supply the lacunas w[hich] he himself had made. I question whether the former be not the loftier thought.

Kingsley to Darwin, 18 November 1859, in *The Correspondence of Charles Darwin*, ed. Frederick Burkhardt et al., 16 vols. to date (Cambridge: Cambridge University Press, 1985–), 7:379–80.

He [Kingsley] talked much of Darwin's new book on species, expressing great admiration for it, but saying that it was so startling that he had not yet been able to make up his mind as to its soundness.

Charles Bunbury, recording his visit to Kingsley from 6 to 8 December 1859, in *Memorials of Sir C. J. F. Bunbury,* ed. Frances Joanna Bunbury, 3 vols. (Mildenhall, Suffolk, UK: S. R. Simpson, 1891–92), 3:201.

I am very busy working out points of Natural Theology, by the strange light of Huxley*, Darwin, and Lyell*. [. . .] Darwin is conquering everywhere, and rushing in like a flood, by the mere force of truth and fact. The one or two who hold out are forced to try all sorts of subterfuges as to fact, or else by evoking the *odium theologicum.*

Kingsley to Frederick Denison Maurice, 1861, in *Charles Kingsley, His Letters and Memories of His Life,* by Charles Kingsley and Frances Eliza Grenfell Kingsley (New York: J. F. Taylor, 1900), 175.

"But surely if there were water-babies, somebody would have caught one at least?"

"Well. How do you know that somebody has not?"

"But they would have put it into spirits, or into the Illustrated News, or perhaps cut it into two halves, poor dear little thing, and sent one to Professor [Richard] Owen*, and one to Professor Huxley*, to see what they would each say about it."

"Ah, my dear little man! that does not follow at all, as you will see before the end of the story."

"But a water-baby is contrary to nature."

"Well, but, my dear little man, you must learn to talk about such things, when you grow older, in a very different way from that. You must not talk about 'ain't' and 'can't' when you speak of this great wonderful world round you, of which the wisest man knows only the very smallest corner, and is, as the great Sir Isaac Newton said, only a child picking up pebbles on the shore of a boundless ocean."

"You must not say that this cannot be, or that that is contrary to nature. You do not know what Nature is, or what she can do; and nobody knows; not even Sir Roderick [Impey] Murchison*, or Professor Owen, or Professor [Adam] Sedgwick*, or Professor Huxley, or Mr. Darwin, or Professor [Michael] Faraday*, or Mr. Grove,† or any other of the great men whom good boys are taught to respect. They are very wise men; and you must listen respectfully to all they say: but even if they should say, which I am sure they never would, 'That cannot exist. That is contrary to nature,' you must wait a little, and see; for perhaps even they may be wrong. It is only children who read Aunt Agitate's Arguments, or Cousin Cramchild's Conversations; or lads who go to popular lectures, and see a man pointing at a few big ugly pictures on the wall, or making nasty smells with bottles and squirts, for an hour or two, and calling that anatomy or chemistry— who talk about 'cannot exist,' and 'contrary to nature.' Wise men are afraid to say that there is anything contrary to nature, except what is contrary to mathematical truth; for two and two cannot make five, and two straight lines cannot join twice, and a part cannot be as great as the whole, and so on (at least, so it seems at present): but the wiser men are, the less they talk about 'cannot.' That is a very rash, dangerous word, that 'cannot'; and if people use it too often, the Queen of all the Fairies, who makes the clouds thunder and the fleas bite, and takes just as much trouble about one as about the other, is apt to astonish them suddenly by showing them, that though they say she cannot, yet she can, and what is more, will, whether they approve or not."

Charles Kingsley, *The Water-Babies* (Boston: T. O. H. P. Burnham, 1864), 65–67.
†William Robert Grove (1811–96), English physicist and electrochemist.

[Does] the doctrine of evolution, by doing away with the theory of creation, [do] away with that of final causes?—let us answer, boldly: Not in the least. We might accept all that Mr. Darwin, all that Professor Huxley*, has so learnedly and so acutely written on physical science, and yet preserve our natural theology on exactly the same basis as that on which Butler[†] and Paley left it. That we should have to develop it, I do not deny. That we should have to relinquish it, I do.

I entreat you to weigh these words, which have not been written in haste; and I entreat you also, if you wish to see how little the new theory, that species may have been gradually created by variation, natural selection, and so forth, interferes with the old theory of design, contrivance, and adaptation, nay, with the fullest admission of benevolent final causes —I entreat you, I say, to study Darwin's *Fertilisation of Orchids*—a book which (whether his main theory be true or not) will still remain a most valuable addition to natural theology.

Charles Kingsley, "The Natural Theology of the Future," lecture delivered at Sion College, London, 10 January 1871, in Kingsley, *Scientific Lectures and Essays* (London: Macmillan, 1893), 329–30. [†]Joseph Butler (1692–1752), English theologian, bishop, and philosopher.

Secondary Literature: Jan Klaver, *The Apostle of the Flesh: A Critical Life of Charles Kingsley* (Leiden: Brill, 2006); Owen Chadwick, *The Victorian Church*, 2 vols. (New York: Oxford University Press, 1966–70).

HENRY KINGSLEY (1830–1876)
English Novelist

About physical science he was absolutely and perfectly ignorant. For this we can scarcely blame him. Mr. [George Henry] Lewes*, and another, whom family reasons prevent my naming, had not then brought science to our doors. Darwin and Huxley* were watching the wonders of God in the deep sea, and had not got epitomized.

Henry Kingsley, *Stretton* (1869; new ed., London: Ward, Lock & Bowden, 1893), 116.

A fly—a blue-bottle fly; for he buzzes, and is difficult to catch, and bangs his idiotic head against the glass; in all respects a blue-bottle, save, oh wonderful fact! that he is brown. Yes, he is the first instance of those parallel types, reproduced in different colors, and with trifling differences—so small as to barely constitute a fresh species—and the origin of which is such a deep deep wonder and mystery to me to this day. Tell me, O Darwin, shall we know on this side of the grave why or how the Adiantum Nigrum and Asplenium capillis Veneris, have reproduced themselves, or, to be more correct, have produced ghosts and fetches of themselves at the antipodes?

Henry Kinglsey, *The Hillyars and the Burtons: A Story of Two Families* (1865; new ed., London: Ward, Lock & Bowden, 1893), 248.

Secondary Literature: J. S. D. Mellick, *The Passing Guest: A Life of Henry Kingsley* (St. Lucia, Queensland, Australia: University of Queensland Press, 1983).

RUDYARD KIPLING (1865–1936)
English Author

I've been trying once more to plough through *The Descent of Man* and every fiber [...] of my body revolted against it. To believe in it that it is necessary never to have—Hullo! Where is this one-idea'd pen going off to anew?

Kipling to Edmonia Hall, 15 May 1888, in *The Letters of Rudyard Kipling*, ed. Thomas Pinney, 6 vols. (Houndsmill, Basingstoke, Hampshire, UK: Macmillan, 1990–2004), 2:216.

Secondary Literature: Reinaldo F. Silva, "Social Darwinism, Race, and the Discourse of Empire in Rudyard Kipling's *Captains Courageous*," *Excavatio* (Edmonton) 19 (2004): 223–35.

ALEXANDER O. KOVALEVSKI (1840–1901)
Russian Biologist

Darwin's theory was received in Russia with profound sympathy. While in Western Europe it met firmly established old traditions which it had first to overcome, in Russia its appearance coincided with the wakening of our society after the Crimean War and here it immediately received the status of full citizenship and ever since has enjoyed widespread popularity.

Alexander O. Kovalevski, quoted in "Russia: Biological Sciences," by Alexander Vucinich, in *The Comparative Reception of Darwinism*, ed. Thomas F. Glick (Austin: University of Texas Press, 1972), 229–30.

Secondary Literature: Alexander Vucinich, *Darwin in Russian Thought* (Berkeley and Los Angeles: University of California Press, 1988).

SERGEI MIKHAILOVICH KRAVCHINSKY
(1852–1895)
Russian Author and Revolutionary

One day there fell into my hands an "open letter" of B. Zaizeff, one of the contributors to the *Russkoi i Slovo,* a widely popular paper of that period. In this "letter" intended for the secret press, the author, speaking of that time, and of the charges brought against the Nihilists of those days by the Nihilists of the present day, says, "I swear to you by everything which I hold sacred, that we were not egotists as you call us. It was an error, I admit, but we were profoundly convinced that we were fighting for the happiness of human nature, and every one of us would have gone to the scaffold and would have laid down his life for Moleschott or Darwin." The remark made me smile. The reader, also, will perhaps smile at it, but it is profoundly sincere and truthful. Had things reached such an extremity, the world would perhaps have seen a spectacle at once tragic and comical; martyrdom to prove that Darwin was right and Cuvier wrong, as two centuries previously the priest Abbaco and his disciples went to the stake, and mounted the scaffold, in support of their view, that Jesus should be written with one J instead of two, as in Greek; and that the Halleluiah should be sung three times and not twice, as in the State Church. It is a fact, highly characteristic of the Russian mind, this tendency to become excited even to fanaticism, about certain things which would simply meet with approval or disapproval from a man of Western Europe.

Stepniak [Sergei Mikhailovich Kravchinsky], *Underground Russia: Revolutionary Profiles and Sketches from Life,* 2nd ed. (New York: Charles Scribner's Sons, 1885), 5–6.

Secondary Literature: Derek Offord, *The Russian Revolutionary Movement in the 1880s* (Cambridge: Cambridge University Press, 1986).

PETER KROPOTKIN (1842–1921)
RUSSIAN REVOLUTIONARY AND GEOGRAPHER

Two aspects of animal life impressed me most during the journeys which I made in my youth in Eastern Siberia and Northern Manchuria. One of them was the extreme severity of the struggle for existence which most species of animals have to carry on against an inclement Nature; the enormous destruction of life which periodically results from natural agencies; and the consequent paucity of life over the vast territory which fell under my observation. And the other was, that even in those few spots where animal life teemed in abundance, I failed to find—although I was eagerly looking for it—that bitter struggle for the means of existence, among animals belonging to the same species, which was considered by most Darwinists (though not always by Darwin himself) as the dominant characteristic of the struggle for life, and the main factor of evolution.

Peter Kropotkin, *Mutual Aid: A Factor of Evolution* (London: Heinemann, 1902), 1.

Natural sciences—that is, mathematics, physics, and astronomy—were my chief studies. In the year 1858, before Darwin had brought out his immortal work, a professor of zoology at the Moscow University, Boulier, published three lectures on transformism, and my brother took up at once his ideas concerning the variability of species. He was not satisfied, however, with approximate proofs only, and began to study a number of special books on heredity and the like; communicating to me in his letters the main facts, as well as his ideas and his doubts. The appearance of *The Origin of Species* did not settle his doubts on several special points, but only raised new questions and gave him the impulse for further studies. We afterward discussed—and that discussion lasted for many years—various questions relative to the origin of variations, their chances of being transmitted and being accentuated; in short, those questions which have been raised quite lately in the [August] Weismann*-Spencer* controversy, in [Francis] Galton's* researches, and in the works of the modern Neo-

Lamarckians. Owing to his philosophical and critical mind, Alexander[†] had noticed at once the fundamental importance of these questions for the theory of variability of species, even though they were so often overlooked then by many naturalists.

Peter Kropotkin, *Memoirs of a Revolutionist* (Boston: Houghton Mifflin, 1899), 97–98. [†]Alexander Kropotkin (d. 1886), Peter Kropotkin's brother.

Another great question also engrossed my attention [ca. 1886]. It is known to what conclusions Darwin's formula, the "struggle for existence," had been developed by his followers generally, even the most intelligent of them, such as Huxley[*]. There is no infamy in civilized society, or in the relations of the whites towards the so-called lower races, or of the strong towards the weak, which would not have found its excuse in this formula.

Even during my stay at Clairvaux I saw the necessity of completely revising the formula itself and its applications to human affairs. The attempts which had been made by a few socialists in this direction did not satisfy me, but I found in a lecture by a Russian zoologist, Professor Kessler,[†] a true expression of the law of struggle for life. "Mutual aid," he said in that lecture, "is as much a law of nature as mutual struggle; but for the progressive evolution of the species the former is far more important than the latter." These few words—confirmed unfortunately by only a couple of illustrations [. . .]—contained for me the key of the whole problem. When Huxley published in 1888 his atrocious article, "The Struggle for Existence; a Program," I decided to put in a readable form my objections to his way of understanding the struggle for life, among animals as well as among men, the materials for which I had been accumulating for two years. I spoke of it to my friends. However, I found that the interpretation of "struggle for life" in the sense of a war-cry of "Woe to the Weak," raised to the height of a commandment of nature revealed by science, was so deeply rooted in this country that it had become almost a matter of religion. Two persons only supported me in my revolt against this misinterpretation of the facts of nature. The editor of the *Nineteenth Century,* Mr. James Knowles, with his admirable perspicacity, at once seized the gist of the matter, and with a truly youthful energy encouraged me to take it in hand. The other supporter was the regretted H. [Henry] W. Bates[*], whom Darwin, in his *Autobiography,* described as one of the most intelligent men he ever met. He was secretary of the Geographical Society, and I knew him; so I spoke to him of my intention. He was delighted with it. "Yes, most assuredly write it," he said. "That is true Darwinism. It is a shame to think of what they have made of Darwin's ideas. Write it, and when you have published it, I will write you a letter of commendation which you may publish."

Peter Kropotkin, *Memoirs of a Revolutionist* (Boston: Houghton Mifflin, 1899), 498–99. †Karl Federovich Kessler (1815–81), Russian zoologist.

Secondary Literature: Brian Morris, *Kropotkin: The Politics of Community* (Amherst, NY: Humanity Books, 2003); Alexander Vucinich, *Darwin in Russian Thought* (Berkeley and Los Angeles: University of California Press, 1988).

JACQUES LACAN (1901–1981)
French Psychoanalyst

Darwin's success seems to derive from the fact that he projected the predations of Victorian society and the economic euphoria that sanctioned for that society the social devastation that it initiated on a planetary scale, and to the fact that it justified its predations by the image of a laissez-faire of the strongest predators in competition for their natural prey.

Before Darwin, however, Hegel had provided the ultimate theory of the proper function of agressivity in human ontology, seeming to prophecy the iron law of our time. [. . .] Here the natural individual is regarded as nothingness, since the human subject is nothingness, in effect, before the absolute Master that is given to him in death. The satisfaction of human desire is possible only when mediated by the desire and the labor of the other. If, in the conflict of Master and Slave, it is the recognition of man by man that is involved, it is also promulgated on a radical negation of natural values, whether expressed in the sterile tyranny of the master or in the productive tyranny of labor.

We all know what an armature this profound doctrine has given to the constructive Spartacism of the Slave recreated by the barbarism of the Darwinian century.

Jacques Lacan, *Écrits: A Selection,* trans. Alan Sheridan (New York: Norton, 1977), 26.

Secondary Literature: Elisabeth Roudinescu, *Jacques Lacan,* trans. Barbary Bray (New York: Columbia University Press, 1997); idem, *Jacques Lacan & Co.,* trans. Jeffrey Mehlman (Chicago: University of Chicago Press, 1990).

PAUL LAFARGUE (1842–1911)
FRENCH MARXIST AUTHOR AND REVOLUTIONARY ACTIVIST

We, who have been accused of sowing disorder, we who want an end of the civil war among classes which exists in our society and which will not end until classes have disappeared; we who want an end to this war between person and person that you and your economists call competition, but which even before Darwin the pitiless logician Hobbes[†] called the war of all against all. But we know that this war of class against class, of individual against individual will not end unless the economic evolution which we see with our very eyes, which centralizes property in fewer and fewer hands [. . .]

[Shouts: No! The Contrary!]

Lafargue, on the floor of the French Chamber of Deputies, 8 December 1891, *Annals de la Chambre des Députés, 5e Législature, Débats parlementaires, Session extraordinaire de 1891* (Paris: Imprimerie des Journaux Officiels, 1892), 758, trans. TFG. Lafargue was Karl Marx's son-in-law. [†]Thomas Hobbes (1588–1679), English philosopher whose depiction of the "war of each against all" reappears in Darwin's account of warring groups of "savages" (*Descent of Man*, chap. 4).

The scientists have not only sold themselves to the governments and the financiers; they have also sold science itself to the capitalist-bourgeoisie. [. . .]

This backward movement still continues; when Darwin published his *Origin of Species*, which took away from God his role of creator in the organic world, as Franklin had despoiled him of his thunderbolt, we saw the scientists, big and little, university professors and members of the Institute, enrolling themselves under the orders of Flourens,[†] who for his own part had at least his eighty years for an excuse, that they might demolish the Darwinian theory, which was displeasing to the government and hurtful to religious beliefs. The intellectuals exhibited that painful spectacle in the fatherland of Lamarck and of [Étienne] Geoffroy Saint-Hilaire, the creators of the evolution theory, which Darwin completed and made proof against criticism.

Today, now that the clerical anxiety is somewhat appeased, the scientists venture to profess the evolution theory, which they never opposed without a protest from their scientific conscience, but they turn it against socialism so as to keep in the good graces of the capitalists. Herbert Spencer*, Haeckel* and the greatest men in the school of Darwinism demonstrate that the classification of individuals into rich and poor, idlers and laborers, capitalists and wage-earners, is the necessary result of the inevitable laws of nature, instead of being the fulfillment of the will and the justice of God. Natural selection, they say, which has differentiated the organs of the human body, has forever fixed the ranks and the functions of the social body. They have, through servility, even lost the logical spirit. They are indignant against Aristotle because he, being unable to conceive of the abolition of slavery, declared that the slave was marked off by nature; but they fail to see that they are saying something equally monstrous when they affirm that natural selection assigns to each one his place in society.

Paul Lafargue, "Socialism and the Intellectuals," in *The Right to be Lazy and Other Studies,* trans. Charles H. Kerr (Chicago: Charles H. Kerr, 1907), 83–85. †Jean Pierre Flourens (1794–1867), French physiologist scornful of Darwin.

FRIEDRICH ALBERT LANGE (1828–1875)
German Philosopher

When the first edition of [my] *History of Materialism* appeared [1866], Darwinism was still new; the parties were just taking up their positions, or, more strictly, the rapidly growing party of "German Darwinians" was still in process of forming, and the reaction, which at present sees here the most threatened point in the old theory of things, was not yet properly in harness, because it had not yet properly appreciated the range of the great problem and the inward power of the new doctrine.

Since then, the interest of friends and foes has been so much concentrated on this point, that not only an extensive literature has sprung up on Darwin and Darwinism, but that we may say that the Darwinian controversy is to-day what the general Materialistic controversy was formerly. [Ludwig] Büchner* is still ever finding new readers for *Kraft und Stoff* [*Force and Matter*], but we no longer hear a literary outcry of indignation when a new edition appears. [Jacob] Moleschott,† the true author of the Materialistic movement, is almost forgotten by the great public, and even Carl Vogt* is now seldom mentioned except in reference to some special question in anthropology or some isolated and immortal utterance of his drastic humor. Instead of this, every periodical takes sides for or against Darwin; there appear almost daily larger or smaller treatises on the theory of descent, natural selection, and especially, as we may expect, on the descent of man, since there are many members of this particular species who lose their wits, if any doubt is raised of the genuineness of their ancestral tree.

Despite this great movement, we may still maintain unchanged nearly everything that we wrote on Darwinism eight years ago, though we can no longer leave the matter where it was then. The material has grown, even if the scientific material has not grown quite in proportion to the printed paper: the questions have been specialized. Formerly Darwin was the only influential champion, not only of the theory of descent, but we may almost say of the natural explanation of organic forms in general. At present it happens that bitter attacks are directed against Darwin and Darwinism by people who confine themselves to the theory of natural selection, as though everything else would have existed even without Darwin.

Friedrich Albert Lange, *History of Materialism,* trans. Ernest Chester Thomas, 3 vols. (Boston: Houghton Mifflin, 1881), 3:26–27. [†]See Friedrich Engels entry.

Secondary Literature: Arnulf Zweig, "Friedrich Albert Lange," in *Encyclopedia of Philosophy,* ed. Paul Edwards, 8 vols. (New York: Macmillan, 1967), 4:383–84.

SIDNEY LANIER (1842–1881)
AMERICAN POET AND PROFESSOR

Lanier's Annotations to *Origin of Species*

Opposite a passage on domestic pigeons: "Man, moving up out of the dark of time, attended by his friends—the pigeon, the dog, the horse, the cow."

On a discussion of adaptation: "Darwin's hearty use of such adjectives as 'exquisite' and 'beautiful' in these connections, is very suggestive of pleasant relations between science and poetry."

Sidney Lanier, *Poems and Poem Outlines,* ed. Charles R. Anderson, Centennial Edition, vol. 1 (Baltimore: Johns Hopkins Press, 1945), lxxiii.

In short, I find that early thought everywhere, whether dealing with physical facts or metaphysical problems, is lacking in what I may call the intellectual conscience—the conscience, for example, which makes Mr. Darwin spend long and patient years in investigating small facts before daring to reason upon them; and which makes him state the facts adverse to his theory with as much care as the facts which make for it.

Sidney Lanier, *The English Novel and the Principle of its Development* (New York: Charles Scribner's Sons, 1883), 126.

Our Darwin boldly takes hold of Nature as if it were a rose, pulls it to pieces, puts it under the microscope, and reports to us what he saw without fear or favor. Beethoven, on the other hand, approaching the same good Nature—from a different direction, with different motives—and looking upon it with the artist's—not the scientist's—eye, finds it a beautiful whole; he does not analyze, but, pursuing the synthetic process, shows it to us as a perfect rose, reporting his observations to us in terms of harmony.

Sidney Lanier, *Music and Poetry: Essays upon some Aspects and Inter-Relations of the Two Arts* (New York: Charles Scribner's Sons, 1899), 71.

I cannot think of the uprising of this sad face of death before our dear Master Shakspere in *Hamlet* from beneath the kingcups and clover and cowslips of the dream, as being the inevitable opposition into which Physical Nature resolves itself, and which every man must grapple with and manage at some time or other of his spiritual career, here—I cannot think of this dual form of Nature without recalling some memorable words in Mr. Darwin's *Origin of Species*. I am fond of bringing together people and books that never dreamed of being side by side: often I find nothing more instructive; and so permit me to quote some words here and there in Mr. Darwin's book which seem to me to give a very precise and scientific account of the very opposition which I have here been trying to bring out as between Nature, the mother of life, in *Midsummer Night's Dream,* and Nature, the mother of death, in *Hamlet.*

Sidney Lanier, *Shakspere and his Forerunners* (New York: Doubleday Page, 1908), 305.

Secondary Literature: Aubrey Harrison Starke, *Sidney Lanier: A Biographical and Critical Study* (New York: Russell & Russell, 1964).

E. RAY LANKESTER (1847–1929)
ENGLISH BIOLOGIST

Darwin by his discovery of the mechanical principle of organic evolution, namely, the survival of the fittest in the struggle for existence, completed the doctrine of evolution, and gave it that unity and authority which was necessary in order that it should reform the whole range of philosophy. The detailed consequences of that new departure in philosophy have yet to be worked out. Its most important initial conception is the derivation of man by natural processes from ape-like ancestors, and the consequent derivation of his mental and moral qualities by the operation of the struggle for existence and natural selection from the mental and moral qualities of animals. Not the least important of the studies thus initiated is that of the evolution of philosophy itself. Zoology thus finally arrives through Darwin at its crowning development: it touches and may be even said to comprise the history of man, Sociology, and Psychology.

E. Ray Lankester, *The Advancement of Science* (London: Macmillan, 1890), 381–82.

A more objectionable misinterpretation of the naturalists' doctrine of the survival of the fittest in the struggle for existence is made by journalists and literary politicians, who declare, according to their political bias, either that science rightly teaches that the gross quality measured by wealth and strength alone can survive and should therefore alone be cultivated, or that science (and especially Darwinism) has done serious injury to the progress of mankind by authorizing their teaching. Both are wrong, and owe their error to self-satisfied flippancy and traditional ignorance in regard to nature-knowledge and the teaching of Darwin. The "fittest" does not mean the "strongest." The causes of survival under Natural Selection are very far indeed from being rightly described as mere strength, nor are they baldly similar to the power of accumulating wealth. Frequently in Nature the more obscure and feeble survive in the struggle because of their modesty and suitability to given conditions, whilst the rich are sent away empty and the mighty perish by hunger.

Edwin Ray Lankester, *Nature and Man,* The Romanes Lecture (Oxford: Clarendon, 1905), 55.

Secondary Literature: Joseph Lester, *E. R. Lankester and the Making of Modern British Biology* (Oxford: British Society for the History of Science, 1995).

HAROLD LASKI (1893–1950)
ENGLISH POLITICAL THEORIST

I have always liked Samuel Butler*. [...] Outside *The Way of All Flesh,* he struck me as taking himself a little seriously, and his controversy with Darwin always reminds me, quite irresistibly, of an angry terrier jumping round a mastiff which refuses to be disturbed.

Laski to Oliver Wendell Holmes Jr., 11 September 1924, in *Holmes-Laski Letters,* ed. Mark DeWolfe Howe (Cambridge, MA: Harvard University Press, 1953), 656.

Then I read and greatly enjoyed the whole of Darwin's correspondence. I lay my hand on my heart and say there never was a more loveable great man—always modest, never aggressive, simple and kindly, and permanently open to new ideas. When you compare him to Descartes or Newton or Leibniz or Goethe he simply overtops them altogether. Really it is impossible to rate him too highly.

Laski to Oliver Wendell Holmes, 15 January 1929, in *Holmes-Laski Letters,* ed. Mark DeWolfe Howe (Cambridge, MA: Harvard University Press, 1953), 1125.

I was intensely interested by what you said about your scientists' [Austin H. Clark, Gerrit S. Miller][†] attitude to Darwinism. I speak of course with ignorance and humility, but I have the sense that the reaction against it is a little exaggerated and secondly that it remains profoundly unsatisfactory that it should not be able to explain (I) the origin of variation or (II) how a variation presented can, often enough, be of any utility for survival in its original stages. But granted all that, the fact that natural selection takes place seems to me to be solidly proved enough, and also that evolution is real, even though the details of the actual pedigree are much thinner and more uncertain than the original enthusiasts thought. And at any rate the supreme result of the seventy years since 1859 has been a body-blow to the Eternal from which he will find it difficult to recover. That is what really matters most. I remain permanently and impenitently anti-clerical.

Laski to Oliver Wendell Holmes, 12 November 1929, in *Holmes-Laski Letters,* ed. Mark DeWolfe Howe (Cambridge, MA: Harvard University Press, 1953), 1130.
†Austin Hobart Clark (1880–1954), American zoologist, entomologist, and evolutionist, and Gerrit S. Miller (1869–1956), American zoologist known for his checklists of North American mammals.

Secondary Literature: David Henry Burton, "The Intellectual Kinship of Oliver Wendell Holmes, Jr., Frederick E. Pollock, and Harold Laski," *Proceedings of the American Philosophical Society* 199, no. 2 (1975): 133–42.

D. H. LAWRENCE (1885–1930)
ENGLISH NOVELIST

I should be very glad to hear your treatment of some of the great religious topics of the day, and regret I cannot attend an evening class. As it is my nights are all too short for the amount of work I ought to put into them.

Reading of Darwin, Herbert Spencer*, [Ernest] Renan*, J. M. Robertson, Blatchford and Vivian† in his *Churches and Modern Thought* [London: Watts, 1908] has serious modified my religious beliefs. [. . .] And I would like to know, because I am absolutely in ignorance, what is precisely the orthodox attitude—or say the attitude of the nonconformist Churches to such questions as Evolution, with that of the Origin of Sin, and as Heaven and Hell.

I know these are tremendous issues, and somehow we hear of them almost exclusively from writers against Christianity.

Lawrence to Reverend Robert Reid, 15 October 1907, in *The Letters of D. H. Lawrence,* ed. James T. Boulton (Cambridge: Cambridge University Press, 2002), 36.
†John Mackinnon Robertson (1856–1933), English journalist and advocate of rationalism and secularism; Robert Blatchford (1851–1943), English socialist journalist and author; and Harry Vivian Majendie Phelips (pseud. Philip Vivian, 1860–1939), critic of Christianity on the grounds that modern science proves it false.

Secondary Literature: Ronald Granofsky, *D. H. Lawrence and Survival: Darwinism in the Fiction of the Transitional Period* (Montreal: McGill-Queens University Press, 2003).

JOSEPH LE CONTE (1823–1901)
AMERICAN GEOLOGIST

I take this opportunity to do justice to the brilliancy and originality of Langdon Chevis, a planter on the coast of South Carolina, near the Savannah River, by recording some views of his expressed to me in a conversation at Flat Rock on the origin of species. We had both read that remarkable book *Vestiges of the Natural History of Creation,* published in 1844, and he had cordially embraced the idea of origin of species by transmutation of previous species, while I contrarily held to Agassiz'* views of creation according to a preordained plan. We had it hot and heavy. When I brought forward the apparently unanswerable objection drawn from the geographical distribution of species and the manner in which contiguous fauna pass into one another, i.e., by substitution instead of transmutation, his answer was exactly what an evolutionist would give to-day—*viz.,* that intermediate links would be killed off in the struggle for life as less suited to the environment; in other words that only the fittest would survive. It must be remembered that this was before the publication of Darwin's book, and the answer was wholly new to me and struck me very forcibly.

I look back with especial pleasure on my writings on evolution. I lay no claim to the discovery of new facts bearing on the theory of evolution, but only to have cleared up its nature and scope and especially to have shown its true relation to religious thought. It is well to stop a moment to show the roles of different thinkers in the advance on this subject. Leaving out of consideration mere vague philosophic speculations, like those of ancient philosophers and of Swedenborg in more modern times, I would say that the role of Lamarck was to introduce evolution as a scientific theory; that of Darwin to present the theory in such wise as to make it acceptable to and accepted by the scientific mind; that of Huxley* to fight the battles of evolution and to win its acceptance by the intelligent popular mind;

that of Spencer* to generalize it into a universal law of nature, thereby making it a philosophy as well as a scientific theory. Finally, it was left to American thinkers to show that a materialistic implication is wholly unwarranted, that evolution is entirely consistent with a rational theism and with other fundamental religious beliefs. My own work has been chiefly in this direction. In my lectures in 1872 on Religion and Science, I might be called a reluctant evolutionist, yet even then, in the sixteenth chapter of the book, I tried to show the mode of origin of the spirit of man from the psyche of animals by a process of evolution. In a few years, however, I was an evolutionist, thorough and enthusiastic. Enthusiastic, not only because it is true, and all truth is the image of God in the human reason, but also because of all the laws of nature it is by far the most religious, that is, the most in accord with religious philosophic thought. It is, indeed, glad tidings of great joy which shall be to all peoples. Woe is me, if I preach not the Gospel. Literally, it can be shown that all the apparent irreligious and materialistic implications of science are reversed by this last child of science, or rather this daughter of the marriage of science and philosophy. During all my life I have striven earnestly to show this. My book on *Evolution and its Relation to Religious Thought* is the embodiment of the result of these strivings, although I believe that if I wrote it again I could add much to the argument. I began this line of thought in 1871, and believe, and therefore claim, that I was the pioneer in this reaction against the materialistic and irreligious implication of the doctrine of evolution. I look with greater pleasure on this than on anything else that I have done. At first I suffered some, not much, obloquy on the part of the extreme orthodox people, but I have lived to see this pass away, and all intelligent clergymen coming to my position.

The Autobiography of Joseph Le Conte, ed. William Sallam Armes (New York: Appleton, 1903), 174–75, 335–37.

Not only did Agassiz* establish the essential identity of the geologic and embryonic succession, the general similarity of the two series, phylogenic and ontogenic, but he also announced and enforced all the formal laws of geologic succession (i.e., of evolution), as we now know them. These, as already stated and illustrated, are the law of differentiation, the law of progress of the whole, and the law of cyclical movement, although he did not formulate them in these words. No true inductive evidence of evolution was possible without the knowledge of these laws, and for this knowledge we are mainly indebted to Agassiz. He well knew also that they were the laws of embryonic development and therefore of evolution; but he avoided the word evolution, as implying the derivative origin of species, and used instead the word development, though it is hard to see in what the words differ. Thus, it is evident that Agassiz laid the whole foundation of evolution, solid and broad, but refused to build any scientific structure on it; he refused to recognize the legitimate, the scientifically necessary outcome of his own work. Nevertheless, without his work a scientific theory of evolution would have been impossible. Without Agassiz (or his equivalent), there would have been no Darwin. There is something to us supremely grand in this refusal of Agassiz to accept the theory of evolution. The opportunity to become the leader of modern thought, the foremost man of the century, was in his hands, and he refused, because his religious, or, perhaps better, his philosophic intuitions, forbade. To Agassiz, and, indeed, to all men of that time, too many, alas! even now, evolution is materialism. But materialism is Atheism. Will some one say, the genuine Truth-seeker follows where she seems to lead whatever be the consequences? Yes; whatever be the consequences to one's self, to one's opinions, prejudices, theories, philosophies, but not to still more certain truth. Now, to Agassiz, as to all genuine thinkers, the existence of God, like our own existence, is more certain than any scientific theory, than anything can possibly be made by proof. From his standpoint, therefore, he was right in rejecting evolution as conflicting with still more certain truth. The mistake which he made was in imagining that there was any such conflict at all. But this was the universal mistake of the age. [. . .]

What, then, is the place of Agassiz in biological science? What is the relation of Agassiz to Darwin—of Agassizian development to Darwinian evolution? I answer, it is the relation of formal science to physical or causal science. Agassiz advanced biology to the formal stage; Darwin carried it forward, to some extent at least, to the physical stage. All true inductive sciences in their complete development pass through these two stages. Science in the one stage treats of the laws of phenomena; in the other, of the causes or explanation of these laws. The former must precede the latter, and form its foundation; the latter must follow the former, and constitute its completion. The change from the one to the other is always attended with prodigious impulse to science.

Joseph Le Conte, *Evolution: Its Nature, its Evidences, and its Relation to Religious Thought,* 2nd rev. ed. (New York: Appleton, 1897), 44–48.

Secondary Literature: Lester D. Stephens, *Joseph Le Conte, Gentle Prophet of Evolution* (Baton Rouge: Louisiana State University Press, 1982).

VLADIMIR LENIN (1870–1924)
RUSSIAN REVOLUTIONARY

It will now be clear that the comparison with Darwin is perfectly accurate: *Capital* is nothing but "certain closely interconnected generalizing ideas crowning a veritable Mont Blanc of factual material." And if anybody has read *Capital* and contrived not to notice these generalizing ideas, it is not the fault of [Karl] Marx*, who, as we have seen, pointed to these ideas even in the preface. And that is not all; such a comparison is correct not only from the external aspect [. . .] but also from the internal aspect. Just as Darwin put an end to the view of animal and plant species being unconnected, fortuitous, "created by God" and immutable, and was the first to put biology on an absolutely scientific basis by establishing the mutability and the succession of species, so Marx put an end to the view of society being a mechanical aggregation of individuals which allows of all sorts of modification at the will of the authorities (or, if you like, at the will of society and the government) and which emerges and changes casually, and was the first to put sociology on a scientific basis by establishing the concept of the economic formation of society as the sum-total of given production relations, by establishing the fact that the development of such formations is a process of natural history.

V. I. Lenin, "What the 'Friends of the People' Are and How They Fight the Social-Democrats (1894)," in *Collected Works,* 4th ed., 45 vols. (Moscow: Progress Publishers, 1960–72), 1:142.

Secondary Literature: Louis Althusser, *Lenin and Philosophy and Other Essays* (New York: Monthly Review Press, 2001).

GEORGE HENRY LEWES (1817–1878)
English Philosopher

Did you not get the letters from me and Mrs. Lewes in reply to your application & with it Darwin's photograph? We have been expecting to hear from you to send Darwin the Professor's photograph, and the weeks have padded by in silence, till I can no longer remain in doubt as to whether our letters to you had miscarried or arrived safely.

Lewes to Frau Karl von Siebold, 2 November 1871, in *The Letters of George Henry Lewes,* ed. William Baker, 2 vols. (Victoria, BC: University of Victoria, 1995), 2:175.

It is impossible to treat of Evolution without taking notice of that luminous hypothesis by which Mr. Darwin has revolutionized Zoology. There are two points needful to be clearly apprehended before the question is entered upon. The first point relates to the lax use of the phrase "conditions," sometimes more instructively replaced by "conditions of existence." Inasmuch as Life is only possible under definite relations of the organism and its medium, the "conditions of existence" will be those physical, chemical, and physiological changes, which *in* the organism, and *out* of it, co-operate to produce the result. There are myriads of changes in the external medium which have no corresponding changes in the organism, not being in any direct relation to it. These, not being co-operant conditions, must be left out of the account; they are not conditions of existence *for* the organism, and therefore the organism does not vary with their variations. On the other hand, what seem very slight changes in the medium are often responded to by important changes in the vital chemistry, and consequently in the structure of the organism. Now the structure of the organism at the time being, that is to say, its structure and the physico-chemical state of its tissues and plasmids, is the main *condition* of this response; the same external agent will be powerful, or powerless, over slightly different organisms, or over the same organism at different times. Usually, and for convenience, when biologists speak of conditions, they only refer to external changes. This usage has been the source of no little confusion in discussing the Development Hypothesis. Mr. Darwin, however, while following the established usage, is careful in several places to declare that of the two factors in Variation—the nature of the organism and the nature of the conditions—the former is by far the more important.

George Henry Lewes, *The Physical Basis of Mind* (London: Trubner, 1877), 101–2.

Secondary Literature: Rosemary Ashton, *G. H. Lewes: A Life* (Oxford: Clarendon, 1991); Alice B. Kaminsky, "George Henry Lewes," in *Encyclopedia of Philosophy*, ed. Paul Edwards, 8 vols. (New York: Macmillan, 1967), 4:451–54.

C. S. LEWIS (1898–1963)
English Author

The Bergsonian critique of orthodox Darwinism is not easy to answer. More disquieting still is Professor D. M. S. Watson's[†] defense. "Evolution itself," he wrote, "is accepted by zoologists not because it has been observed to occur or [. . .] can be proved by logically coherent evidence to be true, but because the only alternative, special creation, is clearly incredible." Has it come to that? Does the whole vast structure of modern naturalism depend not on positive evidence but simply on an a priori metaphysical prejudice. Was it devised not to get in facts but to keep out God. Even, however, if Evolution in the strict biological sense has some better grounds than Professor Watson suggests—and I can't help thinking it must—we should distinguish Evolution in this strict sense from what may be called the universal evolutionism of modern thought. By universal evolutionism I mean the belief that the very formula of universal process is from imperfect to perfect, from small beginnings to great endings, from the rudimentary to the elaborate: the belief which makes people find it natural to think that morality springs from savage taboos, adult sentiment from infantile sexual maladjustments, thought from instinct, mind from matter, organic from inorganic, cosmos from chaos. This is perhaps the deepest habit of mind in the contemporary world. It seems to me immensely implausible, because it makes the general course of nature so very unlike those parts of nature we can observe. You remember the old puzzle as to whether the owl came from the egg or the egg from the owl. The modern acquiescence in universal evolutionism is a kind of optical illusion, produced by attending exclusively to the owl's emergence from the egg. We are taught from childhood to notice how the perfect oak grows from the acorn and to forget that the acorn itself was dropped by a perfect oak. We are reminded constantly that the adult human being was an embryo, never that the life of the embryo came from two adult human beings. We love to notice that the express engine of today is the descendant of the "rocket"; we do not equally remember that the "Rocket" springs not from some even more rudimentary engine, but from something much more perfect and complicated than itself—namely, a man of genius. The obviousness or naturalness which most people seem to find in the idea of emergent evolution thus seems to be a pure hallucination.

C. S. Lewis, "Is Theology Poetry?" address to Oxford Socratic Club, 1944, in *They Asked for a Paper* (London: Geoffrey Bles, 1962), 164–65. †D. M. S. Watson (1886–1973), professor of zoology and comparative anatomy at the University of London.

No, I'm afraid. I shd. lose much and you wd. gain almost nothing by my writing you a preface. No one who is in doubt about your views of Darwin wd. be impressed by testimony from me, who am known to be no scientist. Many who have been or are being moved towards Christianity by my books wd. be deterred by finding that I was connected with anti-Darwinism. I hope (but who knows himself!) that I wd. not allow myself to be influenced by this consideration if it were only my personal concerns as an author that were endangered. But the cause I stand for wd. be endangered too. When a man has become a popular Apologist he must watch his step. Everyone is on the look out for things that might discredit him. Sorry.

Lewis to Bernard Acworth, 4 October 1951, in "C. S. Lewis on Creation and Evolution: The Acworth Letters, 1944–1960," *Perspectives on Science and Christian Faith* 48 (March 1996): 28–33. Acworth (1885–1963), an English submariner, was an evangelical Christian and anti-evolution activist.

Secondary Literature: Michel White, *C. S. Lewis: A Life* (New York: Carroll & Graf, 2004).

GEORGE CORNEWALL LEWIS (1806–1863)

ENGLISH STATESMAN AND MAN OF LETTERS

I return with many thanks the book [*Origin of Species*] which you had the kindness to lend me. I have read the whole of it with much interest, but the author has entirely failed in convincing me of the truth of his opinions, so far as they are new and intelligible. I regard the subject of his enquiry, the origin of species, as unphilosophical and impenetrable. He writes about species, but never determines what a species is; he objects to the received definition, but substitutes none of his own. He uses the phrase "Natural Selection" in half a dozen different senses. Sometimes he applies it to a case when the animals themselves make a selection; sometimes to cases where they are passive, or even reluctant, and are operated on by external causes. Whatever value the book has is confined to the light which it sheds upon the causes which limit animal population. But this light is shed incidentally, for the author impairs the value of his own remarks by confounding the causes which kill individuals with those which exterminate an entire species. A species may be kept constantly within the limits indicated by its potential capacity of multiplication, without being in danger of extinction:—Because birds eat worms, it does not follow that worms will be annihilated. In my opinion he has entirely failed in showing that the various causes which he calls "Natural Selection" have operated upon the animal kingdom, or determines the number of species within the period over which the exact knowledge of man extends. The writer is a man of talent and ingenuity, with a turn for bold speculation and a great command of facts. He is, however, deficient in clearness and soundness; he may suggest to others, but he cannot discover and prove. His mind is of the German type, speculative, laborious, and unsound.

Lewis to Mrs. Robert Lowe, 27 February 1860, in *Life and Letters of the Right Honourable Robert Lowe, Viscount Sherbrooke,* ed. A. Patchett Martin, 2 vols. (London: Longmans, Green, 1893), 2:203.

Secondary Literature: R. W. D. Fenn, *The Life and Times of Sir George Cornewall Lewis* (Almeley, Herefordshire, UK: Logaston, 2005).

HENRY PARRY LIDDON (1829–1890)
English Theologian

April 22.—In afternoon, preached [at St. Paul's] on Professor Darwin, but with discomfort and misgiving.

April 25.—A quantity of letters: one from the Duke of Argyll*, referring to my Sermon in the kindest terms; another, anonymous, abusing me as a Rationalist for what I said about Darwin.

May 3.—Correcting proof of my Sermon on April 22. Dined with Mr. [William] Gladstone* at 10, Downing Street, to meet the Prince of Wales. Sat between Miss Gladstone and Mr. Goschen.[†] A talk with Lord Rosebery,[‡] and a great deal of talk with Bright.[§] Certainly he is a man for whom I can understand a warm affection.

May 4.—Saw Dr. [Edward Bouverie] Pusey*. He said that he should not join the Committee of a Darwin Memorial; but that they would not apply to him, and that my responsibilities might be different. He struck me as being weaker than last Term, and much less able to seize my meaning, and more wandering in what he said. In evening, dined with the Duke of Argyll, whose conversation about Darwin most interesting. He thought the Darwinian view of the world too mechanical—not room enough for the action of a living mind. There is much in Nature not otherwise explainable.

May 10.—Spent some hours in writing a Preface to my Sermon on April 22 on Mr. Darwin. Wrote to Mr. J. Spottiswoode, declining to join the Darwin Memorial Committee; put it on the true ground—a wish not to vex Dr. Pusey, at his age, to whom I owe such respect and affection. If I had only to think of my own convictions, I think I should join, as we owe Darwin much for his courageous adherence to Theistic truths under a great deal of pressure, as I cannot doubt.

Diary entries, 1882, in *Life and Letters of Henry Parry Liddon,* ed. John Octavius Johnston (New York: Longmans, Green, 1904), 275–76. [†]George Goschen (1831–1907), English statesman and businessman. [‡]Archibald Primrose, Lord Roseberry (1847–1929), English statesman and prime minister. [§]John Bright (1811–89), English statesman. April 22, 1882, was a Saturday.

These reflections may naturally lead us to think of the eminent man, whose death during the past week is an event of European importance; since he has been the author of nothing less than a revolution in the modern way of treating a large district of thought, while his works have shed high distinction upon English science. It may be admitted that when the well-known books on the *Origin of Species* and on the *Descent of Man* first appeared, they were largely regarded by religious men as containing a theory necessarily hostile to the fundamental truths of religion. A closer study has generally modified any such impression. If the theory of "natural selection" has given a powerful impulse to the general doctrine of evolution, it is seen that whether the creative activity of God is manifested through catastrophes [...] or by way of a progressive evolution, it is still His creative activity, and that the real questions beyond remain untouched. [...]

Mr. Darwin's greatness is not least conspicuous in the patience and care with which he observed and registered minute single facts, while engaged in arranging groups of facts. Who that has read his book on Earth Worms can forget the experiments by which he set himself to discover whether a worm possessed the faculty of hearing?

H. P. Liddon, *The Recovery of St. Thomas: A Sermon Preached in St. Paul's Cathedral on the Second Sunday after Easter, April 23, 1882, with a Prefatory Note on the Late Mr. Darwin,* 2nd ed. (London: Rivingtons, 1882), 28–29. Persons in the aura of Liddon referred to this presentation as "The Famous Sermon."

Secondary Literature: Michael Chandler, *The Life and Work of Henry Parry Liddon (1829–1890)* (Leominster, Herefordshire, UK: Gracewing, 2000); Owen Chadwick, *The Victorian Church,* 2 vols. (New York: Oxford University Press, 1966–70).

FRANCIS LIEBER (1800–1872)
German-American Jurist

Darwinism is, to my mind and my taste, wayward and repulsive dogmatism, and unintelligible besides. I should like to see the Darwinian who can tell me what he thinks selection is. An action without agent? The most remarkable thing is that these people—Darwin, [Ludwig] Büchner*, &c.—not only prove to you that your great-grandmother was a hideous gorilla, but they do it with enthusiasm, and treat you almost like a heretic if you will not agree with their flimsy and visionary materialism. Give me, rather, the cosmogony of the unethical Greek mythology.

Lieber to Judge Martin Russell Thayer, 26 March 1871, in *The Life and Letters of Francis Lieber,* ed. Thomas Sergeant Perry (Boston: James R. Osgood, 1882), 409.

Secondary Literature: Frank Freidel, *Francis Lieber: Nineteenth-Century Liberal* (Baton Rouge: Louisiana State University Press, 1947).

ABRAHAM LINCOLN (1809–1865)
American President

For many years I subscribed for and kept on our office table the *Westminster* and *Edinburgh Review* and a number of other English periodicals. Besides them I purchased the works of Spencer*, Darwin, and the utterances of other English scientists, all of which I devoured with great relish. I endeavored, but had little success in inducing Lincoln to read them. Occasionally he would snatch one up and peruse it for a little while, but he soon threw it down with the suggestion that it was entirely too heavy for an ordinary mind to digest. A gentleman in Springfield gave him a book called, I believe, *Vestiges of Creation,* which interested him so much that he read it through. The volume was published in Edinburgh, and undertook to demonstrate the doctrine of development or evolution. The treatise interested him greatly, and he was deeply impressed with the notion of the so-called "universal law"—evolution; he did not extend greatly his researches, but by continued thinking in a single channel seemed to grow into a warm advocate of the new doctrine. Beyond what I have stated he made no further investigation into the realm of philosophy.

William H. Herndon, in *Abraham Lincoln,* by Herndon and Jesse Weik, 2 vols. (New York: Appleton, 1895), 2:146–48.

It is interesting to learn that Lincoln, having read the first edition [of *Vestiges*], later procured and read the sixth, in which the religious spirit of the author was made still more apparent.

William E. Barton, *The Soul of Abraham Lincoln* (New York: George H. Doran, 1920), 171.

Often Lincoln would stretch himself on the office cot, weary of his toil, and say, "Now, Billy, tell me about the books"; and Herndon would discourse by the hour, ranging over history, literature, philosophy, and science. [...] With characteristic zest Herndon plunged into Darwin's *Origin of Species* when it appeared, but Lincoln refused to follow on the plea that the water was too deep. He was, however, interested in *Vestiges of Creation*, whose dogma of the universal reign of law fitted into his philosophy in which there were no accidents. He frequently perused the *Westminster* and *Edinburgh Reviews*, which Herndon kept on the office table, but he could not enthuse over Herbert Spencer*. Occasionally, when meditating an important speech, he would ask his partner for books, and Herndon, besides furnishing the books, would sometimes make a brief of his own reading on the subject, especially if it were a question of history. After this manner they worked together, comrades and friends, totally unlike but with the utmost good feeling, until Fame drove her chariot through the back office.

Joseph Fort Newton, *Lincoln and Herndon* (Cedar Rapids, IA: Torch, 1910), 254–55.

Secondary Literature: David Donald, *Lincoln's Herndon* (New York: Knopf, 1948).

CESARE LOMBROSO (1835–1909)
ITALIAN CRIMINOLOGIST

Vaccaro and Grassi believe that they need to attack the new school [of criminology] for its supposed foundation in Darwinism. They want to fight Darwinism and make it disappear. This is typical of the academic world, where professors combat any new theory. But the positivist school is not founded on a theoretical system. As a psychiatrist rather than a natural scientist, I simply replaced the abstract approach of the past with clinical and academic methods in the individual study of the madman and the criminal. If, after applying these new methods, I became aware that they smacked of Darwinism, I had not regrets. Instead, I made use of

Darwinian theory in studying the median occipital fossetta, as well as crimes among animals, children, and savages. But far from being one of Darwin's acolytes, I never mentioned him in the first and second editions of *Criminal Man*. Even in the last edition, I added *disease* to atavism as an explanation of crime—and disease has no place in Darwinian theory. Similarly, [Giuseppe] Sergi*, Garofalo, and Ferri[†] incorporated into our theory the notion of occasional crime, which is not at all Darwinian.

Cesare Lombroso, *Criminal Man* (1876; reprint, Durham, NC: Duke University Press, 2006), 235–36. [†]Raffaelo Garofalo (1851–1934) and Enrico Ferri (1856–1929), criminologists, disciples of Lombroso.

Secondary Literature: Mary Gibson, *Born to Crime: Cesare Lombroso and the Origins of Biological Criminology* (New York: Praeger, 2002); Daniel Pick, *Faces of Degeneration: A European Disorder* (Cambridge: Cambridge University Press, 1993).

HENRY WADSWORTH LONGFELLOW
(1807–1882)
American Poet

One of the things I wished most to say, and which I say first is the delight with which I found your memory so beloved in England. At Cambridge Professor [Adam] Sedgwick* said: "Give my best to Agassiz*. Give him the blessing of an old man." In London, Sir Rod. [Roderick Impey] Murchison* said: "I have known a great many men that I liked; but I *love* Agassiz." In the Isle of Wight, Darwin said, "What a set of men you have in Cambridge! Both our Universities put together cannot furnish the like. Why, there is Agassiz, he counts for three!"[†] [1868, to Agassiz]

I have read both your Historic Sketch and the Scientific paper with much interest. In both you make out a very good case. If Darwinian readers say, "Thou almost persuadest me to be a monkey"; yours can say "Plato, thou reasonest well." [1871, to R. G. Haliburton]

I have given up smoking for a time; and a more ridiculous pre-Adamite, absurd, Darwinian old fool, you never saw. [1872, to G. W. Greene]

The Letters of Henry Wadsworth Longfellow, ed. Andre Hilen, 6 vols. (Cambridge, MA: Harvard University Press, 1967–82), 5:270, 456, 498. †Longfellow had met Darwin in the company of Alfred Lord Tennyson and Julia M. Cameron on the Isle of Wight in July 1868.

Secondary Literature: Charles C. Calhoun, *Longfellow: A Rediscovered Life* (Boston: Beacon, 2004).

JAMES RUSSELL LOWELL (1819–1891)
American Poet and Diplomat

In short, I have been translating into Spanish a sketch of Mr. Darwin's life—no, not *your* Mr. Darwin [i.e., William], certainly, you foolish little person, but his father. Not that I like science any better than I ever did. I hate it as a savage does writing, because he fears it will hurt him somehow; but I have a great respect for Mr. Darwin, as almost the only perfectly disinterested lover of truth I ever encountered. I mean, of course, in his books, for I never had the pleasure of seeing him. So I volunteered my services as dragoman, and when the opuscule is printed (which will not be for some time yet), I shall ask permission to lay a copy at your feet, as we say here.

Lowell to Mrs. W. E. Darwin, 1 September 1878, in *Letters of James Russell Lowell,* ed. Charles Eliot Norton, 3 vols. (Boston: Houghton Mifflin, 1904), 3:55–56.

from "Credidimus Jovem Regnare"

I don't object, not I, to know
My sires were monkeys, if 'twas so;
I touch my ear's collusive tip
And own the poor-relationship.
That apes of various shapes and sizes
Contained their germs that all the prizes
Of senate, pulpit, camp, and bar win
May give us hopes that sweeten Darwin.
Who knows but from our loins may spring
(Long hence) some winged sweet-throated thing
As much superior to us
As we to Cynocephalus?

The Complete Poetical Works of James Russell Lowell, ed. Horace E. Scudder (Cambridge, MA: Houghton Mifflin, 1896), 425.

Lowell's religious belief at the end of his life was of a piece with his political,—doubt subdued by hope but not killed. Despite the warm friendship of his middle age with men of science, the doctrines that were advanced by Darwin and his followers had always repelled him—perhaps because with that affinity of certain of his moods for pessimism which we have noticed, he perceived the temperamental danger for him of evolutionary doctrines. He had written to Leslie Stephen* in 1876:—

"I continue to shut my eyes resolutely in certain speculative directions, and am willing to find solace in certain intimations that seem to me from a region higher than my reason";—and to Miss Grace Norton in 1879 he wrote of science: "I hate it as a savage does writing, because I fear it will hurt me somehow"; and he goes on to say: "I think the evolutionists will have to make a fetich of their protoplasm before long. Such a mush seems to me a poor substitute for the Rock of Ages, by which I understand a certain set of higher instincts which mankind have found solid under their feet in all weathers."

Ferris Greenslet, *James Russell Lowell: His Life and Work* (Boston: Houghton Mifflin, 1905), 224–25.

The protoplasmic germ to which it was incalculable promotion to become a stomach, has it not, out of the resources with which God had endowed it, been able to develop the brain of Darwin, who should write its biography? Even Theology is showing signs that she is getting ready to exchange a man who fell in Adam for a man risen out of nonentity and still rising through that aspiring virtue in his veins which is spurred onwards and upwards by the very inaccessibility of what he sees above him.

James Russell Lowell, *Latest Literary Essays and Addresses* (Boston: Houghton Mifflin, 1892), 168–69.

Secondary Literature: Martin Duberman, *James Russell Lowell* (Boston: Houghton Mifflin, 1966).

JOHN LUBBOCK (1834–1913)
English Prehistorian and Naturalist

Like [Huxley*], but, of course, far less effectively, from the date of the appearance of *The Origin of Species*, I stood by Darwin and did my best to fight the battle of truth against the torrent of ignorance and abuse which was directed against him. Sir J. Hooker* and I stood by Huxley's side and spoke up for Natural Selection in the great Oxford debate of 1860.

John Lubbock, *Essays and Addresses, 1900–1903* (Leipzig: Bernhard Tauchnitz, 1904), 8.

Many, I believe, are deterred from attempting what are called stiff books for fear they should not understand them; but there are few who need complain of the narrowness of their minds, if only they would do their best with them. In reading, however, it is most important to select subjects in which one is interested. I remember years ago consulting Mr. Darwin as to the selection of a course of study. He asked me what interested me most, and advised me to choose that subject. This, indeed, applies to the work of life generally.

John Lubbock, *The Pleasures of Life*, pt. 1 (London: Macmillan, 1903), 72.

It is still possible for a boy to pass though the school without any real scientific training, and to leave Eton without ever having heard of Darwin or Newton.

John Lubbock, "On the Study of Science," *Contemporary Review* 50 (1886): 209–20, on 213, quoting F. W. Cornish, "Eton Reform," *Nineteenth Century*, 1884, 587.

Secondary Literature: Mark Patton, *Science, Politics and Business in the Work of Sir John Lubbock* (Aldershot: Ashgate, 2007).

CHARLES LYELL (1797–1875)
Scottish Geologist

[October 1859]
My dear Darwin: I have just finished your volume, and right glad I am that I did my best with Hooker* to persuade you to publish it without

waiting for a time which probably could never have arrived, though you lived to the age of a hundred, when you had prepared all your facts on which you ground so many grand generalizations.

It is a splendid case of close reasoning and long sustained argument throughout so many pages, the condensation immense, too great perhaps for the uninitiated, but an effective and important preliminary statement, which will admit, even before your detailed proofs appear, of some occasional useful exemplifications, such as your pigeons and cirripedes, of which you make such excellent use. [...]

It is this which has made me so long hesitate, always feeling that the case of Man and his Races and of other animals, and that of plants, is one and the same, and that if a *Vera causa* be admitted for one instant, of a purely unknown and imaginary one, such as the word "creation," all the consequences must follow.

Lyell to Darwin, 3 October 1859, in *Life, Letters, and Journals of Sir Charles Lyell, Bart,* 2 vols. (London: John Murray, 1881), 2:325.

[January 1860]
I have been so absorbed in preparation for a new edition of my *[Principles of] Geology* that I have really had no ideas to exchange, except on those matters which the initiated are discussing, or that question which my friend Charles Darwin's book has brought before the British reading public, both scientific, literary, and theological; whether, as Dean [Henry Hart] Milman* expresses it, Lyell and his friend have come from tadpoles, against which the Dean, after reading the book on the *Origin of Species,* vehemently protests, saying that the production of such a book is in itself enough to refute the possibility of such an origin.

Nevertheless it is easier to say and feel this, than to gainsay the continually increasing body of evidence to which I shall try to add some arguments not yet advanced.

Of all the small books of a readable kind which have come out in our time, you will find it, I expect, the one which takes the longest to read and digest. [...]

To return to Darwin's book, Twistleton,[†] who has called since I wrote the above, being up here for Lord [Thomas Babington] Macaulay's* funeral, told me he had been much taken with the new theory, and stated some objections, and ended with asking me what I thought Agassiz* would say to it, after he had nailed his colors to the mast in his recent work on *Classification*. Now I should like much if you will learn what Agassiz does think and say, and if he has already written anything, please send it to me. Asa Gray*, among your scientific men of note, is, I think, the one who comes nearest in his opinions to Darwin. I confess that Agassiz's last work drove me far over into Darwin's camp, or the Lamarckian view, for when he attributed the original of every race of man to an independent starting point, or act of creation, and not satisfied with that, created whole "nations" at a time, every individual out of "earth, air, and water" as Hooker* styles it, the miracles really became to me so much in the way of S. Antonio of Padua, or that Spanish saint whose name I forget, that I could not help thinking Lamarck must be right, for the rejection of his system led to such license in the cutting of knots.

Lyell to George Ticknor, 9 January 1860, in *Life, Letters, and Journals of Sir Charles Lyell, Bart,* 2 vols. (London: John Murray, 1881), 2:329–31. †Edward Twistleton (1809–74) was an English civil servant and poor law commissioner. Through his wife, who was the aunt of Henry Parkman (1850–74), a Boston banker and lawyer, Twistleton was acquainted with a broad circle of Boston intellectuals.

I was glad to get the news in Mrs. Ticknor's last letter of the opening of the Museum and of Agassiz's* doings. I fully expect it will soon be a model collection. I shall be curious to see a second reply of Agassiz to Darwin, which I understand is coming. Murray has sold all that remained, and more, of my friend's *Origin of Species;* 4,250 copies printed, and only out about a year, and he must now prepare a new edition.

Whatever faith we may settle down into, opinions can never go back exactly to what they were before Darwin came out. The Oxford Professor of Geology, J. Phillips,† has fought Darwin by citing me in pages out of my *Principles,* but I must modify what I said in a new edition. Agassiz helped Darwin and the Lamarckians by going so far in his *Classification,* not hesitating to call in the creative power to make new species out of nothing whenever the slightest difficulty occurs of making out how a variety got to some distant part of the globe. Asa Gray's* articles, all of which I have procured, appear to me the ablest, and on the whole grappling with the subject, both as a naturalist and metaphysician, better than anyone else on either side of the Atlantic.

Lyell to George Ticknor, 29 November 1860, in *Life, Letters, and Journals of Sir Charles Lyell, Bart,* 2 vols. (London: John Murray, 1881), 2:340–41. †John Phillips (1800–1874), English geologist.

When in 1874 I spent some time with Lyell in his Forfarshire home, a communication from Darwin was always an event which made a "red-letter day," as Lyell used to say; and he gave me many indications in his conversation of how strongly he relied on the opinion of Darwin—more indeed than on the judgment of any other man—this confidence not being confined to questions of science, but extending to those of morals, politics, and religion.

J. W. Judd, "Darwin and Geology," in *Darwin and Modern Science: Essays in Commemoration of the Birth of Charles Darwin and of the Fiftieth Anniversary of the Publication of the Origin of Species* (Cambridge: Cambridge University Press, 1909), 337–84, on 383.

Secondary Literature: Sir Charles Lyell's Scientific Journals of the Species Question, ed. Leonard G. Wilson (New Haven, CT: Yale University Press, 1970).

M

THOMAS BABINGTON MACAULAY (1800–1859)
English Historian and Politician

Chevening, July 16, 1856.—After breakfast Lord Stanhope very kindly and sensibly left me to rummage his library. A fine old library it is, of, I should guess, fifteen thousand volumes: much resembling a college library both in appearance and in the character of the books. I was very agreeably entertained till two in the afternoon. Then we set off for Mountstuart Elphinstone's,[†] six miles off. I saw him probably for the last time; still himself, though very old and infirm. A great and accomplished man as any that I have known. In the evening Darwin, a geologist and traveler, came to dinner.

The Life and Letters of Lord Macaulay, ed. George Otto Trevelyan, 2 vols. (London: Longmans, Green, 1876), 1:403–4. [†]Mountstuart Elphinstone (1779–1859), Scottish statesman and historian, governor of Bombay.

Secondary Literature: Owen Dudley Edwards, *Macaulay* (London: Weidenfeld & Nicolson, 1988); Sheridan Gilley, "Thomas Babington Macaulay," in *Encyclopedia of Historians and Historical Writing,* ed. Kelly Boyd, 2 vols. (London: Fitzroy Dearborn, 1999), 2:746–47.

RAMSAY MacDONALD (1866–1937)
English Prime Minister

The scientific method employs the processes of both induction and deduction. [. . .] Darwin's work consisted not so much in proving the theory of evolution from a series of grouped facts (though he did that more than any of his predecessors), but in using that theory to explain facts, and so to this day we hear occasional disputes as to whether the Darwinian method was inductive or deductive, whereas, as a matter of fact, it was a scientific blending of both.

The Socialist method is the Darwinian method. It begins with social phenomena, with the rational desire to group them in systems, and with the equally rational desire to discover their causes and visualize their complete fulfilment. Its interest consists in the whence and whither of society.

James Ramsay MacDonald, *The Socialist Movement* (London: Williams & Norgate, 1911), 114–15.

Secondary Literature: D. A. Stack, "The First Darwinian Left: Radical and Socialist Response to Darwin, 1859–1914," *History of Political Thought* 21 (2000): 682–710.

ANNE LUMB MACDONELL (b. 1850)
English Aristocrat

My father entertained the great naturalist Darwin, when he came to Buenos Ayres in 1837 in H.M.S. *Beagle*. He stayed with my parents, as there were few hotels and those were poor and uncomfortable. A story is told of how, on his return from one of his expeditions, he brought a little species of mole that belonged to that part of South America and was almost extinct. It is called the *touca-touca* from the noise it makes. Darwin rolled it up in one of his fine cambric handkerchiefs, wanting very much to boil it down to preserve its skeleton, and for this purpose put it carefully in his chest of drawers. He went away again for a three weeks' excursion, as he had heard of the mammoth shell at Quilmes, near to Uncle John Yates' place. During his absence the housemaid complained that in the Señor Professor's room there was such a bad smell that she could not go into it. My mother went, and soon finding the cause of the trouble, promptly threw the handkerchief and *touca-touca* into the fire.

On Mr. Darwin's return he inquired for the specimen, and was much upset at not finding it. My mother did not dare acknowledge her fault, and so begged my father [Edward Lumb] to tell Darwin what she had done, whereupon he said, "I will forgive Mrs. Lumb, for she is nearly as beautiful as the *touca-touca*." I wonder whether my beautiful mother appreciated the compliment?

Lady Anne Lumb Macdonell, *Reminiscences of Diplomatic Life* (London: Adam & Charles Black, 1913), 27–29.

Secondary Literature: Adriana Novoa and Alex Levine, *From Man to Ape: Darwinism in Argentina—Central Analogies in Peripheral Science* (Chicago: University of Chicago Press, 2010).

ERNST MACH (1838–1916)
German Physicist and Philosopher

The fundamental theorems of dynamical science, of optics, of heat, and of electricity were all disclosed in the century that followed Galileo. Of scarcely less importance, it seems, was that movement which was prepared for by the illustrious biologists of the hundred years just past, and formally begun by the late Mr. Darwin. Galileo quickened the sense for the simpler phenomena of inorganic nature. And with the same simplicity and frankness that marked the efforts of Galileo, and without the aid of technical or scientific instruments, without physical or chemical experiment, but solely by the power of thought and observation, Darwin grasps a new property of organic nature—which we may briefly call its plasticity. With the same directness of purpose, Darwin, too, pursues his way. With the same candor and love of truth, he points out the strength and the weakness of his demonstrations. With masterly equanimity he holds aloof from the discussion of irrelevant subjects and wins alike the admiration of his adherents and of his adversaries.

Ernst Mach, *Popular Scientific Lectures,* trans. Thomas J. McCormack (Chicago: Open Court, 1894), 215–16.

As we recognize no real gulf between the physical and the psychical, it is a matter of course that, in the study of the sense-organs, general physical as well as special biological observations may be employed. Much that appears to us difficult of comprehension when we draw a parallel between a sense-organ and a physical apparatus, is rendered quite obvious in the light of the theory of evolution, simply by assuming that we are concerned with a living organism with particular memories, particular habits and manners, which owe their origin to a long and eventful race-history. [. . .] Even teleological conceptions, as aids to investigation, are not to be shunned. It is true, our comprehension of the facts of reality is not enhanced by referring them to an unknown World-Purpose, itself problematic. Nevertheless, the question as to the value that a given function has for the existence of an organism, or as to what are its actual contributions to the existence of the same, may be of great assistance in the comprehension of this function. Of course, we must not suppose, on this account, as many Darwinians have done, that we have "mechanically explained" a function, when we discover that it is necessary for the survival of the species. Darwin himself is doubtless quite free from this short-sighted conception. The physical means by which a function is developed still remains a physical problem; while the mode and reason of an organism's voluntary adaptation continues to be a psychological problem. The preservation of the species is only one, though an actual and very valuable, point of departure for inquiry, but it is by no means the last and the highest. Species have certainly been destroyed, and new ones have as certainly arisen. The pleasure-seeking and pain-avoiding will, therefore, is directed perforce beyond the preservation of the species. It preserves the species when it is advantageous to do so; transforms it when it is advantageous; and destroys it when its continuance would not be advantageous. Were it directed merely to the preservation of the species, it would move aimlessly about in a vicious circle, deceiving both itself and all individuals. This would be the biological counterpart of the notorious "perpetual motion" of physics.

Ernst Mach, *Contributions to the Analysis of the Sensations,* trans. C. M. Williams (Chicago: Open Court, 1897), 35–40.

Secondary Literature: John T. Blackmore, *Ernst Mach: His Life, Work, and Influence* (Berkeley and Los Angeles: University of California Press, 1972); Peter Alexander, "Ernst Mach," in *Encyclopedia of Philosophy,* ed. Paul Edwards, 8 vols. (New York: Macmillan, 1967), 5:115–19.

WILLIAM SHARP MACLEAY (1792–1865)
ENGLISH-AUSTRALIAN ENTOMOLOGIST

[May 1860]

It is lucky for me, therefore, that both you and Mrs. Lowe have given me the subject of this letter in asking me for my opinion of Darwin's book. To me, now on the verge of the tomb, I must confess the subject of it is more interesting than either the extension of British commerce or even the progress of national education. This question is no less than "What am I?" "What is man?" a created being under the direct government of his Creator, or only an accidental sprout of some primordial type that was the common progenitor of both animals and vegetables. The theologian has no doubt answered those questions, but leaving the Mosaic account of the Creation to Doctors of Divinity, the naturalist finds himself on the horns of a dilemma. For, either from the facts he observes, he must believe in a special creation of organized species, which creation has been progressive and is now in full operation, or he must adopt some such view as that of Darwin, viz. that the primordial material cell of life has been constantly sprouting forth of itself by "natural selection" into all the various forms of animals and vegetables. Darwin, indeed, for no reason that I can perceive, except his fear of alarming the clergy, speaks of a Creator of the original material cell. But there appears to be very little necessity for His existence, if it be true, as Darwin says, that this material cell can go on by itself eternally sprouting into all the animals and vegetables that have existed or will exist. Again, if this primordial cell had a Creator, as Darwin seems to admit, I do not see what we gain by denying the Creator, as Darwin does, all management of it after its creation. Lamarck was more logical in supposing it to have existed of itself from all eternity—indeed this is the principal difference that I see between this theory of Darwin's and that of Lamarck, who propounded everything essential in the former theory, in a work now rather rare—his *Philosophie zoologique*. [...] It is the system also, with some small alterations, of the *Vestiges of Creation*, a work which I recollect telling you at the time is more incorrect as to facts and therefore valueless, however attractive it may be in style. Darwin, on the other hand, like his predecessor Lamarck, is a most able naturalist; and though I agree with what Mrs. Lowe tells me is the opinion of Sir R. [Roderick Impey] Murchison*, via. that his facts are not always sound; still, quite enough of them are so far unexceptionable as to entitle his lucubrations—however preposterous—to our respect, if not to our assent. Natural selection (sometimes called "struggles" by Darwin) is identical with the *"Besoins des choses"* of Lamarck, who, by means of his hypothesis, for instance, assigns the constant stretching of the neck to reach the acacia leaves as the cause of the extreme length of it in the giraffe; much in the same way the black bear, according to Darwin, became a whale, which I believe as little as his other assertion that our progenitors anciently had gills—only they had dropped off by want of use in the course of myriads of generations. As for his sexual selection, it is the only original invention of the three. It is truly Darwin's own, and if anyone can believe, that the sexes of every animal were originally alike: that the cock, for instance, owes his comb, wattle, and other distinguishing marks to the taste of the hens who have constantly sought such a type to breed from—why, all I can say is that such a believer must have a very wide swallow. I can only assume from the favor which this book has received from the English public, that either they do not understand the tenor of it, or that what is termed Revealed Religion, and particularly the Mosaic account of the Creation, sits somewhat uneasily on the minds of a great many thinking persons. The theory is almost a materialist one—nay, even so far atheistic that, if it allows of a deity at all, He has been ever since the institution of the primordial type of life fast asleep. [...] All special interference of a Creator with it Darwin repeatedly denies. I am myself so far a Pantheist that I see God in everything: but then I believe in His special Providence, and that He is the constant and active sole Creator and all-wise Administrator of the Universe. Darwin seems at times to have been led to his most wild conclusions by his anxiety to avoid the constant and special interference of a Creator. He parodies your legal axiom, and says, *De minimis non curat Deus* [God is not interested in trivia]. But there can be nothing great or small in respect to the absolute. [...]

To conclude: It is far easier for me to believe in the direct and constant government of the Creation by God, than that He should have created the world and then left it to manage itself, which is Darwin's theory in a few words. Nevertheless, Charles Darwin is an old friend of mine and I feel grateful to him for his work. I hope it will make people attend to such matters, and to be no longer prevented by the first chapter of Genesis from asking for themselves what the Book of Nature says on the subject of the Creation. I have now complied with your and Mrs. Lowe's requests. I could say much more if I entered into the examination of Darwin's facts, or rather the facts on which he founds his theory, but I must have already tired you—I only wish I had your gift of writing tersely—*sed non cuivis* [but it is not everyone's lot]. This letter is for Mrs. Lowe as well as you, so pray tell her, with my affectionate regards, that I trust she will tell me whether she agrees with Darwin's notions or with mine, i.e. saving intact her own saving faith in the Bible story of six days and the concluding apple.

W. S. Macleay to Robert Lowe, May 1860, in *Life and Letters of the Right Honourable Robert Lowe, Viscount Sherbrooke,* ed. A. Patchett Martin, 2 vols. (London: Longmans, Green, 1893), 2:204–7.

Secondary Literature: Dov Ospovat, *The Development of Darwin's Theory* (Cambridge: Cambridge University Press, 1981).

ALEXANDER MACMILLAN (1818–1896)
English Publisher

[26 November 1859]
Pray look at Huxley's* article, "Time and Life." Darwin's book, which it mentions, is remarkable certainly. I thought of "Nature acts in tooth and claw" as I was glancing over it. I wish someone could bring out the other side. But surely the scientific men ought on no account to be hindered from saying what they find are facts.

Macmillan to Alfred Tennyson, 26 November 1859, in *Life and Letters of Alexander Macmillan,* ed. Charles L. Graves (London: Macmillan, 1910), 140. The reference is to Huxley's "Time and Life: Mr Darwin's *Origin of Species,*" *Macmillan's Magazine* 1 (1859–60): 142–48.

[January 1860]
Meanwhile the Thursday evenings at Henrietta Street[†] were continued with increasing vigor and prestige. In January [1860] they had "much fine talk" from Huxley*, [Charles] Kingsley*, [Frederick Denison] Maurice*, [Thomas] Hughes*, [David] Masson* and Henry Kingsley* on Darwin's *Origin of Species.*

Life and Letters of Alexander Macmillan, ed. Charles L. Graves (London: Macmillan, 1910), 147. [†]Macmillan's second branch office was located at 23 Henrietta Street, London.

[25 July 1860]
To Daniel Wilson[†]
Your letter of July 8 is indeed very grateful to us. Of course we shall most gladly undertake your book. [. . .]

The sooner you write and let me know the exact title and that I may announce it the better I shall be pleased. The subject is one that is sure to be interesting, especially at present. That Darwin controversy has awakened interest in races and origins to a large extent. Every original investigation as to the relics of bygone races is looked for with the keenest interest. Unfortunately a good deal of theological asperity has mixed up with the controversy on both sides. But I think a true scientific spirit is on the whole gaining ground. Huxley* on the one side, and Hopkins‡ on the other, are dealing with the matter in the true spirit of truth-seeking, I do believe. Between [Richard] Owen* and Huxley some personal soreness has hindered the enquiry from going on in a smooth, orderly, enquiring spirit, but I think this is likely to lessen. There was a wonderful passage of arms between Huxley and the Bishop of Oxford [Samuel Wilberforce*] at the last British Association, in which I suspect the Bishop came off second best from the simple fact that he was suspected of using the *odium theologicum* unfairly. [. . .]

Huxley says definitely that he is not a Darwinian, but has the greatest anxiety in the interests of science that all justice should be done to the theory and perfect fair-play given to its discussion untrammelled by the yelping of the curs of orthodoxy.

Hopkins here, who is one of the greatest authorities on certain departments of Geology, and has been President of the Geological Society and of the British Association, has written long papers in *Fraser* [*Fraser's Magazine*] on the anti-Darwinian side, which Huxley says are valuable contributions to the subject. I have had a good deal of talk with Mr. Hopkins on the subject, and he too is very strong on the importance of keeping clear of all pseudo-religious dealing with scientific questions. I have been telling you this mainly to show you that the fullest and fairest consideration is likely to be given to anything you may bring forward, and that the minds of men are running in that very direction that will make them look with interest to your forthcoming work.

Letters of Alexander Macmillan, ed. George A. Macmillan (Glasgow: Glasgow University Press for private circulation, 1908), 57. †Daniel Wilson (1816–92), archaeologist, coined the term *prehistory.* ‡William Hopkins (1793–1866), English mathematician and geologist.

Secondary Literature: Charles Morgan, *The House of Macmillan* (New York: Macmillan, 1944).

HENRY SUMNER MAINE (1822–1888)
ENGLISH LEGAL HISTORIAN

I have never myself imagined that any amount of evidence of law or usage, written or observed, would by itself solve the problems which cluster round the beginnings of human society. "The imperfection of the geological record" is a mere trifle to the imperfection of the archaeological record. "What were the motives" I asked in my *Ancient Law* (p. 270), "which originally prompted men to hold together in the family union?" "To such a question," I answered, "Jurisprudence unassisted by other sciences is not competent to give a reply." This anticipation of aid to be expected from biological science has been fulfilled, and it is remarkable that, while the greatest luminary of ancient science invented or adopted the Patriarchal theory, the greatest name in the science of our day is associated with it. Mr. Darwin appears to me to have been conducted by his own observations and studies to a view of the primitive condition of mankind, which cannot be distinguished from this theory. "We may conclude (*Descent of Man*, ii. 362) from what we know of the passions of all male quadrupeds that promiscuous intercourse in a state of nature is extremely improbable. [...] If we look far enough back in the stream of time, it is exceedingly improbable that primeval men and women lived promiscuously together. Judging from the social habits of man as he now exists and from most savages being polygamists, the most probable view is that primeval men aboriginals lived in small communities, each with as many wives as he could support or obtain, whom he would have jealously guarded against all other men. [...] In primeval times men [...] would probably have lived as polygamists or temporarily as monogamists. [...] They would not at that period have lost one of the strongest of all instincts, common to all the lower animals, the love of their young offspring" (p. 367). With his usual candor Mr. Darwin admits, though with some hesitation, the conclusions of writers who have followed a different path of inquiry from his, but he thinks that the licentiousness attributed to savages belonged to a "later period when man had advanced in his intellectual powers but retrograded in his instincts." It must be remembered that a difference in the nature of the sexual union, answering to the difference of view separating the Patriarchal theory from its opposite, runs through the whole animal world; and, under such circumstances, considering the extreme scantiness of the archaeological evidence, it would seem reasonable to call in the testimony of those who have made the animal world their study. When man had most of the animal in him, he belonged to the highest animals; and this is the consideration which gives such importance to Mr. Darwin's opinion.

Henry Maine, *Dissertations on Early Law and Custom,* new ed. (London: John Murray, 1891), 205–7.

Secondary Literature: Alan Diamond, ed., *The Victorian Achievement of Henry Maine* (Cambridge: Cambridge University Press, 1991).

OSIP MANDELSTAM (1891–1938)
RUSSIAN POET

I would compose similes to convey an idea of your nature and got more and more accustomed to your anti-Darwinian essence; I studied the living language of your long ungainly arms, created to provide a handshake in a moment of peril, and given to passionate protestation, on the run, against natural selection.

Osip Mandelstam, "Journey to Armenia," in *The Noise of Time,* trans. Clarence Brown (Harmondsworth: Penguin, 1993), 191–225, on 205.

I have signed an armistice with Darwin and placed him on my imaginary bookstand next to Dickens. If they should dine together the third member of the party would be Mr. Pickwick. One can't help being charmed by Darwin's good nature. He is an unintentional humorist. The humor of the situation is habitual to him, follows him everywhere.

Osip Mandelstam, "Journey to Armenia," in *The Noise of Time,* trans. Clarence Brown (Harmondsworth: Penguin, 1993), 191–225, on 214–15.

Secondary Literature: Clarence Brown, *Mandelstam* (Cambridge: Cambridge University Press, 1973).

MAO TSE-TUNG (1893–1976)
CHINESE COMMUNIST LEADER

Throughout history, at the outset new and correct things often failed to win recognition from the majority of people and had to develop by twists and turns through struggle. Often, correct and good things were first regarded not as fragrant flowers but as poisonous weeds. Copernicus' theory of the solar system and Darwin's theory of evolution were once dismissed as erroneous and had to win out over bitter opposition. Chinese history offers many similar examples. In a socialist society, the conditions for the growth of the new are radically different from and far superior to those in the old society. Nevertheless, it often happens that new, rising forces are held back and sound ideas stifled. Besides, even in the absence of their deliberate suppression, the growth of new things may be hindered simply through lack of discernment. It is therefore necessary to be careful about questions of right and wrong in the arts and sciences, to encourage free discussion and avoid hasty conclusions. We believe that such an attitude will help ensure a relatively smooth development of the arts and sciences.

Mao Tse-tung, "On the Correct Handling of Contradictions among the People," in *Selected Works of Mao Tse-tung*, 5 vols. (1965; reprint, Peking: Foreign Languages Press, 1977), 5:384–421, on 408–9.

Secondary Literature: Nick Knight, *Marxist Philosophy in China* (New York: Springer, 2005).

OTHNIEL C. MARSH (1831–1899)
American Paleontologist

I was passing a memorable day with Darwin, during which he spoke freely of many scientific men. Referring to Huxley*, he said with more than usual earnestness, "Huxley is the king of men!" A few days later I mentioned this to Huxley, and he was deeply moved by it. His reply I shall never forget: "Now you can understand why we who know Darwin all have such an affection for him, and when his enemies reviled the noble man, why my right arm was so heavy in his defense."

Othniel C. Marsh, "Thomas Henry Huxley," *American Journal of Science*, 3rd ser., 50 (1895): 177–83, on 182.

Secondary Literature: Charles Schuchert and Clara Mae LeVene, *O. C. Marsh: Pioneer in Paleontology* (New Haven, CT: Yale University Press, 1940); Mark Jaffe, *The Gilded Dinosaur: The Fossil War between E. D. Cope and O. C. Marsh and the Rise of American Science* (New York: Three Rivers, 2000).

ALFRED MARSHALL (1842–1924)
English Economist

Before Adam Smith's book had yet found many readers, biologists were already beginning to make great advances towards understanding the real nature of the differences in organization which separate the higher from the lower animals; and before two more generations had elapsed, Malthus' historical account of man's struggle for existence started Darwin on that inquiry as to the effects of the struggle for existence in the animal and vegetable world, which issued in his discovery of the selective influence constantly played by it. Since that time biology has more than repaid her debt; and economists have in their turn owed much to the many profound analogies which have been discovered between social and especially industrial organization on the one side and the physical organization of the higher animals on the other. [...] This central unity is set forth in the general rule [...] that the development of the organism, whether social or physical, involves an increasing subdivision of functions between its separate parts on the one hand, and on the other a more intimate connection between them. Each part gets to be less and less self-sufficient, to depend for its wellbeing more and more on other parts, so that any disorder in any part of a highly-developed organism will affect other parts also.

This increased subdivision of functions, or "differentiation" as it is called, manifests itself with regard to industry in such forms as the division of labor, and the development of specialized skill, knowledge and machinery: while "integration," that is, a growing intimacy and firmness of the connection between the separate parts of the industrial organism, shows itself in such forms as the increase of security of commercial credit, and of the means and habits of communication by sea and road, by railway and telegraph, by post and printing-press.

The doctrine that those organisms which are the most highly developed, in the sense in which we have just used the phrase, are those which are likely to survive in the struggle for existence, is itself in process of development. It is not yet completely thought out either in its biological or its economic relations. [...]

[...] the struggle for survival may fail to bring into existence organisms that would be highly beneficial: and in the economic world the demand for any industrial arrangement is not certain to call forth a supply, unless it is something more than a mere desire for the arrangement or a need for it. It must be an efficient demand; that is, it must take effect by offering adequate payment or some other benefit to those who supply it. A mere desire on the part of employees for a share in the management and the profits of the factory in which they work, or the need on the part of clever youths for a good technical education, is not a demand in the sense in which the term is used when it is said that supply naturally and surely follows demand. This seems a hard truth: but some of its harshest features are softened down by the fact that those races, of which the members render unrequited services to other members, are not only the most likely to flourish for the time, but most likely to rear a large number of descendants who inherit their beneficial habits.

Alfred Marshall, *Principles of Economics*, 2 vols. (London: Macmillan, 1898), 1: 319–21.

Secondary Literature: Justin Cunningham Wood, *Alfred Marshall* (London: Routledge, 1998), esp. 183–87; Joseph Schumpeter, *History of Economic Analysis* (New York: Oxford University Press, 1954).

HARRIET MARTINEAU (1802–1876)
English Philosopher and Feminist

[26 December 1859]
I think when [Mr A] has finished Charles Darwin's book he will see that the theory has *not* a theological basis. I think it a pity that 2 or 3 expressions wd seem to warrant the notion: but I believe them to be like the same expressions in the translation (not the original) of [Alexander von] Humboldt's *Cosmos,*—used as ordinary current expressions, without reference to their primitive meaning. If so, they ought not to have been used: but the theory does not require the notion of a creation; & my conviction is that Charles D. does not hold it. What a work it is!—overthrowing (if true) revealed religion on the one hand, & Natural [theology] (as far as Final Causes and Design are concerned) on the other. The range & mass of knowledge takes away one's breath.

Martineau to George Holyoake, 26 [December 1859], in *The Collected Letters of Harriet Martineau*, vol. 4, *Letters, 1856–1862* (London: Pickering & Chatto, 2007), 208–9.

[W]hat I write for is to thank you again for sending me your brother's book [*Origin of Species*]. As for thanking *him* for the book itself, one might say "thank you" all one's life without giving any idea of one's sense of obligation. [. . .] I believed, and have often described, the quality and conduct of your brother's mind; but it is an unspeakable satisfaction to see here the full manifestation of its earnestness and simplicity, its sagacity, its industry, and the patient power by which it has collected such a mass of facts, to transmute them by such sagacious treatment into such portentous knowledge. I should much like to know how large a proportion of our scientific men believe that he has found a sound road to the upper ranges of the history of organized existence. It does not very much matter; for it is the next generation that effectively profits by such works: but it would be pleasant to know that a good many remain open-minded.

Martineau to Erasmus Darwin, 2 February 1860, in *Harriet Martineau's Letters to Fanny Wedgwood*, ed. Elisabeth Sanders Arbuckle (Stanford, CA: Stanford University Press, 1983), 185–86.

It seemed to me, after I had written to [Erasmus Darwin], that I ought to have said one thing more about C.D.'s book, for honesty's sake: and the notices I have seen reminded me of this since, more than once. I rather regret that C.D. went out of his way two or three times (I think not more) to speak of "the Creator" in the popular sense of the First Cause; and also *once* of the "final cause" of certain cuckoo affairs [cf. *Origin of Species*, 1st ed., 488]. This latter is sure to be misunderstood, in the face of all the rest of the book: and the other gives occasion for people to ride off from the argument in a way which need not have been granted to them. It is curious to see how those who would otherwise agree with him turn away because his view is "derived from," or "based on," "theology," while he admits critics, by this opening, who would otherwise have no business with the book at all. It seems to me that having carried us up to the earliest group of forms, or to the single primitive one, he and we have nothing to do with those few forms, or that one, came there. His subject is the "Origin of Species," and not the origin of Organization; and it seems a needless mischief to have opened the latter speculation at all.

Martineau to Fanny Wedgwood, 13 March 1860, in *Harriet Martineau's Letters to Fanny Wedgwood*, ed. Elisabeth Sanders Arbuckle (Stanford, CA: Stanford University Press, 1983), 189.

Lest C. Darwin should mind his critics I will console him with a bit of remarkable patronage. Our cook is a sort of rude genius—a devouring reader of tales, and a greedy listener to new things. In going after the fowls with her, and discussing some phenomena, M[†] told her about *Origin of Species,* (gospel according to Darwin) and Mary[‡] observed she "shouldn't wonder it was true." She will never forget it again.

Martineau to Fanny Wedgwood, 20 October 1860, in *Harriet Martineau's Letters to Fanny Wedgwood,* ed. Elisabeth Sanders Arbuckle (Stanford, CA: Stanford University Press, 1983), 189. [†]Maria Martineau, niece of Harriet Martineau. [‡]Mary Martineau, Harriet's aunt.

Secondary Literature: Deborah Anna Logan, *The Hour and the Woman: Harriet Martineau's "Somewhat Remarkable" Life* (DeKalb: Northern Illinois University Press, 2002).

KARL MARX (1818–1883)
German Political Theorist

Darwin's work is most important and suits my purpose in that it provides a basis in natural science for the historical class struggle. One does, of course, have to put up with the clumsy English style of argument. Despite all shortcomings, it is here that, for the first time, "teleology" in natural science is not only dealt a mortal blow but its rational meaning is empirically explained.

Marx to Ferdinand Lasalle, 16 January 1861, in Marx and Engels, *Collected Works,* 50 vols. (Moscow: Progress Publishers, 1975–2005), 41:245.

I'm amused that Darwin, at whom I've been taking another look, should say that he also applies the "Malthusian" theory to plants and animals, as though in Mr Malthus's case the whole thing didn't lie in its not being applied to plants and animals, but only—with its geometric progression— to humans as against plants and animals. It is remarkable how Darwin rediscovers, among the beasts and plants, the society of England with its division of labor, competition, opening up of new markets, "inventions" and Malthusian "struggle for existence." It is Hobbes'[†] *bellum omnium contra omnes* and is reminiscent of Hegel's Phenomenology, in which civil society figures as an "intellectual animal kingdom," whereas, in Darwin, the animal kingdom figures as civil society.

Marx to [Friedrich] Engels, 18 June 1862, in Marx and Engels, *Collected Works*, 50 vols. (Moscow: Progress Publishers, 1975–2005), 41:380. †See Paul Lafargue entry.

As to Paul [Lafargue]'s* lively narration of his adventure with Mlle [Clémence] Royer* [Darwin's Lamarckian French translator], it has tickled [Friedrich] Engels* and my humble self. I was not at all astonished at his failure. He will remember that, having read her preface to Darwin, I told him at once she was a bourgeois. Darwin was led by the struggle for life in English society—the competition of all with all, bellum omnium contra omnes—to discover competition to [. . .] [illegible] as the ruling law of "bestial" and vegetative life. Darwinism, conversely, considers this a conclusive reason for human society never to emancipate itself from its bestiality.

Marx to Paul and Laura Lafargue, 15 February 1869, in Marx and Engels, *Collected Works*, 50 vols. (Moscow: Progress Publishers, 1975–2005), 43:216.

"Herr [Friedrich Albert] Lange* (*On the Labour Question*, etc., 2nd ed.)," Marx wrote to Kugelmann† on June 27, 1870, "sings my praises loudly, but with the object of making himself important. Herr Lange, you see, has made a great discovery. The whole of history can be brought under a single great natural law. This natural law is the *phrase* (in this application Darwin's expression becomes nothing but a phrase) 'struggle for life,' and the content of this phrase is the Malthusian law of population or, rather, over-population. So, instead of analyzing the 'struggle for life' as represented historically in various definite forms of society, all that has to be done is to translate every concrete struggle into the phrase 'struggle for life,' and this phrase itself into the Malthusian 'population fantasy.' One must admit that this is a very impressive method—for swaggering, sham-scientific, bombastic ignorance and intellectual laziness."

V. I. Lenin, *Materialism and Empirio-Criticism*, pt. 3 (Peking: Foreign Languages Press, 1972), 397–98. †Ludwig Kugelmann (1828–1902), German gynecologist, confidant of Engels and Marx.

Secondary Literature: Ralph Colp Jr., "The Myth of the Darwin-Marx Letter," *History of Political Economy* 14 (1982): 461–82; Giuliano Pancaldi, "The Technology of Nature: Marx's Thoughts on Darwin," in *The Natural Sciences and the Social Sciences*, ed. I. Bernard Cohen (Dordrecht, Netherlands: Kluwer, 1994), 257–94.

DAVID MASSON (1822–1907)
Scottish Author and Literary Historian

Rumours there were and ominous whisperings
Of the great bomb-shell book from Murray's press,
Threatening flagration in the common mind,
And scientific warring not less loud
Than that which followed the "Vestiges"—I mean
Charles Darwin's "Origin of Species"
By Natural Selection; in which words
The author indicates this principle
As principle-in-chief of all his work,
That Nature is a progress of conditions
Still tightening, tightening, tightening through the ages,
Growing more strict and difficult, and thus
Strangling the worser breeds, and letting through
The finer, subtler, stronger, who escape
Minus their clumsier parts, and modified
On and still on for ever.

"Colloquy of the Round Table," *Macmillan's Magazine* 1 (1869–60): 148–60, on 157.
The unsigned "colloquy" was presumed to be the work of Masson and Thomas
Hughes. Darwin mentions this verse in a letter to Asa Gray of 8 March 1860. See
Darwin Correspondence Project, letter 4843, http://www.darwinproject.ac.uk/
darwinletters/calendar/entry-4843.html (accessed 21 May 2009).

Secondary Literature: Henry James, "David Masson," in *Literary Criticism: Essays
on Literature, American Writers, English Writers* (New York: Library of America,
1984), 1169–74.

MAXWELL T. MASTERS (1833–1907)
English Botanist and Horticulturist

Let any one who knows what was the state of botany in this country even
so recently as fifteen or twenty years ago, compare the feeling between
botanists and horticulturists at that time with what it is now. What sympathy had the one for the pursuits of the other? The botanist looked down
on the varieties, the races, and strains, raised with so much pride by the

patient skill of the florist as on things unworthy of his notice and study. The horticulturist, on his side, knowing how very imperfectly plants could be studied from the mummified specimens in herbaria, which then constituted in most cases all the material that the botanist of this country considered necessary for the study of plants, naturally looked on the botanist somewhat in the light of a laborious trifler. [...] Darwin altered all this. He made the dry bones live; he invested plants and animals with a history, a biography, a genealogy, which at once conferred an interest and a dignity on them. Before, they were as the stuffed skin of a beast in the glass case of a museum; now they are living beings, each in their degree affected by the same circumstances that affect ourselves, and swayed, *mutatis mutandis,* by like feelings and like passions. If he had done nothing more than this we might still have claimed Darwin as a horticulturist; but as we shall see, he has more direct claims on our gratitude. The apparently trifling variations, the variations which it was once the fashion for botanists to overlook, have become, as it were, the keystone of a great theory.

Gardeners' Chronicle, 22 April 1882, quoted in *Life of Charles Darwin,* by George Thomas Bettany (London: Walter Scott, 1887), 167–68.

Secondary Literature: Robert Olby, "Mendelism: From Hybrids and Trade to a Science," *Comptes Rendus de l'Académie des Sciences,* 3rd ser., *Sciences de la Vie,* 323 (2000): 1043–51.

HENRY MAUDSLEY (1835–1918)
English Psychiatrist

The attention of the philosophic and scientific world has been so much fixed on the theory of evolution, ever since Darwin set forth the main manner of the process by means of the survival of the fittest through natural selection, that there has been a proneness to overlook the fact that all we see and feel around us is not progress—in the sense we understand progress. Survival of the fittest does not mean always survival of the best in the sense of the highest organism; it means only the survival of that which is best suited to the circumstances, good or bad, in which it is placed—the survival of a savage in a savage social medium, of a rogue among rogues, of a parasite where a parasite alone can live. A decline from a higher to a lower level of being, a process, that is to say, of degeneration, is an integrant and active part of the economy of nature.

Henry Maudsley, *Body and Will* (New York: Appleton, 1884), 237.

We learn, then, that when man is born with a brain no higher—indeed, lower—than that of an ape, he may have the convolutions fewer in number, and individually less complex, than they are in the brain of a chimpanzee and an orang; the human brain may revert to, or fall below, that type of development from which, if the theory of Darwin be true, it has gradually ascended by evolution through the ages.

Henry Maudsley, *Body and Mind: An Inquiry into their Connection and Mutual Influence, Specially in Reference to Mental Disorders* (New York: Appleton, 1886), 46.

Secondary Literature: Michael Collie, *Henry Maudsley, Victorian Psychiatrist: A Bibliographical Study* (Winchester, Hampshire, UK: St. Paul's Bibliographies, 1988).

FREDERICK DENISON MAURICE (1805–1872)
English Theologian

But of all modern schools of thought, the purely scientific, represented by Mr. Darwin (who, I believe returned fully the warm admiration which Mr. Maurice felt for him) has most troubled the minds of simple English Christians. A passage or two from Mr. Maurice's writings may, perhaps, lead any such who may read this book to take courage, and look the *Origin of Species* squarely in the face—at any rate it will show them that he could do so:

"It has been our wont to speak of man as formed in the image of God, and yet as made out of the dust of the earth. I think those who have used the words have been aware—if not at the same moment, yet at certain moments of their lives—of both the facts to which the words point, and have been trying to learn how they are compatible.

"I have myself little hope that we shall become fully aware of our relation to One who is above us, if from any cowardly self-glorification we shrink from confessing these baser affinities. The more thoroughly we accept the facts which attest our humiliation, the more overwhelming will be the force of the facts which attest the glory of our human parentage. If Mr. Darwin has added new strength to the one kind of evidence—whether he has or not, as I told you before, I have no right to affirm, or even to guess—I can have no doubt whence the discoveries have come, or by whom that search has been prompted. I perceive that in his last book he speaks with much reverence of the moral elevation which the belief in a one omnipotent ruler of the universe is likely to produce in those who cherish it. I am afraid that in me such a belief would cause more depression than elevation. Mere omnipotence is crushing. Whereas any one whose heart confesses that every step in the apprehension of nature or man, or the archetype of man, is due to the education of a loving parent, must be sure that no diligence, such as that of Mr. Darwin, in studying the meanest insect or flower, can be wasted; but will also be sure that the processes in the student himself—the springs of his zeal and patience—must have a far deeper interest, must carry us into another region altogether."

Frederick Denison Maurice, *The Friendship of Books and Other Lectures,* ed. Thomas Hughes (London: Macmillan, 1889), xxii–xxiii.

Secondary Literature: Jeremy N. Morris, *F. D. Maurice and the Crisis of Christian Authority* (New York: Oxford University Press, 2005); Owen Chadwick, *The Victorian Church,* 2 vols. (New York: Oxford University Press, 1966–70).

JUSTIN HUNTLY McCARTHY (1859–1936)
IRISH POLITICIAN

In the meantime that keen struggle between science and orthodoxy, to which I have already alluded, had set in, and the younger men were beginning to regard [Richard] Owen* as one who had been pushed aside by the greater energy and boldness of the newer school which was represented by Darwin, Huxley*, and [John] Tyndall*. I am myself entirely lacking in all culture of the fields of science, and I never was able to understand why there should have been any antagonism between the followers of Owen and the followers of Darwin, or, indeed, whether there was any actual antagonism at all. But I found that most of the people whom I knew talked of Owen as representing the old school and Darwin as representing the new, and I could only accept the supposition that there was some sort of struggle going on, and that Owen was to me a living man because I had known him, and Darwin only a book or the head of a school because I had not known him.

Justin McCarthy, *Reminiscences,* 2 vols. (New York: Harper, 1899), 2:267.

Secondary Literature: Paul A. Townsend, "'No Imperial Privilege': Justin McCarthy, Home Rule, and Empire," *Éire—Ireland: Journal of Irish Studies* 42 (2007): 201–8.

JAMES McCOSH (1811–1894)
Scottish-American Philosopher and Theologian

During my student-days the great work on "Typical forms and Special Ends in Creation" [New York: Carter, 1856] was published under the joint authorship of Dr. McCosh and Dr. George Dickie.[†] [...] The book, though presenting what I regard as the best summary of the old argument for Natural Theology, would not apply in our time without some readjustment. Its "Typical forms," borrowed from Goethe and other Nature-Philosophers of the last century, would need to be transformed into the Types or Phyla by heredity of our day; and its "special ends" are very like Darwin's "Survival of the fittest," but giving prominence to the principle of Design, which Darwin so carefully eliminated, and which is now forcing its way back even into Evolution-Biology. Dickie's method of argument by marshalling long hosts of carefully observed facts, which point towards the goal, is so strangely suggestive of Darwin's method, that if the relative date of their works were reversed, one might imagine that Dickie copied Darwin. [...]

What really ruined the run of this book was the appearance soon after it publication of Darwin's *Origin of Species,* which carried the whole controversy into new regions. This may explain, in part, the hostility to Darwinism of my revered friend, Dr. Dickie, whose carefully drawn and really sound lines of argument were overwhelmed by the new theory; just as Louis Agassiz*, in the New World, was annoyed to find all the speculations which had lifted him to eminence buried by the same influence.

Memoir by George MacLoskie of his student days in Scotland, in *The Life of James McCosh: A Record Chiefly Autobiographical,* ed. William Milligan Sloane

(New York: Scribner's, 1897), 122; MacLoskie (1834–1920) was a professor of biology at Princeton. [†]George Dickie (1812–82), Scottish botanist, professor of natural history at Queen's College, Belfast.

Greatly to the disappointment of some of his followers, Darwin is obliged to postulate three or four germs of life created by God. To explain the continuance of life, he is obliged to call in a pangenesis, or universal life, which is just a vague phrase for that inexplicable thing life, and life is just a mode of God's action. Plants, the first life that appeared, have no sensation. How did sensation come in? Whence animal instinct? [...] It is felt by all students of mental science that Darwin is weak when he seeks to account for these high ideas or sentiments. Careful, as being so trained, in noticing the minutest peculiarities of plants and animals, he seems utterly incapable of understanding man's higher capacities and noble aspirations—of seeing how much is involved in consciousness, in personal identity, in necessary truth, in unbending rectitude; he explains them only by overlooking their essential peculiarities. [...] But these inquires have brought us face to face with a remarkable body of facts. [...] How curious, should it turn out that these scientific inquirers, so laboriously digging in the earth, have, all unknown to themselves, come upon the missing link which is partially to reconcile nature and revealed religion.

James McCosh, "Religious Aspects of the Doctrine of Development," in *History, Essays, Orations, and other Documents of the Sixth General Conference of the Evangelical Alliance (New York, October 2–12, 1873),* ed. Philip Schaff and S. Irenaeus Prime (New York: Harper Brothers, 1874), 264–71, on 267–71.

Secondary Literature: J. David Hoeveler Jr., *James McCosh and the Scottish Intellectual Tradition: From Glasgow to Princeton* (Princeton, NJ: Princeton University Press, 1981); Paul Charles Kemeny, *Princeton in the Nation's Service: Religious Ideals and Educational Practice, 1868–1928* (New York: Oxford University Press, 1998).

WILLIAM McDOUGALL (1871–1938)
English Psychologist

I read widely and before my first graduation in general science at seventeen years of age, I had read nearly all of Spencer*, Darwin, and Huxley*,

Lyell's* *Principles of Geology*, and other standard works of science. The great controversy between evolutionary theory and religions was still raging, and I delighted in Huxley's attacks on [William] Gladstone* and all the orthodoxies. [. . .] In those days the word "agnostic," recently popularized by Huxley, seemed to me the best banner under which to sail. But my agnosticism was not militant, aggressively negative, or hostile to religion. I said *ignoramus*, I could not follow [Emil] Dubois-Reymond* in adding *ignorabimus*.

Carl Murchison, ed., *A History of Psychology in Autobiography*, vol. 1 (New York: Russell & Russell, 1930), 194.

Secondary Literature: Raymond van Over and Laura Oteri, eds., *William McDougall: Explorer of the Mind* (New York: Garrett, 1967).

MARGARET MEAD (1901–1978)
American Anthropologist

Family beliefs about the sources of genius of its most conspicuously gifted members themselves create an atmosphere in which talent can actively flourish or may suffer special eclipse. In the Darwin family, the belief that Darwins are slow learners provides protection for Darwin descendants so that they are not pressured into premature school performances. The emphasis on the tremendous amount of hard work that Charles did has also provided an atmosphere within which choice of level of accomplishment is possible.

Margaret Mead, *Continuities in Cultural Evolution* (New York: Transaction, 1999), 183–85.

[Mead's] family are, she says, "rationalistic, agnostic, Spencer-reading, New England Puritans," and when I boasted of my five generations of atheism, she capped the tale with a statement that her great-grandmother was "read out of" the Unitarian Church for heresy. [. . . She has] a good sound plain intelligent—almost female Darwin face.

Gregory Bateson to his mother (Caroline Beatrice Bateson), 31 March 1936, in *Gregory Bateson: The Legacy of a Scientist*, by David Lipset (Englewood Cliffs, NJ: Prentice-Hall, 1980), 150–51.

Secondary Literature: Nancy Lutkehaus, *Margaret Mead: The Making of an American Icon* (Princeton, NJ: Princeton University Press, 2008).

H. L. MENCKEN (1880–1956)
AMERICAN JOURNALIST

A bishop in his robes, playing his part in the solemn ceremonial of the mass, is a dignified spectacle; the same bishop, bawling against Darwin half an hour later, is seen to be simply an elderly Irishman with a bald head, the son of a respectable police sergeant in South Bend, Indiana.

H. L. Mencken's Smart Set Criticism, ed. William H. Nolte (Washington, DC: Regnery Gateway, 2001), 94.

I think [...] of Charles Darwin and his associates, and of how they were reviled in their time. This reviling, of course, is less vociferous than it used to be, chiefly because later victims are in the arena, but the underlying hostility remains. Within the past two years the principal Great Thinker of Britain, George Bernard Shaw*, has denounced the hypothesis of natural selection to great applause, and a three-time candidate for the American Presidency, William Jennings Bryan*, has publicly advocated prohibiting the teaching of it by law. The great majority of Christian ecclesiastics in both English-speaking countries, and with them the great majority of their catechumens, are still committed to the doctrine that Darwin was a scoundrel, and Herbert Spencer* another, and Huxley* a third—and [Friedrich] Nietzsche* is to the three of them what Beelzebub himself is to a trio of bad boys.

H. L. Mencken, *Prejudices: Third Series* (New York: Knopf, 1924; [Whitefish, MT?]: Kessinger, 2006), 129.

Secondary Literature: Terry Teachout, *The Skeptic: A Life of H. L. Mencken* (New York: HarperCollins, 2002).

MIKHAIL ALEKSANDROVICH MENZBIR
(1855–1935)
RUSSIAN ZOOLOGIST

The broad sweep of his work, particularly of the *Origin of Species,* made Darwin a target of massive attacks. The ideas he presented in this book were not exactly new: [for example,] earlier writers had referred to the genealogical unity of organisms. What Darwin gave was a new explanation of this unity. In a relatively small volume he collected such a mass of masterly arranged facts that his conclusions stood out before our eyes. Earlier, it was customary to talk about the limitless variation of organisms only in jest; Darwin was the first to introduce serious talk about this matter. Earlier, most questions relating to the life of animals were not recognized as pertinent questions and were ignored; he showed that ignoring these questions did not mean proving that they were not significant. Briefly, Darwin gave much attention to questions of this kind and he raised others that could no longer be ignored.

M. A. Menzbir, "Charlz Darvin i sovremennoe sosttoianie evoliutsionnogo ucheniia" [Russian thought], *Russkaia Mysl',* no. 7 (1882): 53–79, on 53–54, quoted in *Darwin in Russian Thought,* by Alexander Vucinich (Berkeley and Los Angeles: University of California Press, 1988), 87.

Secondary Literature: Alexander Vucinich, *Darwin in Russian Thought* (Berkeley and Los Angeles: University of California Press, 1988).

JOHN EDWARD MERCER (1857–1922)
ANGLICAN CLERGYMAN

The curious, and still obscure, history of totemism supplies abundant evidence on this point; and not less so that modern sympathy with all living things, which is largely based on what may be termed the new totemism of the Darwinian theory. But while attention will thus be focused on the sphere of the inorganic, seemingly so remote from human modes of experience, some attempt will nevertheless be made to suggest the inner harmonies which link together all modes of existence. A further limitation to be noted is that "nature" will be taken to cover only such natural objects as remain in what is generally called their "natural" condition—that is, which are independent of, and unaffected by, human activities.

As for the animal world in the widest sense, it is plain that its study, from the mystical point of view, forms a department to itself. Granted that the transition from the mineral to the organism is gradual, and that from the vegetable to the animal still more gradual, the broad fact remains that, when we reach the higher forms of the realm of living matter, we definitely recognize many of the characteristics which are found in the human soul—will, emotion, impulse, even intellectual activities. Not only primitive man, but those also who are often far advanced in mental development, attribute souls to animals, and find it difficult to believe otherwise—as witness the totemistic systems followed by theories of metempsychosis. And Darwinism, far from destroying these old ideas, has simply furnished a scientific basis for a new totemism.

John Edward Mercer, *Nature Mysticism* (London: George Allen, 1913), 9, 248.

Secondary Literature: Richard P. Davis, *Bishop John Edward Mercer: A Christian Socialist in Tasmania* (Hobart: Printing Section, University of Tasmania, 1982).

HERMAN MERIVALE (1806–1874)
English Civil Servant and Historian

Power to thine elbow, thou newest of sciences,
All the old landmarks are ripe for decay;
Wars are but shadows, and so are alliances,
Darwin the great is the man of the day.

All other 'ologies want an apology;
Bread's a mistake—Science offers a stone;
Nothing is true but Anthropobiology—
Darwin the great understands it alone.

Mighty the great evolutionist teacher is,
Licking Morphology clean into shape;
Lord! what an ape the Professor or Preacher is,
Ever to doubt his descent from an ape.

Herman Merivale, "Darwinity," in *A Nonsense Anthology*, ed. Carolyn Wells (New York: Charles Scribner's Sons, 1919), 31–32.

Secondary Literature: David T. McNab, "Herman Merivale and Colonial Office Indian Policy in the Mid-Nineteenth Century," *Canadian Journal of Native Studies* 1 (1981): 277–302.

ELIE METCHNIKOFF (1845–1916)
Russian Microbiologist

Meanwhile, Elie was yearning for independent and more general study. During his unsuccessful journey, he had acquired in Leipzig many recently published scientific books, and, among them, Darwin's *Origin of Species*. The theory of evolution deeply struck the boy's mind and his thoughts immediately turned in that direction. He said to himself that isolated forms which had found no place in definite animal or vegetable orders might perhaps serve as a bond between those orders and elucidate their genetic relationships.

Olga Metchnikoff, *Life of Elie Metchnikoff* (Boston: Houghton Mifflin, 1921), 41–42.

At Giessen, Elie had read Fritz Müller's* *For Darwin,* a book which had a decisive influence on the future direction of his researches. Fritz Müller, in his embryological works on certain crustaceans, had been the first to confirm in a concrete manner Darwin's evolutionist theories; he had thus demonstrated that it was chiefly in embryology that precious indications were to be found concerning the genealogy of organisms. [In later years Metchnikoff often dwelt on the fact that Fritz Müller was not fully appreciated and that it was he who had most efficaciously contributed to the confirmation of Darwinian theories.] Under the influence of this work, Elie, who until now had limited himself to introductory researches, resolved to concentrate all his efforts on the comparative embryology of animals. He started to work in that direction, and his researches confirmed him more and more in the opinion that the key of animal evolution and genealogy was to be sought for in the most primitive stages, in those simple phases of development where no secondary element has yet been introduced from external conditions. In those primordial stages, essential characters, common to all, reveal the analogy and connections between animals from different groups.

Olga Metchnikoff, *Life of Elie Metchnikoff* (Boston: Houghton Mifflin, 1921), 50, brackets in the original.

Secondary Literature: Alfred I. Tauber and Leon Chernyak, *Metchnikoff and the Origin of Immunology: From Metaphor to Theory* (New York: Oxford University Press, 1991).

LÉON METCHNIKOFF (1838–1888)
RUSSIAN ORIENTALIST

The shining merit of Darwin resides especially in the amazing perspicacity with which his genius transformed that worn-out politico-economical thesis into the very principle of regeneration, not only for the biological science of our days, but also for modern philosophy altogether. Such a miracle could be performed only by his clear perception of the fact that the great law of competition or struggle for life, unduly applied by the Malthusian politico-economy to a series of phenomena for which it cannot account, is really a capital principle pervading the individual life throughout. Since the Malthusian law, stating that the number of competitors always exceeds the means of subsistence, is true with animals, we might logically foresee that it would not do for human societies; because the animals, being far more prolific than men, simply consume the food they find ready in Nature, whilst the lowest human tribes—provided that they possess some social organization—generally produce a large part of what they consume; and slavery, appearing at a very low degree of social evolution, yields us a sufficient proof that, even in those destitute conditions, men united into a society produce more food than is strictly required for the subsistence of them all.

Léon Metchnikoff, "Revolution and Evolution," *Contemporary Review* 50 (1886), 412–37, on 431.

Secondary Literature: Anne M. Bailey and Josep R. Llobera, *The Asiatic Mode of Production: Science and Politics* (London: Routledge, 1981).

JOHN STUART MILL (1806–1873)
ENGLISH PHILOSOPHER

[April 1860]

I have read since my return here several things which have interested me, above all Darwin's book. It far surpasses my expectation. Though he cannot be said to have proved the truth of his doctrine, he does seem to have proved that it *may* be true, which I take to be as great a triumph as knowledge and ingenuity could possibly achieve on such a question. Certainly nothing can be at first sight more entirely unplausible than his theory, and yet after beginning by thinking it impossible, one arrives at something like an actual belief in it, and one certainly does not relapse into complete disbelief.

Mill to Alexander Bain, 11 April 1860, in *The Letters of John Stuart Mill*, 2 vols. (London: Longmans, Green, 1910), 1:236.

I was spending an evening last week with my friend Mr. John Stuart Mill, and I am sure you will be pleased to hear from such an authority that he considers your reasoning throughout is in the most exact accordance with the strict principles of logic. He also says the method of investigation you have followed is the only one proper to such a subject.

Henry Fawcett to Darwin, 16 July 1861, *More Letters of Charles Darwin*, ed. Francis Darwin, 2 vols. (New York: Appleton, 1903), 1:189–90.

In regard to the Darwinian hypothesis, I occupy nearly the same position as you do. Darwin has found (to speak Newtonically) a *vera causa*, and has shown that it is capable of accounting for vastly more than had been supposed; beyond that it is but the indication of what may have been though it is not proved to be, the origin of the organic world we now see. I do not think it an objection that it does not, even hypothetically, resolve the question of the first origin of life, any more than it is an objection to chemistry that it cannot analyze beyond a certain number of simple or elementary substances. Your remark that the development theory naturally leads to convergences as well as divergences is just, striking, and as far as I know, has not been made before. But does not this very fact resolve one of your difficulties, viz., that species are not, by divergence, multiplied to infinity?

since the variety is kept down by frequent blending. The difficulty is also met by the fact that the law of natural selection must cause all forms to perish except those which are superior to others in power of keeping themselves alive in circumstances actually realized on earth.

Mill to Hewett C. Watson, 30 January 1869, in *The Letters of John Stuart Mill*, 2 vols. (London: Longmans, Green, 1910), 2:181.

Secondary Literature: Joseph Schumpeter, *History of Economic Analysis* (New York: Oxford University Press, 1954).

HENRY HART MILMAN (1791–1868)
English Historian and Cleric

[Milman to Lyell*, 4 April 1862]
What say you to this dilemma as to the Scriptural geologists? It seems to me rather a puzzler. Many writers, from Mr. Granville Penn[†] to Dr. Mc-Caul,[‡] have asserted and endeavored to prove the cosmogony in the Book of Genesis to be anticipative of and, rightly interpreted, in perfect accordance with modern discoveries in astronomy and geology. The difficulty seems to me insuperable. Either the writer (suppose Moses) understood the full signification of all these words and images, or he did not. If he did, he was consciously or unconsciously a premature Newton, Cuvier, Lyell*; and this without any advantage to mankind, for no one for four or five thousand years after was the wiser for his prophetic knowledge. If he did not, the Almighty inspired into his mind words utterly without meaning, which he himself wanted knowledge to interpret even to himself—an enigmatic yet pregnant oracle, of which the key was not to be discovered till the nineteenth century after Christ. Down to that time it was not and could not be understood, or rather must have been misunderstood, by all the successive generations of mankind. [. . .]

Sir Charles Lyell answers:—I see no escape from the dilemma you have put, and hope you will force some of them to acknowledge it; for it is too bad, after what was gone through by the astronomers, that in geology we should have the same stand made in favor of a revelation in science. But the pre-Adamite heresy[§] will soon cast into the shade all difficulties about the age of the world before the advent of man; while, on the other hand, there is a monster of more hideous mien looming in the future, in comparison with which the doctrine of pre-Adamites is orthodoxy itself. I allude, of course, to the Lamarckian and Darwinian theory of the origin of species, which, so long as I thought it visionary, I thought amusing enough, but which, now that I have grown more familiar with its face and think it so probable that it must be endured, I by no means find more welcome. Extreme and unauthorized views upon the literal inspiration of the Bible on the one hand, the intolerant and contemptuous dogmatism of the modern German school of criticism on the other hand, being, as my father believed, responsible for much of the assumed oppugnancy between religion and science, it seemed to him that it would be well to set them face to face, to examine carefully and dispassionately the facts upon which they were at issue, the facts upon which it might at the end appear that there was internecine, unapproachable enmity. [. . .] It is clear that, as to the moral truths of religion, there is no such collision, no such antagonism. The profoundest men of science may be the most virtuous as well as the wisest of mankind. A [John] Herschel* or a [Michael] Faraday*, while ranging the heavens to unfold the wonders of astronomy, or tracing the most subtle law of chemistry, finds no impediment to the unimpeachable practice of all the domestic and social excellencies.

Arthur Milman, *Henry Hart Milman, Dean of St. Paul's: A Biographical Sketch* (London: John Murray, 1900), 270–72. †Granville Penn (1761–1844), English biblical geologist. ‡James McCaul (1841–1906), Irish-Canadian Presbyterian minister who lectured on Agassiz's geology. §Pre-Adamites held that human beings existed before Adam.

Secondary Literature: "Milman, Henry Hart," in *The Oxford Dictionary of the Christian Church*, ed. F. L. Cross and E. A. Livingstone (Oxford: Oxford University Press, 1997).

PETER CHALMERS MITCHELL (1864–1945)
English Zoologist

Let me sum up my argument. It is asserted that war is just, necessary and admirable, and that this proposition is a deduction from biology. In the words of von Bernhardi:[†] "Wherever we look in nature, we find that war is a fundamental law of development. This great verity, which has been recognized in past ages, has been convincingly demonstrated in modern times by Charles Darwin." I hope to have succeeded in showing:

1. That even if the struggle for existence were a scientific law, it does not necessarily apply to human affairs.

2. That modern nations are not units of the same order as the units of the animal and vegetable kingdom from which the law of struggle for existence is a supposed inference.

3. That the struggle for existence as propounded by Charles Darwin, and as it can be followed in nature, has no resemblance with human warfare.

4. That man is not subject to the laws of the unconscious and that his conduct is to be judged not by them, but by its harmony with a real and external not-self that man has built up through the ages.

Peter Chalmers Mitchell, *Evolution and the War* (New York: Dutton, 1915), 108.
[†]Friedrich Adam Julius Bernhardi (1849–1930) was a militaristic Prussian general, author of *England as Germany's Vassal* (London: Doran, 1914).

Secondary Literature: D. P. Crook, "Peter Chalmers Mitchell and Antiwar Evolutionism in Britain during the Great War," *Journal of the History of Biology* 22 (1989): 325–56.

ST. GEORGE JACKSON MIVART (1827–1900)
English Biologist

It is easy to complain of onesidedness in the views of many who oppose Darwinism in the interest of orthodoxy; but not at all less patent is the intolerance and narrow-mindedness of some of those who advocate it, avowedly or covertly, in the interest of heterodoxy. This hastiness of rejection or acceptance, determined by ulterior consequences believed to attach to "Natural Selection," is unfortunately in part to be accounted for by some expressions and a certain tone to be found in Mr. Darwin's writings. That his expressions, however, are not always to be construed literally is manifest. His frequent use metaphorically of the theistic expressions, "contrivance," for example, and "purpose," has elicited, from the Duke of Argyll* and others, criticisms which fail to tell against their opponent, solely because such expressions are, in Mr. Darwin's writings, merely figurative—metaphors, and nothing more.

St. George Jackson Mivart, *On the Genesis of Species*, 2nd ed. (London: Macmillan, 1871), 16–17.

Since the first edition of this work appeared, Mr. Darwin has published his *Descent of Man*. Therein he shows elaborately the resemblances which exist both in structure and mode of development between man's body and the bodies of inferior forms. He also calls attention to similarity in diseases, parasites, the effects of medicines, stimulants, &c. All this, however, merely amounts to a proof of what no one denies, namely, that man *is* an animal; and consequently to the establishment of an *a priori* probability that *if* other animals have arisen by "Natural Selection," the animal man has also arisen in like manner, unless a valid objection can be raised from some other part of his nature. It is patent that an objection can be raised from his intellectual and moral faculties, and accordingly Mr. Darwin endeavors to show that there is no difference of *kind* between these faculties and the psychical powers of brutes. In this endeavor he fails utterly. The result is that Man (the totality of his being and not his anatomy only being considered) is seen, yet more clearly from this very failure, to differ from every other animal by a distinction far more profound than any which separates each irrational animal from every other.

St. George Jackson Mivart, *On the Genesis of Species*, 2nd ed. (London: Macmillan, 1871), 319n1.

One of the most grotesque conceptions suggested by Mr. Darwin is that of the nakedness of man, and especially of woman, having been produced by the gradual extension over the body (through the persistent choice of more and more hairless spouses) of an incipient local nakedness like that now existing in certain apes. No zoological facts known to the author afford the slightest basis for this bizarre hypothesis.

St. George Mivart, "Man and Apes," *Eclectic Magazine of Foreign Literature, Science and Art*, new ser., 17 (1873): 698–707, on 702.

Secondary Literature: Mariano Artigas, Thomas F. Glick, and Rafael A. Martínez, *Negotiating Darwin: The Vatican Confronts Evolution, 1877–1902* (Baltimore: Johns Hopkins University Press, 2006).

CECIL JAMES MONRO (1833–1882)
ENGLISH MATHEMATICIAN

I can easily believe, as Darwin would say, that before we were tidal ascidians we were a slimy sheet of cells floating on the surface of the sea. Well, in those days, the missing dimension, and the two forthcoming ones respectively, kept changing with the rotation of the earth—we now know how, but could not guess then. So, now, the missing dimension or dimensions, if any, might be determined by circumstances which we could not tell unless we knew all about the said dimension or dimensions.

Monro to James Clerk Maxwell, 10 September 1871, in *The Life of James Clerk Maxwell,* by Lewis Campbell and William Garnet (London: Macmillan, 1884), 292.

Secondary Literature: Simon Schaffer, "Metrology, Metrication, and Victorian Values," in *Victorian Science in Context,* ed. Bernard V. Lightman (Chicago: University of Chicago Press, 1997), 438–75.

MARIA MONTESSORI (1870–1952)
ITALIAN EDUCATOR

But inasmuch as Quetelet's[†] *homme moyen* was, so to speak, at once a mathematical and philosophical reconstruction of the *non-existent perfect man,* who furthermore could not possibly exist, this classical and masterly study by the great statistician was strenuously combatted and then forgotten, so far as its fundamental concepts were concerned, and remembered only as a scientific absurdity. The thought of that period was too analytical to linger over the great, the supreme synthesis expounded by Quetelet.

Mankind must needs grow weary of anatomizing bodies and tracing back to origins, before returning to an observation of the whole rather than the parts, and to a contemplation of the future. In fact, the thought of the nineteenth century was so imbued with the evolutionary theories as set forth by Charles Darwin, that it believed the reconstruction of the *Pithecanthropus erectus* from a doubtful bone a more positive achievement than that of the *medial man* from the study of millions of living men.

Maria Montessori, *Pedagogical Anthropology*, trans. Frederic Taber Cooper (New York: Frederick A. Stokes, 1913), 455. †Adolphe Quetelet (1796–1874), Belgian statistician and astronomer.

Secondary Literature: Marion O'Donnell, *Maria Montessori* (London: Continuum, 2007).

AUBREY MOORE (1848–1900)
ENGLISH CLERIC AND CHRISTIAN DARWINIAN

Whatever were the steps by which Israel was led to that doctrine of God which constituted its mission and its message to the world, as we look back from the point of view of Christianity we see that the religion of Israel stands to the teaching of Christ in a relation in which no Pagan religion stands. It is strange that Mr. Darwin should have failed to see that this was the answer to his difficulty. It appeared to him, he tells us (*Autobiography*, 308), "utterly incredible that if God were now to make a revelation to the Hindoos, he would permit it to be connected with the belief in Vishnu, Siva, &c., as Christianity is connected with the Old Testament." Incredible, no doubt. But why? For the very reason which makes it "incredible" that man should be evolved directly from a fish, as Anaximander is said to have taught, and not incredible that he should be evolved, as Darwin teaches, from one of the vertebrates. He very idea of development, whether in species or religions, implies a law, and order in the development.

The conception of creation out of nothing was of course unknown to Anaxagoras. Intelligence is only the arranger of materials already given in a chaotic condition. With Aristotle too it is reason which makes everything what it is. But the reason is in things, not outside them. Nature is rational from end to end. In spite of failures and mistakes, due to her materials, nature does the best she can and always aims at a good end. She works like an artist with an ideal in view. Only there is this marked difference—Nature has the principle of growth within herself, while the artist is external to his materials. Here we have a clear and consistent statement of the doctrine of immanent reason as against the Anaxagorean doctrine of a transcendent intelligence. If we translate both into the theological language of our own day, we should call the latter the deistic, the former of science, we might say that we have here, face to face, the mechanical and the organic view of nature. Both were teleological, but to the one, reason was an extra-mundane cause, to the other, an internal principle. It was the contrast between external and inner design, as we know it in Kant and Hegel; between the teleology of Paley and the "wider teleology" of Darwin and Huxley* and [John] Fiske*; between the transcendent and immanent views of God, when so held as to be mutually exclusive.

Science had pushed the deist's God farther and farther away, and at the moment when it seemed as if He would be thrust out altogether, Darwinism appeared, and, under the disguise of a foe, did the work of a friend. It has conferred upon philosophy and religion an inestimable benefit, by showing us that we must choose between two alternatives. Either God is everywhere present in nature, or He is nowhere. He cannot be here and not there. He cannot delegate His power to demigods called "second causes." In nature everything must be His work or nothing.

Aubrey Moore, "The Christian Doctrine of God," in *Lux Mundi: A Series of Studies in the Religion of the Incarnation*, ed. Charles Gore, 10th ed. (London: John Murray, 1890), 71, 93–94, 99.

Secondary Literature: Richard England, "Natural Selection, Teleology, and the Logos," *Osiris* 16 (2001): 270–87.

G. E. MOORE (1873–1958)
ENGLISH PHILOSOPHER

The modern vogue of Evolution is chiefly owing to Darwin's investigations as to the origin of species. Darwin formed a strictly biological hypothesis as to the manner in which certain forms of animal life became established, while others died out and disappeared. His theory was that this might be accounted for, partly at least, in the following way. When certain varieties occurred (the cause of their occurrence is still, in the main, unknown), it might be that some of the points, in which they have varied from their parent species or from other species then existing, made them better able to persist in the environment in which they found themselves less liable to be killed off. They might, for instance, be better able to endure the cold or heat or changes of the climate; better able to find nourishment from what surrounded them; better able to escape from or resist other species which fed upon them; better fitted to attract or master the other sex. Being thus liable to die, their numbers relative to other species would increase; and that very increase in their numbers might tend towards the extinction of those other species. This theory, to which Darwin gave the name Natural Selection, was also called the theory of survival of the fittest. The natural process which it thus described was called evolution. It was very natural to suppose that evolution meant evolution from what was lower into what was higher; in fact it was observed that at least one species, commonly called higher—the species man—had so survived, and among men again it was supposed that the higher races, ourselves for example, had shewn a tendency to survive the lower, such as the North American Indians. We can kill them more easily than they can kill us. The doctrine of evolution was then represented as an explanation of how the higher species survives the lower. Spencer*, for example, constantly uses more evolved as equivalent to higher. But it is to be noted that this forms no part of Darwin's scientific theory. That theory will explain, equally well, how by an alteration in the environment (the gradual cooling of the earth, for example), quite a different species from man, a species which we think infinitely lower, might survive us. The survival of the fittest does not mean, as one might suppose, the survival of what is fittest to fulfill a good purpose—best adapted to a good end: at the last, it means merely the survival of the fittest to survive; and the value of the scientific theory, and it is a theory of great value, just consists in shewing what are the causes which produce certain biological effects. Whether these effects are good or bad, it cannot pretend to judge.

G. E. Moore, *Principia Ethica* (1903; facsimile, New York: Dover Philosophical Classics, 2004), 47–48.

Secondary Literature: Paul Levy, *Moore: G. E. Moore and the Cambridge Apostles* (New York: Holt, Rinehart & Winston, 1980); John O. Nelson, "George Edward Moore," in *Encyclopedia of Philosophy*, ed. Paul Edwards, 8 vols. (New York: Macmillan, 1967), 5:372–81.

ALEXANDER GOODMAN MORE (1830–1895)
English Botanist

Besides his work for [*Annals of Natural History*], [More] began a review of *the* book of the season, Darwin's *Origin of Species;* and some correspondence on Darwinism and the current criticisms of it, passed between him and Mr. [Hewett Cottrell] Watson*; but this article was never finished; indeed the subject, though fascinating, did not suit him; nor after the first six months of 1860 does he seem ever to have contemplated taking part in the great Darwinian controversy.

Life and Letters of Alexander Goodman More with Selections from his Zoological and Botanical Writings, ed. C. B. Moffat (Dublin: Hodges, Figgis, 1898), 125.

At this time [June 1860] he had an interesting correspondence with Mr. Darwin about orchids. Mr. Darwin, in his study of orchid fertilization, encountered from the first a troublesome stumbling-block in the Bee Orchis[†] *(Ophrys apifera).* Of the twelve kinds which grew in the neighborhood of his Kentish home, eleven, in different ways, perfectly fitted his view of the adaptation of their structures to cross-fertilization by insects. But the "Bee," with a structure as beautiful and complicated as any, alone seemed to be visited by no insect, yet to form seed. Unwilling to think that so highly developed a plant could be quite independent of insect agency, Mr. Darwin became very anxious to have specimens examined in other parts of the country where it was more plentiful than at Down, and therefore, he thought, more likely to attract notice from nectar-seeking insects. This desire led him to consult Mr. [Hewett Cottrell] Watson*, by whose advice Mr. More was applied to.

Mr. Darwin's first letter to Mr. More is dated June 24th, 1860:

"Dear Sir: I hope that you will forgive the liberty which I take in writing to you and requesting a favor. Mr. H. C. Watson has given me your address, and has told me that he thought that you would be willing to oblige me. Will you please to read the enclosed; and then you will understand what I *wish observed* with respect to the *bee orchis.* What I especially wish, from information which I have received since publishing the enclosed, is that the state of the pollen-masses should be noted in flowers just beginning to wither, in a district *where the bee orchis is extremely com-*

mon. I have been assured that in part of the Isle of Wight, viz. Freshwater Gate, *numbers* occur, almost crowded together. Whether anything of this kind occurs in your vicinity I know not; but if in your power, I should be infinitely obliged for any information. [. . .] Could you oblige me by taking the great trouble to send me, in an old tin canister, any of the orchids; it would be a great kindness, but perhaps I am unreasonable to make such a request."

The point which Mr. Darwin here thought would take too much explanation was probably his idea that in some of the orchids mentioned the labella might be irritable. He was quite taken by surprise with the actual arrangement of the first species sent him in response to his application. [. . .] "You can hardly imagine (he wrote) what an interesting morning's work you have given me" [. . .] (August 3rd, 1860). The results of that "interesting morning's work" are set forth at page 99 of Mr. Darwin's book on the "Fertilisation of Orchids," where it may be seen how largely the domestic economy of the flower is thought by the author to depend on a little circumstance described for him from growing specimens by Mr. More—the very delicate hinging of the flap (or "distal portion") of the segment of the flower called the labellum. "So flexible and elastic is the hinge (between the two halves of the labellum) that the weight of even a fly, as Mr. More informs me, depresses the distal portion; but when the weight is removed it instantly springs up to its former and ordinary position, and with its curious medial ridge partly closes the entrance into the flower." The use of this mechanism, Mr. Darwin at once concluded, was to enable an insect readily to crawl in, *via* the yielding labellum, and then, since the door behind it instantly closes, to cause the insect to crawl out another way upwards, and thereby detach the pollen-masses for the benefit of the next flower visited.

Life and Letters of Alexander Goodman More with Selections from his Zoological and Botanical Writings, ed. C. B. Moffat (Dublin: Hodges, Figgis, 1898), 153–54. [†]*Orchis* is a genus in the orchid family, Orchidaceae.

Secondary Literature: Donall Symmott, "Botany in Ireland," in *Nature in Ireland,* ed. John Wilson Foster (Montreal: McGill-Queen's University Press, 1998), 157–83.

LEWIS HENRY MORGAN (1818–1881)
AMERICAN ANTHROPOLOGIST

Finally, it will be perceived that the state of society indicated by the consanguine family points with logical directness to an anterior condition of
promiscuous intercourse. There seems to be no escape from this conclusion, although questioned by so eminent a writer as Mr. Darwin. It is not
probable that promiscuity in the primitive period was long continued even
in the horde; because the latter would break up into smaller groups for
subsistence, and fall into consanguine families. The most that can safely
be claimed upon this difficult question is, that the consanguine family was
the first organized form of society, and that it was necessarily an improvement upon the previous unorganized state, whatever that state may have
been. It found mankind at the bottom of the scale, from which, as a starting point, and the lowest known, we may take up the history of human
progress, and trace it through the growth of domestic institutions inventions, and discoveries, from savagery to civilization. By no chain of events
can it be shown more conspicuously than in the growth of the idea of the
family through successive forms. With the existence of the consanguine
family established, of which the proofs adduced seem to be sufficient, the
remaining families are easily demonstrated.

Lewis H. Morgan, *Ancient Society* (1877; reprint, Chicago: Charles H. Kerr, 1910),
427.

Morgan's Library

Morgan's first inventory entry on Darwin is in September, 1863, and it
refers to T. H. Huxley's* popular exposition of Darwin's doctrine; the *Origin* itself appears in the inventory as late as December 1865. [. . .] He now
added to his library Darwin's *Descent of Man;* the works of natural selection's co-discoverer A. R. Wallace* *(The Malay Archipelago*, and *Contributions to the Theory of Natural Selection);* further works of Darwin's most
ardent publicist, T. H. Huxley [. . .] and *On the Genesis of Species* by St.
George [Jackson] Mivart*, a biologist and a Catholic who supported evolutionism but was critical of natural selection.

Thomas R. Trautman and Karl Sanford Kebelac, *The Library of Lewis Henry
Morgan* (Philadelphia: American Philosophical Society, 1994), 50–51.

Secondary Literature: Meyer Fortes, *Kinship and Social Order: The Legacy of Lewis Henry Morgan* (London: Routledge, 2004).

THOMAS HUNT MORGAN (1866–1945)
AMERICAN GENETICIST

Darwin appealed to chance variations as supplying evolution with the material on which natural selection works. If we accept, for the moment, this statement as the cardinal doctrine of natural selection it may appear that evolution is due, (1) not to an orderly response of the organism to its environment, (2) not in the main to the activities of the animal through the use or disuse of its parts, (3) not to any innate principle of living material itself, and (4) above all not to purpose either from within or from without. Darwin made quite clear what he meant by chance. By chance he did not mean that the variations were not causal. On the contrary he taught that in Science we mean by chance only that the particular combination of causes that bring about a variation are not known. They are accidents, it is true, but they are causal accidents. In his famous book on *Animals and Plants under Domestication,* Darwin dwells at great length on the nature of the conditions that bring about variations. If his views seem to us today at times vague, at times problematical, and often without a secure basis, nevertheless we find in every instance, that Darwin was searching for the physical causes of variation. He brought, in consequence, conviction to many minds that there are abundant indications, even if certain proof is lacking, that the causes of variation are to be found in natural processes. Today the belief that evolution takes place by means of natural processes is generally accepted. It does not seem probable that we shall ever again have to renew the old contest between evolution and special creation. But this is not enough. We can never remain satisfied with a negative conclusion of this kind. We must find out what natural causes bring about variations in animals and plants; and we must also find out what kinds of variations are inherited, and how they are inherited. If the circumstantial evidence for organic evolution, furnished by comparative anatomy, embryology and paleontology is cogent, we should be able to observe evolution going on at the present time, i.e. we should be able to observe the occurrence of variations and their transmission. This has actually been done by the geneticist in the study of mutations and Mendelian heredity.

Thomas Hunt Morgan, *A Critique of the Theory of Evolution* (Princeton, NJ: Princeton University Press, 1916), 37–39.

Secondary Literature: Robert E. Kohler, *Lords of the Fly:* Drosophila *Genetics and the Experimental Life* (Chicago: University of Chicago Press, 1994).

EDWARD S. MORSE (1838–1925)
American Zoologist

I gave my first [regular] lecture [in Tokyo] September 12 [1877]. [. . .] it is a delight to teach such good boys, all greedy to learn. Their attention, their courtesy, and their respectful demeanor is an inspiration. Most of them are rationalists and a few may be Buddhists, so with these conditions I anticipate a delightful experience in presenting Darwinism pure and simple.

Saturday, October 6, 1877. I gave my first lecture in a course of three on Evolution to-night in the large college hall. A number of professors and their wives and from five hundred to six hundred students were present, and nearly all of them were taking notes. It was an interesting and inspiring sight. [. . .] The audience seemed to be keenly interested, and it was delightful to explain Darwinian theory without running up against theological prejudice as I often did at home. The moment I finished there was a rousing and nervous clapping of the hands which made my cheeks tingle. One of the Japanese professors told me that it was the first lecture ever given in Japan on Darwinism or evolution. I am looking forward with interest to the other lectures, for I shall have objects to illustrate the points, though the Japanese are quick as a flash to interpret my blackboard drawings.

Edward S. Morse, *Japan Day by Day,* 2 vols. (Boston: Houghton Mifflin, 1917), 1:284, 339–40.

[The *Origin of Species*] effect on zoological literature was striking. The papers were at first tinged with the new doctrine, then saturated, and now, without reference to the theory, derivation is taken for granted. [...] To a naturalist it may seem well-nigh profitless to discuss the question of evolution since the battle has been won; and if there be any discussion, it is as to the relative merits and force of the various factors involved. The public, however, are greatly interested in the matter, as may be seen by a renewal of the fight in the English reviews; and the agitation is still kept up by well-meaning though ignorant advisers, who insist that science has not accepted the doctrine; and great church organizations meet to condemn and expel their teachers of science from certain schools of learning because their teachings are imbued with the heresy.

Edward S. Morse, "What American Zoologists Have Done for the Evolution," *Science* 10 (1887): 73–75, on 73.

Suddenly [Walt Whitman*] asked: "What have you over at the Club tonight, anyway?" "Professor Edward S. Morse is to be there: he is to talk on Evolution." "Evolution? Well, that's a big enough job to give Morse plenty to do: it's like starting at the beginning—at the root or the seed before the root. Is this Morse the Japanese man? Tell them after they have had Morse upon Evolution they should have him for Japan—to describe what he saw there, what Japanese life signifies, to an American, in this nineteenth century."

Horace Traubel, *With Walt Whitman in Camden (November 1, 1888–January 20, 1889)* (New York: Appleton, 1914), 87.

Secondary Literature: Ronald L. Numbers, *Darwinism Comes to America* (Cambridge, MA: Harvard University Press, 1998).

ANGELO MOSSO (1846–1910)
Italian Physiologist

Cicero says that "even the best orators, those who speak with the greatest ease and elegance, feel some apprehension when they are preparing to speak, and are nervous during their exordium." In some men this emotion is so powerful that they never succeed in mastering themselves sufficiently to speak in public. Darwin suffered such discomfort on finding himself the object of anyone's attention, that only very rarely could he make up his mind to take part in any public ceremony.

Angelo Mosso, *Fatigue,* trans. Margaret Drummond and W. B. Drummond (London: Swan Sonnenschein, 1906), 240.

Secondary Literature: Camillo di Giulio, Franca Daniele, and Charles M. Tipton, "Angelo Mosso and Muscular Fatigue," *Advances in Physiological Education* 30 (2006): 51–57.

JAMES BOWLING MOZLEY (1813–1878)
ENGLISH THEOLOGIAN

We have only to do, however, with Mr. Darwin's theory with reference to the special purpose before us. For this purpose we need not say that we do, or do not, adopt the theory of the Transmutation of Species. Let us assume it to be true; it cannot be worked without a principle of design. And first, what is the place which natural selection has in it? Does it do everything? If it does, then the theory is as a theory complete without the principle of design. But if natural selection, according as Mr. Darwin himself defines its functions, does not do everything, but leaves a void and chasm in the theory which must he filled up by some other principle, what is this other principle, when we come to examine it, but design?

We know Mr. Darwin's own account of natural selection; and from this very account it allows that natural selection is not an agent at all, but a result. It is the effect which proceeds from a favorable modification, or development of structure in one animal in the struggle for existence with another animal not thus additionally endowed; viz., his survivorship and continuance on the field while the other perishes. [. . .] But natural selection only weeds, and does not plant; it is the drain of Nature carrying off the irregularities, the monstrosities, the abortions; it comes in after and upon the active developments of Nature to prime and thin them; but it does not create a species; it does not possess one productive or generative function.

James Bowling Mozley, *Essays, Historical and Theological,* 2 vols. (New York: Dutton, 1878), 2:395–97.

Secondary Literature: Owen Chadwick, *The Victorian Church,* 2 vols. (New York: Oxford University Press, 1966–70).

JOHN MUIR (1838–1914)
AMERICAN NATURALIST

The vegetation of the Sierra does not slant up against the cold frosty sky of the summits to end in a sharp colorless edge of lichens. There are ten flowering plants of large size that go above all of the pinched blinking dwarfs which almost justify Darwin's ungodly word "struggle," and burst into bloom of purple and yellow as rich and abundant as ever responded to the thick creamy sun-gold of the plain.

It is my faith that every flower enjoys the air it breathes. Wordsworth, Professors [Moritz] Wagner,[†] French, and Darwin claim that plants have minds, are conscious of their existence, feel pain and have memories.

John of the Mountains: The Unpublished Journals of John Muir, ed. Linnie Marsh Wolfe (Boston: Houghton Mifflin, 1938), 118, 436–37. [†]See John Thomas Gulick entry.

Libby sent me [John] Tyndall's* new book. [. . .] Runkle[†] is going to send me Darwin [*Descent of Man,* pub. February 1871]. These, with my notes and maps, will fill my winter hours, if my eyes do not fail.

Muir to Mrs. Ezra S. Carr, 8 September 1871, in *Life and Letters of John Muir,* ed. William Frederic Badè, 2 vols. (Boston: Houghton Mifflin, 1924), 1:197–98. [†]John David Runkle (1822–1902), American educator and mathematician.

Yours announcing Dr. [Asa] Gray* is received. I have great longing for Gray whom I feel to be a great, progressive, unlimited man like Darwin and Huxley* and [John] Tyndall*. I will be most glad to meet him.

Muir to Mrs. Carr, 14 July 1872, in *Life and Letters of John Muir,* ed. William Frederic Badè, 2 vols. (Boston: Houghton Mifflin, 1924), 1:335.

Secondary Literature: Donald Worcester, *A Passion for Nature: The Life of John Muir* (New York: Oxford University Press, 2008).

FRIEDRICH MAX MÜLLER (1823–1900)
German Philologist and Orientalist

We have lately been very much occupied, partly with the Oriental Congress, partly by visits to us and visits we have paid. Happily I had finished my Veda, and have only a little to do to the Preface, Index, &c., which must be finished this week, and then I shall not grudge myself a little freedom. We had Lepsius, father and son,[†] on a visit. Then we went to stay with Mr. [Mountstuart E.] Grant Duff*, formerly Under-Secretary for India, and with Sir John Lubbock*, where we met Darwin, [John] Tyndall*, Spencer*, and the two Lepsius. Lectures will soon begin here, and though I have been here through the Vacation and worked hard, it has suited me very well. Perhaps I shall take a holiday in winter, if I need it, and a visit to Rome has long been hoped for by us both. My principal work is finished, and I can at last give myself a treat.

Müller to his mother, 9 October 1874, in *The Life and Letters of the Right Honourable Friedrich Max Müller,* ed. Georgina A. Müller, 2 vols. (London: Longmans, Green, 1902), 1:494. [†]Karl Richard Lepsius (1810–84), German Egyptologist, and his son Richard (1851–1915), geologist.

This visit was the only time Max Müller and Darwin met. The conversation turning on apes as the progenitors of man, Max Müller asserted that if speech were left out of consideration, there was a fatal flaw in the line of facts. "You are a dangerous man," said Darwin, laughingly.

Georgina Müller, in *The Life and Letters of the Right Honourable Friedrich Max Müller,* ed. Georgina A. Muller, 2 vols. (London: Longmans, Green, 1902), 1: 494–95.

In religion, as in language and other intellectual manifestations, what is really important is the germ, not the fruit. "A suspicion of something beyond what is seen," springing naturally from a healthy mind, would be far more important in the early ages of mankind than a ready-made catechism. Man has to gain not only his daily bread, but what is far more important, his thoughts, his words, his faith, in the sweat of his brow. In that sense, I am a thorough Darwinian. Where I differ from Darwin, is when he does not see that nothing can become actual but what was potential, that mere environment explains nothing, because what surrounds and determines is as much given as what is surrounded and determined, that both presuppose each other and are meant for each other. Now I take my stand against Darwin on language, because language is the necessary condition of every other mental activity, religion not excluded, and I am able to prove that this indispensable condition of all mental growth is entirely absent in animals. This is my palpable argument. [. . .]

I have no feeling for or against Darwinism, and I always try in approaching these problems to care for nothing that I may care for in my heart. I am certain that we are led; I am certain we ought to follow; I am certain that, even if we go wrong, as long as we do it because we will not resist the power of facts and arguments, we are right. If Darwin's facts were irresistible, I should accept the ape-theory without a murmur, because I should feel that we were meant to accept it. But I feel with you that never was a theory of such importance put forward with a smaller array of powerful arguments than by Darwin. "What is, is best": these were [Charles] Kingsley's* last words,† used no doubt in a purely ethical sense, but applicable nevertheless to all pursuit of truth.

Müller to the Duke of Argyll, 4 February 1875, in *The Life and Letters of the Right Honourable Friedrich Max Müller*, ed. Georgina A. Müller, 2 vols. (London: Longmans, Green, 1902), 1:508–9. †Kingsley had died on 23 January.

Secondary Literature: Lourens van den Bosch, *Friedrich Max Müller: A Life Devoted to Humanities* (Boston: Brill, 2002).

FRITZ MÜLLER (1821–1897)
German Zoologist

When I had read Charles Darwin's book *On the Origin of Species*, it seemed to me that there was one mode, and that perhaps the most certain, of testing the correctness of the views developed in it, namely, to some particular group of animals. Such an attempt to establish a genealogical tree, whether for the families of a class, the genera of a large family, or for the species of an extensive genus, and to produce pictures as complete and intelligible as possible of the common ancestors of the various smaller and larger circles, might furnish a result in three different ways.

1. In the first place, Darwin's suppositions when thus applied might lead to irreconcilable and contradictory conclusions, from which the erroneousness of the suppositions might be inferred. If Darwin's opinions are false, it was to be expected that contradictions would accompany their detailed application at every step, and that these, by their cumulative force, would entirely destroy the suppositions from which they proceeded, even though the deductions derived from each particular case might possess little of the unconditional nature of mathematical proof.

2. Secondly, the attempt might be successful to a greater or less extent. If it was possible upon the foundation and with the aid of the Darwinian theory, to show in what sequence the various smaller and larger circles had separated from the common fundamental form and from each other, in what sequence they had acquired the peculiarities which now characterize them, and what transformations they had undergone in the lapse of ages—if the establishment of such a genealogical tree, of a primitive history of the group under consideration, free from internal contradictions, was possible—then this conception, the more completely it took up all the species within itself, and the more deeply it enabled us to descend into the details of their structure, must in the same proportion bear in itself the warrant of its truth, and the more convincingly prove that the foundation upon which it is built is no loose sand, and that it is more than merely "an intellectual dream."

3. In the third place, however, it was possible, and this could not but appear, *prima facie,* the most probable case, that the attempt might be frustrated by the difficulties standing in its way, without settling the question, either way, in a perfectly satisfactory manner. But if it were only possible in this way to arrive for oneself at a moderately certain independent judgment upon a matter affecting the highest questions so deeply, even this alone could not but be esteemed a great gain. [. . .]

Among the parasitic Crustacea, especially, everybody has long been accustomed to speak, in a manner scarcely admitting of a figurative meaning, of their arrest of development by parasitism, as if the transformation of species were a matter of course. It would certainly never appear to any one to be a pastime worthy of the Deity, to amuse himself with the contrivance of these marvelous cripplings, and so they were supposed to have fallen by their own fault, like Adam, from their previous state of perfection.

A false supposition, when the consequences proceeding from it are followed further and further, will sooner or later lead to absurdities and palpable contradictions. During the period of tormenting doubt—and this was by no means a short one—when the pointer of the scales oscillated before me in perfect uncertainty between the pro and the con, and when any fact leading to a quick decision would have been most welcome to me, I took no small pains to detect some such contradictions among the inferences as to the class of Crustacean furnished by the Darwinian theory. But I found none, either then, or subsequently. Those which I thought I had found were dispelled on closer consideration, or actually became converted into supports for Darwin's theory. Nor, so far as I am aware, have any of the necessary consequences of Darwin's hypotheses been proved by anyone else, to stand in clear and irreconcilable contradiction. And yet, as the most profound students of the animal kingdom are amongst Darwin's opponents, it would seem that it ought to have been an easy matter for them to crush him long since beneath a mass of absurd and contradictory inferences, if any such were to be drawn from his theory. To this want of demonstrated contradictions I think we may ascribe just the same importance in Darwin's favor, that his opponents have attributed to the absence of demonstrated intermediate forms between the species of the various strata of the earth. Independently of the reasons which Darwin gives for the preservation of such intermediate forms being only exceptional, this last mentioned circumstance will not be regarded as of very great significance by any one who has traced the development of an animal upon larvae fished from the sea, and had to seek in vain for months, and even years, for those transitional forms, which he nevertheless knew to be swarming around him in thousands.

Fritz Müller, *Facts and Arguments for Darwin,* trans. W. S. Dallas (London: John Murray, 1869), 1–4, 7–8.

Secondary Literature: David A. West, *Fritz Müller: A Naturalist in Brazil* (Blacksburg, WV: Pocahontas, 2003).

HERMANN MÜLLER (1829–1883)
German Botanist

As a foundation for the hypothetic natural law [the Knight-Darwin law] that "no organic being fertilizes itself for a perpetuity of generations, but that a cross with another individual is occasionally—perhaps at very long intervals—indispensable," Darwin showed that in all higher and the great majority of lower animals the sexes are separate, and that most hermaphrodite forms pair regularly; that, in the experience of breeders of animals and cultivators of plants, breeding in-and-in diminishes the strength and the productiveness of the offspring, while crossing with another breed, or with another stock of the same breed, increases both [. . .] finally, that in no living organism do the structure or situation of the reproductive organs prevent occasional crossing with another individual of the same species. These statements, taken separately, were neither decisive nor free from objection, but collectively they lent a high degree of probability to Darwin's hypothesis; and so, from its close connection with the question of the origin of species and the fundamental importance that it therefore had for all botanical research, botanists could not help at once taking part for or against it, according to whether they were impelled by the general weight of evidence or deterred by the gaps in the chain.

The most complete collection of all the known facts which contribute to prove Knight's[†] law is given by Darwin in his work on the *Variation of Animals and Plants under Domestication,* in which he suggests many new and fruitful lines of research; but the three methods of investigation which Darwin originally used have been the chief aids in investigating the determining conditions of the forms of flowers. Numerous observers, among whom Friedrich Hildebrand, Federico Delpino, my brother Fritz Müller[*], and Severin Axell[‡] deserve special mention, have pushed forward along these new paths that Darwin opened; they have not only brought to light a mass of new facts, all tending to elucidate floral mechanisms on the basis of the Knight-Darwin law, but they have also disclosed many new general principles.

Hermann Müller, *The Fertilisation of Flowers,* trans. D'Arcy W. Thompson (London: Macmillan, 1883), 5, 11. [†]Thomas Andrew Knight (1759–1838), English horticulturist. Knight's law is that "organic beings shall not fertilize themselves for perpetuity," as Darwin expressed it in *The Variation of Plants and Animals under Domestication,* 2 vols. (1868; reprint, New York: Appleton, 1900), 2:154. ‡Friedrich Hildebrand (1833–1915), professor of botany at Freiberg; Federico Delpino (1833–1905), professor of botany at Genoa; and Severin Axell (1843–92), Swedish botanist known for studies of flower pollination.

Secondary Literature: Charles Darwin, "Prefatory Notice" in *The Fertilisation of Flowers,* by Hermann Müller, trans. D'Arcy W. Thompson (London: Macmillan, 1883).

LEWIS MUMFORD (1895–1990)
AMERICAN HISTORIAN

In his essay on population the Reverend T. R. Malthus shrewdly generalized the actual state of England in the midst of the disorders that attended the new industry. He stated that population tended to expand more rapidly than the food supply, and that it avoided starvation only through a limitation by means of the positive check of continence, or the negative checks of misery, disease, and war. In the course of the struggle for food, the upper classes, with their thrift and foresight and superior mentality emerged from the ruck of mankind. With this image in mind, and with Malthus's *Essay on Population* as the definite stimulus to their thought, two British biologists, Charles Darwin and Alfred Wallace*, projected the intense struggle for the market upon the world of life in general. Another philosopher of industrialism, just as characteristically a railroad engineer by profession as Spinoza had been a lens grinder, coined a phase that touched off the whole process: to the struggle for existence and the process of natural selection Spencer* appended the results: "the survival of the fittest." The phrase itself was a tautology; for survival was taken as the proof of fitness: but that did not decrease its usefulness.

This new ideology arose out of the new social order, not out of Darwin's able biological work. His scientific study of modifications, variations, and the processes of sexual selection were neither furthered nor explained by a theory which accounted not for the occurrence of new organic adaptations, but merely for a possible mechanism whereby certain forms had been weeded out after the survivors had been favorably modified. Moreover, there were the demonstrable facts of commensalism and symbiosis, to say nothing of ecological partnership, of which Darwin himself was fully conscious, to modify the Victorian nightmare of a nature red in tooth and claw.

The point is, however, that in paleotechnic society the weaker were indeed driven to the wall and mutual aid had almost disappeared. The Malthus-Darwin doctrine explained the dominance of the new bourgeoisie, people without taste, imagination, intellect, moral scruples, general culture or even elementary bowels of compassion, who rose to the surface precisely because they fitted an environment that had place and no use for any of these human attributes. Only anti-social qualities had survival value. Only people who valued machines more than men were capable under these conditions of governing men to their own profit and advantage.

Lewis Mumford, *Technics and Civilization* (New York: Harcourt, Brace & World, 1963), 186–87.

Secondary Literature: Lewis Mumford and Patrick Geddes: The Correspondence, ed. Frank Novak Jr. (New York: Routledge, 1995); Christopher MacGregor Scribner, "Lewis Mumford," in *Encyclopedia of Historians and Historical Writing,* ed. Kelly Boyd, 2 vols. (London: Fitzroy Dearborn, 1999), 2:844.

RODERICK IMPEY MURCHISON (1792–1871)
Scottish Geologist

[1 December 1859]
We have been much startled by the apparition of Darwin's book on the *Origin of Species.* Huxley* is quite a believer in this ultra-Lamarckian theory of *Natural Selection* & the *Struggle for Existence,* by which one race has passed according to general laws, into another; & so we all descend form a Monad up thence as from inferior types. *I am decidedly hostile to the whole thing,* & am as firmly a believer as ever in *Successive Creations.* By the bye that view of Creation is not in Darwin's book & I do not therefore think it will suit *Your book.*

Murchison to George Gordon, 1 December 1859, quoted in Michael Collie, *Murchison in Moray: A Geologist on Home Ground* (Philadelphia: American Philosophical Society, 1995), 184.

Secondary Literature: David Roger Oldroyd, *The Highlands Controversy: Constructing Geological Knowledge in Fieldwork in Nineteenth-Century Britain* (Chicago: University of Chicago Press, 1990); M. J. S. Rudwick, "Roderick Impey Murchison," in *Dictionary of Scientific Biography,* ed. Charles Coulston Gillispie, 16 vols. (New York: Scribner's, 1970–80), 9:582–85.

ROBERT MUSIL (1880–1942)
Austrian Novelist

What doesn't quite fit into the picture is the a-Catholicism, indeed the a-religiosity of the family. Perhaps a kind of theism reigned, a kind of adherence to goodness and order; perhaps too a shyness about mentioning God too often; perhaps, too, a kind of treaty between the more gentle mother and the father who was influenced by Darwin.

Robert Musil, *Diaries, 1899–1941,* trans. Philip Payne (New York: Basic Books, 1999), 476.

Secondary Literature: Hannah Hickman, *Robert Musil and the Culture of Vienna* (London: Croom Helm, 1984).

N

JAWAHARLAL NEHRU (1889–1964)
Indian Statesman

There was a great argument and conflict in England and elsewhere in Europe between science and religion. There could be no doubt of the result. The new world of industry and mechanical transport depended on science, and science thus could not be discarded. Science won all along the line and "natural selection" and "survival of the fittest" became part of the ordinary jargon of the people, who used the phrases without fully understanding what they meant. Darwin had suggested in his *Descent of Man* that there might have been a common ancestor of man and certain apes. This could not be proved by examples showing various stages in the process of development. From this there grew the popular joke about the "missing link." And, curiously enough, the ruling classes twisted Darwin's theory to suit their own convenience, and were firmly convinced that it supplied yet another proof of their superiority. They were the fittest to survive in the battle of life, and so by "natural selection" they had come out on top and were the ruling class. This became the justification for one class dominating over another, or one race ruling over another. It became the final argument of imperialism and the supremacy of the white race. And many people in the West thought that the more domineering they were, the more ruthless and strong, the higher up in the scale of human values they were likely to be. It is not a pleasant philosophy, but it explains to some extent the conduct of western imperialist powers in Asia and Africa.

Nehru to his daughter, Indira, 3 February 1933, in *Glimpses of World History, Being Further Letters to His Daughter Written in Prison, and Containing a Rambling Account of History for Young People*, by Jawaharlal Nehru (New York: John Day, 1942), 525–26.

Secondary Literature: Judith Brown, *Nehru: A Political Life* (New Haven, CT: Yale University Press, 2003).

LADY DOROTHY NEVILL (1826–1913)
ENGLISH HORTICULTURALIST AND HOSTESS

[ca. 1875]

Orchids and insectivorous plants were her especial hobby, and so she got in touch with the author of the *Origin of Species*. Writing to a friend at the time she said, "Mr. Darwin has expressed a wish to see me—I dare hardly hope for such happiness." She was able to furnish the great naturalist with many specimens, which she liked to think were of use to him in his wonderful researches.

"I am sending" (she wrote to Lady Airle)† "curious plants to experimentalize upon to Mr. Darwin. I am so pleased to help in any way the labours of such a man—it is quite an excitement for me in my quiet life, my intercourse with him—he promises to pay me a visit when in London. I am sure he will find I am the missing link between man and apes."

A great friend of hers wrote: "Keep me some hour when you come to London and let me see you and talk to you, and tell me about Darwin and the plants, and, if you can, do let me come some day and see them. It is long since I have seen your wonders, and I would like to come again. How far does Darwinianism enter into you—does it disturb your old beliefs or not? I think the mind of the real Naturalist is sometimes so bent upon each fact and each discovery as never to generalize, and so they manage to keep the two things separate in their mind."

Darwin himself used to tell a story of a pious professor who, ever seeking to reconcile biology with the Bible, accounted for the extinction of the mastodon by saying that the door of the ark had been made too small to admit it!

The Life and Letters of Lady Dorothy Nevill, ed. Ralph Neville (London: Methuen, 1919), 56, 58. †Mabel Ogilvy, Countess of Airlie (1866–1956), English courtier and author.

Darwin was a man of the utmost simplicity of life, and his household was a very haven of tranquility. On one occasion, when there was a question of my paying the Darwins a visit of some days, Mrs. Darwin wrote to me, saying that she understood that those who moved much in London society were accustomed to find their country-house visits enlivened by all sorts of sports and practical jokes—she had read that tossing people in blankets had become highly popular as a diversion. "I am afraid," her letter ended, "we should hardly be able to offer you anything of that sort."

I did pay Darwin a visit at Down, but as ill-luck would have it he was just at this time suffering from a violent attack of the malady—for it amounted to that—which he had contracted during his voyage on the *Beagle,* when he had become a martyr to sea-sickness, which never afterwards entirely left him, and throughout his tireless life of investigation intermittently rendered his existence a burden.

Lady Dorothy Nevill, *Under Five Reigns* (New York: John Lane, 1910), 106–7.

Secondary Literature: Philip J. Waller, *Writers, Readers, and Reputations: Literary Life in Britain, 1870–1918* (Oxford: Oxford University Press, 2006).

SIMON NEWCOMB (1835–1909)
CANADIAN-AMERICAN ASTRONOMER AND MATHEMATICIAN

21 February 1860: Prof. [Benjamin] Peirce* got me permission to attend the special meetings of the [American Academy of Arts and Sciences].

22 February: Went into Boston, and heard a discussion at the A.A.S. on the origin of plants and animals. Walked out with Bartlett[†] and Prof. [Francis] Bowen*, whom I attacked on the subject of fatalism.

27 March: Attended the Academy. Darwin's *Origin of Species* was the only subject discussed.

1 May: Attended a special meeting of the Academy called for the purpose of finishing up Darwinism discussion, which I hope was done.

Simon Newcomb, unpublished diary entries quoted in Albert Moyer, *A Scientist's Voice in American Culture: Simon Newcomb and the Rhetoric of Scientific Method* (Berkeley and Los Angeles: University of California Press, 1992), 39. [†]William Francis Bartlett (1840–76), future American general, who attended this lecture with his Harvard classmate Simon Newcomb.

Darwin, Huxley*, Spencer*, [William Kingdon] Clifford*, [John] Tyndall*, and Haeckel* came one by one into note, and I read the discussions of their views with gradually increasing interest, though my occupations left me no time for a careful reading of their books. [...] So much discussion of these men [in religious journals] gradually led to a desire to know what they had to say for themselves, and I began now and then to glance at their shorter essays, and to dip into their books. [...] What I have read of Darwin, Huxley, Tyndall, and Spencer has been in a critical rather than a sympathetic spirit, and I have always had a repugnance to wholesale attacks on Christianity.

Excerpts from two different drafts of Newcomb's unpublished "religious autobiography," quoted in Albert Moyer, *A Scientist's Voice in American Culture: Simon Newcomb and the Rhetoric of Scientific Method* (Berkeley and Los Angeles: University of California Press, 1992), 40.

Secondary Literature: Albert Moyer, *A Scientist's Voice in American Culture: Simon Newcomb and the Rhetoric of Scientific Method* (Berkeley and Los Angeles: University of California Press, 1992).

JOHN HENRY NEWMAN (1840–1902)
English Prelate

[1863] It is strange that monkeys should be so like men with no historical connection between them. [...] I will either go whole hog with Darwin or dispensing with time & history altogether, hold, not only the theory of distinct species but also that of the creation of fossil bearing rocks.

John H. Newman, *The Philosophical Notebook,* ed. Edward J. Sillem, 2 vols. (Louvain, Belgium: Nauwelaerts, 1970), 158.

[1870] I have not fallen in with Darwin's book [*Origin of Species*]. I conceive it to be an advocacy of the theory that that principle of propagation, which we are accustomed to believe began with Adam, and with the patriarchs of the brute species, began in some one common ancestor millions of years before.

1. Is this against the distinct teaching of the inspired text? if it is, then he advocates an antichristian theory. For myself, speaking under correction, I don't see that it does contradict it.

2. Is it against Theism (putting Revelation aside)—I don't see how it

can be. Else, the fact of a propagation from Adam is against Theism. If second causes are conceivable at all, an Almighty Agent was supposed, I don't see why the series should not last for millions of years as well as for thousands.

The former question is the more critical. Does Scripture contradict the theory? Was Adam *not* immediately taken from the dust of the earth? "All are of dust"—Eccles iii, 20—yet *we* never *were* dust—we are from fathers. Why may not the same be the case with Adam? I don't say that it *is* so—but, if the sun does not go round the earth and the earth stand still, as Scripture seems to say, I don't know why Adam needs be immediately out of dust—*Formavit Deus hominem de limo terrae*—i.e. out of what really was dust and mud in its nature, before He made it what it was, living. But I speak under Correction. Darwin does not *profess* to oppose Religion. I think he deserves a degree as much as many others, who have had one.

Newman to E. B. Pusey (about the prospective award to Darwin of an honorary degree at Oxford), in *The Letters and Diaries of John Henry Newman*, vol. 25 (Oxford: Clarendon, 1973), 137–38.

Secondary Literature: James Collins, "John Henry Newman," in *Encyclopedia of Philosophy*, ed. Paul Edwards, 8 vols. (New York: Macmillan, 1967), 5:480–85; Owen Chadwick, *The Victorian Church*, 2 vols. (New York: Oxford University Press, 1966–70).

ALFRED NEWTON (1829–1907)
English Zoologist

Not many days after my return home there reached me the part of the *Journal of the Linnean Society* which bears on its cover the date 20th August, 1858, and contains the papers by Mr. Darwin and Mr. Wallace*, which were communicated to that Society at its special meeting of the first of July preceding, by Sir Charles Lyell*, and Dr. (now Sir Joseph) Hooker*. I think I had been away from home the day this publication arrived, and I found it when I came back in the evening. At all events, I know that I sat up late that night to read it; and never shall I forget the impression it made upon me. Herein was contained a perfectly simple solution of all the difficulties which had been troubling me for months past. I hardly know whether I at first felt more vexed at the solution not having occurred to me than pleased that it had been found at all. However, after reading these papers more than once, I went to bed satisfied that a solution had been found. All personal feeling apart, it came to me like the direct revelation of a higher power; and I awoke next morning with the consciousness that there was an end of all the mystery in the simple phrase, "Natural Selection." I am free to confess that in my joy I did not then perceive, and I cannot say when I did begin to perceive, that though my especial puzzles were thus explained, dozens, scores, nay, hundreds of other difficulties lay in the path, which would require an amount of knowledge, to be derived from experiment, observation, and close reasoning, of which I could form no notion, before this key to the "mystery of mysteries" could be said to be perfected; but I was convinced a *vera causa* had been found, and that by its aid one of the greatest secrets of creation was going to be unlocked. I lost no time in drawing the attention of some of my friends, with whom I happened to be at the time in correspondence, to the discovery of Mr. Darwin and Mr. Wallace; and I must acknowledge that I was somewhat disappointed to find that they did not so readily as I had hoped approve of the new theory. In some quarters I failed to attract notice; in others my efforts received only a qualified approval. But I am sure I was not discouraged in consequence; and I never doubted for one moment, then nor since, that we had one of the grandest discoveries of the age—a discovery all the more grand because it was so simple.

Alfred Newton, "The Early Days of Darwinism," *Macmillan's Magazine,* February 1888, 112–14, reprinted in *Life of Alfred Newton,* by A. F. R. Wollaston (New York: Dutton, 1921), 112–13.

In the Nat. Hist. Section we had another hot Darwinian debate. Mr. F. O. Morris[†] had a paper on the list to be read "On the Permanence of Species," but in the committee we decided it should not be produced (he was not there himself), Babington[‡] treating us to some selections from it and remarking that it would, of course, appear in due course of time in the new series of the *Naturalist.*

The ball was opened by a paper containing diluted Owenism by Dr. Collingwood,[§] followed by a long undiluted atheistical rigmarole by a

Prof. Draper,[††] a Yankee. After this a hot discussion took place. Huxley[*] was called upon by [John Stevens] Henslow[*] to state his views at greater length, and this brought up the Bp. [Bishop] of Oxford [Samuel Wilberforce[*]], who made of course, a wonderfully good speech if the facts had been correct. Referring to what Huxley had said two days before, about after all its not signifying to him whether he was descended from a Gorilla or not, the Bp. chaffed him and asked whether he had a preference for the descent being on the father's or the mother's side? This gave Huxley the opportunity of saying that he would sooner claim kindred with an Ape than with a man like the Bp. who made so ill an use of his wonderful speaking powers to try and burke, by a display of authority, a free discussion on what was, or what was not, a matter of truth, and reminded him that on questions of physical science "authority" had always been bowled out by investigation, as witness astronomy and geology.

He then caught hold of the Bp.'s assertions and showed how contrary they were to facts, and how he knew nothing about what he had been discoursing on. A lot of other people afterwards spoke; Broody [i.e., Benjamin Collins Brodie][‡‡] on the medical view of the thing, which he did very temperately, declaring that at present it was impossible to say what was the truth; Lubbock, a son of Sir John's[*], who is a very clever young fellow, who took a decided Darwinian view, and Admiral FitzRoy, the man who commanded the *Beagle,* and who had better have let it alone.

The feeling of the audience was much against the Bp., and [Richard] Simpson[*], who had been very anti-Darwin, declared that if that was all that could be said in favor of the old idea, he was a convert. Not so [Henry Baker] Tristram[*], who waxed exceedingly wrath as the discussion went on, and declared himself more and more anti-Darwinian. The discussion was adjourned until the Monday, but it was then thought by the leaders on both sides that it had better be dropped, and so the matter rests.

On the Sunday, at the University Church, Temple,[§§] the Master of Rugby, treated his audience to a sermon on Darwinism, in which he espoused Darwin's ideas fully! Nothing very particular occurred during the last few days, and I did a good deal of lionizing. Oxford is no doubt finer than Cambridge, but not to the extent that her sons make out.

From a letter from Newton to his brother, 25 July 1860, quoted in *Life of Alfred Newton,* by A. F. R. Wollaston (New York: Dutton, 1921), 118–20. [†]Francis Orpen

Morris (1810–93), Irish ornithologist and entomologist. ‡Churchill Babington (1821–89), English classicist, archaeologist, and naturalist (conchology expert). §Cuthbert Collingwood (1826–1908), physician naturalist, lecturer on botany and biology in Liverpool. "Owenism" refers to the utopian socialism of Robert Owen. ††See John Richard Green entry. ‡‡See Thomas H. Huxley entry. §§Frederick Temple (1821–1902), headmaster of Rugby School, later archbishop of Canterbury.

The history of opinion on evolution and natural selection, in the years which followed the publication of the *Origin,* can be traced in the titles of the papers and subjects of discussion at successive meetings of the British Association. In the Presidential Address delivered by Professor Newton to the Biological Section of the Manchester meeting in 1887, there is a most interesting account of the struggles which took place:

"The ever-memorable meeting [. . .] at Oxford in the summer of 1860 saw the first open conflict between the professors of the new faith and the adherents of the old one. Far be it from me to blame those among the latter who honestly stuck to the creed in which they had educated themselves; but my admiration is for the few dauntless men who, without flinching from the unpopularity of their cause, flung themselves in the way of obloquy, and impetuously assaulted the ancient citadel in which the sanctity of 'species' was enshrined and worshipped as a palladium. However strongly I myself sympathized with them, I cannot fairly state that the conflict on this occasion was otherwise than a drawn battle; and thus matters stood when in the following year the Association met in this city [Manchester]. That, as I have already said, was a time of 'slack water.' But though the ancient beliefs were not much troubled, it was for the last time that they could be said to prevail; and thus I look upon our meeting in Manchester 1861 as a crisis in the history of biology. All the same, the ancient beliefs were not allowed to pass wholly unchallenged; and one thing is especially to be marked—they were challenged by one who was no naturalist at all, by one who was a severe thinker no less than an active worker; one who was generally right in his logic, and never wrong in his instinct; one who, though a politician, was invariably an honest man—I mean the late Professor Fawcett.† On this occasion he brought the clearness of his mental vision to bear upon Mr. Darwin's theory, with the result that Mr. Darwin's method of investigation was shewn to be strictly in accordance with the rules of deductive philosophy, and to throw light where all was dark before."

Quoted in *Charles Darwin and the Theory of Natural Selection,* by Edward B. Poulton (London: Cassell, 1896), 154. †Henry Fawcett (1833–84), English statesman and economist.

Secondary Literature: Frank James, "An 'Open Clash between Science and the Church'? Wilberforce, Huxley and Hooker on Darwin at the British Association, Oxford, 1980," in *Science and Beliefs: From Natural Philosophy to Natural Science, 1700–1900,* ed. David M. Night and Matthew D. Eddy (Aldershot: Ashgate, 2005), 171–93.

FRIEDRICH NIETZSCHE (1844–1900)
German Philosopher

There are truths which are best recognized by mediocre minds, because they are best adapted for them, there are truths which only possess charms and seductive power for mediocre spirits—one is pushed to this probably unpleasant conclusion, now that the influence of respectable but mediocre Englishmen—I may mention Darwin, John Stuart Mill*, and Herbert Spencer*—begins to gain the ascendancy in the middle-class region of European taste. Indeed, who could doubt that it is a useful thing for *such* minds to have the ascendancy for a time? It would be an error to consider the highly developed and independently soaring minds as specially qualified for determining and collecting many little common facts, and deducing conclusions from them; as exceptions, they are rather from the first in no very favorable position towards those who are "the rules." After all, they have more to do than merely to perceive—in effect, they have to *be* something new, they have to *signify* something new, they have to *represent* new values! The gulf between knowledge and capacity is perhaps greater, and also more mysterious, than one thinks: the capable man in the grand style, the creator, will possibly have to be an ignorant person—while on the other hand, for scientific discoveries like those of Darwin, a certain narrowness, aridity, and industrious carefulness (in short, something English) may not be unfavorable for arriving at them.—Finally, let it not be forgotten that the English, with their profound mediocrity, brought about once before a general depression of European intelligence.

Friedrich Wilhelm Nietzsche, *Beyond Good and Evil: Prelude to a Philosophy of the Future,* trans. Helen Zimmern, vol. 12 of *Complete Works of Friedrich Nietzsche,* ed. Alexander Tille (New York: Macmillan, 1911), 212–13.

Anti-Darwin.—As regards the celebrated "struggle for life," it seems to me, in the meantime, to be more asserted than proved. It occurs, but only as an exception; the general aspect of life is not a state of want or hunger; it is rather a state of opulence, luxuriance, and even absurd prodigality— where there is a struggle, it is a struggle for power.—We must not confound Malthus with nature. Granted, however, that this struggle exists— and in fact it does occur—its results, alas, are the reverse of what the Darwinian school wish, the reverse of what one might perhaps wish, in accordance with them: it is prejudicial to the strong, the privileged, the fortunate exceptions. The species does not grow in perfection: the weak again and again get the upper hand of the strong,—their large number, and their greater cunning are the cause of it. Darwin forgot the intellect (that was English!); the weak have more intellect. [...] One must need intellect in order to acquire it; one loses it when it is no longer necessary. He who has strength rids himself of intellect ("let it go hence!" is what people think in Germany at present, "the Empire will remain" [...]).

Friedrich Nietzsche, *The Twilight of the Idols,* trans. Thomas Common, vol. 11 of *The Works of Friedrich Nietzsche,* ed. Alexander Tille (New York: Macmillan, 1896), 173–74.

Secondary Literature: John Richardson, *Nietzsche's New Darwinism* (Oxford: Oxford University Press, 2004); Walter Kaufmann, "Friedrich Nietzsche," in *Encyclopedia of Philosophy,* ed. Paul Edwards, 8 vols. (New York: Macmillan, 1967), 5:504–14.

MAX NORDAU (1849–1923)
Austrian Social Theorist

Each individual is [...] the result of the operation of these two influences, the primeval law of life and heredity. The former seeks to create fresh forms, adapted to the performance of the business of life, while the latter seeks to repeat a plan already in existence, that, namely, of its parents. I cannot sufficiently impress this truth upon you, namely, that from my point of view an unrestricted liberty of choice among all available forms is the original feature, and that the feature of similarity to parents which has encroached upon that liberty only became developed at a later period; since it is this hypothesis alone which makes the entire Darwinian theory comprehensible, and the neglect of which makes that theory not an explanation, but only a verification of facts within one's observation.

In fact, if, as Darwin and with him the entire troop of his followers and interpreters think, heredity is to be considered the original and more important of the laws which decided the development of the individual, how then would any deviation therefrom or any improvement thereon be conceivable? The product would have under all circumstances to retain a resemblance to the producer, and if its external relations made this impossible for it, it would simply have to fall to the ground. The great phenomenon of adaptation to one's allotted conditions of life, which, according to Darwin, is one of the chief causes of the origin of species, would continue to be a perfectly irresolvable problem. My theory, on the contrary, offers a solution of this problem. The living creature, I repeat, is not limited to any one form as will render the absorption of oxygen and the formation of protoplasm possible to it; it is just this individual absolute liberty which enables it to assume the form which is impressed upon it by its external relationships, just as a floating body at rest will be driven towards that one of all the directions possible to it, to which even the very slightest external impulse urges it. Is it the parent organism which gives it its particular form? Very well, then, the young organism will also then assume the form of the parents. Do the external conditions among which it has to live seek to transform it, to make it dissimilar to its parents? Very well, then, it will resign its inherited form, and, yielding to the new impulse, assume that which the external conditions of life endeavor to force upon it. In this way we can explain adaptation to circumstances, which, if this theory is correct, becomes no longer a contradiction of but an analogy to the law of heredity.

Biology, the science of life, only recognizes the individual, not the species. The former alone is something actually existent, independent, clearly defined, while the latter is much less distinct, nay, is frequently quite incapable of being defined with certainty.

Max Nordau, *Paradoxes* (1885; reprint, London: William Heinemann, 1896), 37–38.

It can be shown by figures that science does not lose, but continually gains ground. But the million does not care about exact statistics. In France it accepts without resistance the suggestion, that science is retreating before religion, from a few newspapers, written mainly for clubmen and gilded courtesans, into the columns of which the pupils of the clerical schools have found an entrance. Of science itself, of its hypotheses, methods, and results, they have never known anything. Science was at one time the fashion. The daily press of that date said, "We live in a scientific age"; the news of the day reported the travels and marriages of scientists; the feuilleton-novels contained witty allusions to Darwin; the inventors of elegant walking-sticks and perfumes called their productions "Evolution Essence" or "Selection Canes"; those who affected culture took themselves seriously for the pioneers of progress and enlightenment.

Max Nordau, *Degeneration* (New York: Appleton, 1895), 114–15.

Secondary Literature: Daniel Pick, *Faces of Degeneration: A European Disorder* (Cambridge: Cambridge University Press, 1993).

MARIANNE NORTH (1830–1890)
English Naturalist

One bright summer day [in 1882] I went down to Bromley Common to see my father's dear old friend, George Norman, then in his ninety-fourth year. [...] They drove me on to Down, the dear old man sitting with his back to the horses with that old-fashioned courtesy towards women, which is now nearly forgotten. Kentish lanes are full of beauty, with their high tangled hedges and fine oak trees. We skirted Hayes Common, and that grand park with the Roman camp in it which old Mr. Brassey is said to have bought and then forgotten its possession for a whole year. When something reminded him of it, he said, "God bless my soul! I forgot all about it," went to see it, did not like it, and sold it again immediately. Down is about six miles from Bromley Common, a pretty village, and a most unpretentious old house with grass plot in front, and a gate upon the road. On the other side the rooms opened on a verandah covered with creepers, under which Mr. Darwin used to walk up and down, wrapped in the great boatman's cloak John Collier[†] has put in his portrait. He seldom went further for exercise, and hardly ever went away from home: all his heart was there and in his work. No man ever had a more perfect home, wife, and children; they loved his work as he did, and shared it with him. He and Mr. Norman had been friends for many years, and it was pretty to see the greater man pet his old neighbor and humor him; for with all his great spirit he was very much of a spoilt child, and proud of his age. Of Charles Darwin's age I never had the smallest idea. He seemed no older than his children, so full of fun and freshness. He sat on the grass under a shady tree, and talked deliciously on every subject to us all for hours together, or turned over and over again the collection of Australian paintings I brought down for him to see, showing in a few words how much more he knew about the subjects than any one else, myself included, though I had seen them and he had not.

Mr. and Mrs. Vernon Lushington were staying there. She had the good art of making others shine. Every one wished to interest her, and to bring out that wondrous smile and look of sympathy on her beautiful face, and I felt that we owed much of the interesting talk of that day to her tact and power of fascination. She also played in her own peculiar way, as if the things she played had been written for her alone by Bach or Handel, while Mr. Darwin rested on the sofa, and made her repeat them over and over, with an enjoyment which was real. When I left he insisted on packing my sketches and putting them even into the carriage with his own hands. He was seventy-four: old enough to be courteous too. Less than eight months after that he died, working till the last among his family, living always the same peaceful life in that quiet house, away from all the petty jealousies and disputes of lesser scientific men.

Marianne North, *Recollections of a Happy Life*, ed. Mrs. John Addington Symonds, 2 vols. (London: Macmillan, 1893), 2:214–15. [†]John Collier (1850–1934), Pre-Raphaelite painter.

Secondary Literature: Suzanne Le-May Sheffield, *Revealing New Worlds: Three Victorian Women Naturalists* (London: Routledge, 2001); M. Jeanne Peterson, *Family, Love, and Work in the Lives of Victorian Gentlewomen* (Bloomington: Indiana University Press, 1989).

CHARLES ELIOT NORTON (1827–1908)
AMERICAN AUTHOR AND CRITIC

On December 26, 1859, Norton describes in the course of a letter to Mrs. [Elizabeth] Gaskell*, a visit to the Physiological Museum of Harvard College, with [James Russell] Lowell*, Torrey,[†] and Jeffries Wyman*, who was in charge of it. "The museum was cold and chilly as the gallery of a Roman palace in February," the letter says, "and we were glad to go down into Wyman's working-room where, round the fire, we grew warm discussing the new book of Mr. Darwin's which is exciting the admiration and the opposition of all our philosophers. Agassiz* is busy in writing a review of the book, in which he intends to refute Mr. Darwin's chief contentions,—for if Darwin is right, Agassiz is wrong. You, I fancy, have not read the book. [...] I admire the patience of Mr. Darwin's research, the wide reach of his knowledge and his thought, and above all the honesty and manliness of his plain speech. His book will help to overthrow many old and cumbrous superstitions even if it establish but few truths in their place. But with what a sense of ignorance such books oppress one,—not merely of one's own ignorance, just as it may be, but of that of the men who know most in any special field."

Norton to Elizabeth Gaskell, 26 December 1859, in *Letters of Charles Eliot Norton*, ed. Sara Norton and M. A. DeWolfe Howe, 2 vols. (Boston: Houghton Mifflin, 1913), 2:201–2. [†]John Torrey (1796–1873), American botanist.

[September 1860]
You will find in the number of the *Atlantic* for October, an excellent and able article by [James Russell] Lowell* on "The Election in November," which gives as fair a view as I have seen of our political conditions and prospects. The first article in the same number is by [Nathaniel] Hawthorne,—the second is mine, and the one on Darwin is by Dr. [Asa] Gray*. The controversy about Darwin's book has been carried on with great activity and animation among our men of science. The best among them seem to be ready to admit that his theory though not proved, and not likely to be proved and accepted in all its parts, is one of those theories which help science by weakening some long-established false notions, and by suggestions leading toward truth if not actually embracing it.

Norton to A. H. Clough, 24 September 1860, in *Letters of Charles Eliot Norton*, 2 vols. (Boston: Houghton Mifflin, 1913), 2:210–11.

Last Sunday, April 5 [1873], Grace and I lunched with the Darwins, who are spending a few weeks in town, in a house in Montagu Street. Mr. Darwin was even more than usually pleasant: his modesty, his simplicity, his geniality of temper, the pleasant unaffected animation of his manners, are always delightful; but on Sunday there was a tenderness in his expression, and he was in better health for the day than common. His talk is not often memorable on account of brilliant or impressive sayings, but it is always the expression of the qualities of mind and heart which combine in such rare excellence in his genius.

Journal entry, in *Letters of Charles Eliot Norton*, 2 vols. (Cambridge, MA: Houghton Mifflin, 1913), 2:476–77.

Secondary Literature: Linda Dowling, *Charles Eliot Norton: The Art of Reform in Nineteenth-Century America* (Hanover, NH: University Press of New England, 2007).

MARGARET OLIPHANT (1828–1897)
ENGLISH NOVELIST AND ESSAYIST

It is difficult to dissociate [Thomas Carlyle* and John Stuart Mill*] from their productions; but when we turn to Charles Darwin, who perhaps is the most influential of all the scientific writers of our epoch, we associate no personality with his work, and feel no temptation to inquire what kind of man he was. This is one drawback which attaches to wealth and comfort, and a quiet life, that there is little attraction for human sympathy in them. But the importance of Darwin in the literary and scientific history of his time is not to be mistaken. His works have been read according to a very unusual formula not so applicable now as in former days—like novels. It would be perhaps a truer form of applause to say of a successful novel that it has been read like Darwin. His works have been discussed in every drawing-room as well as studied in every scientific retirement; but this, we are disposed to be believe, as has been the case with many other scientific works of the period, rather because of the lucidity and interest of the style and the manner of putting these wonderful new doctrines—from their character as literature, in short—than from interest in their subjects or conviction of their truth. It is harder than any philosopher has ever conceived to make ordinary men and women consider in any other light than that of a piquant pleasantry, touching upon the burlesque, the idea that they are themselves the offspring of jellyfish. Notwithstanding this, there can be no doubt that the doctrine of Evolution has had the greatest effect in science, has exercised a considerable influence upon the religious polemics or apologetics of the time, and has been very startling to many minds and very stimulating to many others. [. . .] Darwin's work has the peculiarity that it is unpolemical; his conclusions are worked out with all the calm of scientific research, with nine of the lively pleasure in flinging a challenge to the upholders of religious systems, whose theory of the origin of man is that he was developed from above and not from below, which actuates, for instance, the writing and utterances of Professor Huxley* and [John] Tyndall*, and other philosophers of their class.

Margaret Oliphant, "The Literature of the Last Fifty Years," *Blackwoods Magazine,* June 1887, reprinted in *Prose by Victorian Women,* ed. Andrea Bloomfield and Sally Mitchell (London: Taylor & Francis, 1996), 392–428, on 411–12.

Secondary Literature: Elisabeth Jay, *Mrs. Oliphant: "A Fiction to Herself"; A Literary Life* (New York: Oxford University Press, 1995).

EUGENE O'NEILL (1888–1953)
AMERICAN PLAYWRIGHT

What's the bad news, Pop? Has another Fundamentalist been denying Darwin?

Eugene O'Neill, *Dynamo* (1929), in *The Plays of Eugene O'Neill* (New York: Kessinger, 2004), 431.

Secondary Literature: John P. Diggins, *O'Neill's America: Desire under Democracy* (Chicago: University of Chicago Press, 2007).

ELEANOR ORMEROD (1828–1901)
ENGLISH ENTOMOLOGIST

If you want something very good about the lower creatures up to date I suppose you could not mend *Text Book of Zoology,* by Dr. [Carl] Claus*, translated by Adam Sedgwick*. This is a grand book, but I would not put it in my students' hands without a strong observation that I consider Darwinianism, &c., of this nature perfectly unproved and baseless. I certainly think that presently this view will follow "spontaneous generation."

Ormerod to Robert Wallace, 10 November 1889, *Eleanor Ormerod LLD, Economic Entomologist: Autobiography and Correspondence,* ed. Robert Wallace (New York: Dutton, 1904), 276.

Secondary Literature: Suzanne Le-May Sheffield, *Revealing New Worlds: Three Victorian Women Naturalists* (London: Routledge, 2001).

HENRY FAIRFIELD OSBORN (1857–1935)
AMERICAN PALEONTOLOGIST

Between the appearance of *The Origin of Species*, in 1859, and the present time there have been great waves of faith in one explanation and then in another: each of these waves of confidence has ended in disappointment, until finally we have reached a stage of very general skepticism. Thus the long period of observation, experiment, and reasoning which began with the French natural philosopher Buffon, one hundred and fifty years ago, ends in 1916 with the general feeling that our search for causes, far from being near completion, has only just begun. [...]

For a long period after *The Origin of Species* appeared, Haeckel* and many others believed that Darwin had arisen as the Newton for whom Kant did not dare to hope; but no one now claims for Darwin's law of natural selection a rank equal to that of Newton's law of gravitation.

Henry Fairfield Osborn, *The Origin and Evolution of Life* (New York: Scribner's, 1921), ix–xii.

Chance is the very essence of the original Darwinian selection hypothesis of evolution. William James* and many other eminent philosophers have adopted the "chance" view as if it had been actually demonstrated. Thus James observes: "Absolutely impersonal reasons would be in duty bound to show more general convincingness. Causation is indeed too obscure a principle to bear the weight of the whole structure of theology. As for the argument from design, see how Darwinian ideas have revolutionized it. Conceived as we now conceive them, as so many fortunate escapes from almost limitless processes of destruction, the benevolent adaptations which we find in nature suggest a deity very different from the one who figured in the earlier versions of the argument. The fact is that these arguments do but follow the combined suggestions of the facts and of our feeling. They prove nothing rigorously. They only corroborate our pre-existent partialities." Again, to quote the opinion of a recent biological writer [G. R. Davies]: "And why not? Nature has always preferred to work by the hit-or-miss methods of chance. In biological evolution millions of variations have been produced that one useful one might occur."

I have long maintained that this opinion is a biological dogma; it is one of the string of hypotheses upon which Darwin hung his theory of the origin of adaptations and of species, a hypothesis which has gained credence through constant reiteration, for I do not know that it has ever been demonstrated through the actual observation of any evolutionary series.

Henry Fairfield Osborn, *The Origin and Evolution of Life* (New York: Scribner's, 1921), 7–8.

Secondary Literature: Ronald Rainger, *An Agenda for Antiquity: Henry Fairfield Osborn and Vertebrate Paleontology at the American Museum of Natural History, 1890–1935* (Huntsville: University of Alabama Press, 2004).

WILLIAM OSLER (1849–1919)
Canadian Physician

From John Hunter[†] to Charles Darwin enormous progress was made in every department of zoology and botany, not only in the accumulation of facts relating to structure, but in the knowledge of function, so that the conception of the phenomena of living matter was progressively widened. Then with the *Origin of Species* came the awakening, and the theory of evolution has not only changed the entire aspect of biology, but has revolutionized every department of human thought.

Even the theory itself has come within the law; and to those of us whose biology is ten years old, the new conceptions are, perhaps, a little bewildering. The recent literature shows, however, a remarkable fertility and strength. Around the nature of cell-organization the battle wages most fiercely, and here again the knowledge of structure is sought easily as the basis of explanation of vital phenomena. So radical have been the changes in this direction that a new and complicated terminology has sprung up, and the simple, undifferentiated bit of protoplasm has now its cytosome, cytolymph, carysome, chromosome, with their somacules and biophores. These accurate studies in the vital units have led to material modifications in the theory of descent. [August] Weismann's[*] views, par-

ticularly on the immortality of the unicellular organisms, and of the reproductive cells of the higher forms, and on the transmission or non-
transmission of acquired characters, have been based directly upon studies
of cell-structure and cell-fission.

William Osler, "The Leaven of Science," [21 May 1894], in *A Way of Life & Other
Addresses & Annotations*, ed. Shigeaki Hinohara and Hisae Niki (Durham, NC:
Duke University Press, 2001), 153–70, on 167–68. †See Asa Gray entry.

It is not too much to say that Charles Darwin has so turned man right-
about-face that, no longer looking back with regret upon a Paradise Lost,
he feels already within the gates of a Paradise Regained.

William Osler, "Man's Redemption of Man," *American Museum* 71 (1910): 246–52,
on 248.

Secondary Literature: Michael Bliss, *William Osler: A Life in Medicine* (New York:
Oxford University Press, 1999).

NIKOLAUS OSTERROTH (1875–1933)
German Working-class Leader

I let myself be guided by the titles of works and did not find the truths I
sought. [...] But finally I succeeded. [...] The book was called *Moses or
Darwin?* (1889) and was written by the well known Professor A. Dodel.†
Written in a very popular style, it compared the Mosaic story of creation
with the natural evolutionary history, illuminated the contradictions of
the biblical story, and gave a concise description of the evolution of organic and inorganic nature, interwoven with plenty of striking proofs.

What particularly impressed me was a fact that now became clear to
me: that evolutionary natural history was monopolized by the intuitions
of higher learning; that Newton, Laplace, Kant, Darwin, and Haeckel*
brought enlightenment only to the upper social classes; and that for the
common people in the grammar school the old Moses with his six-day
creation of the world was still the authoritative world view. For the upper
classes there was evolution, for us creation; for them productive liberating
knowledge, for us rigid faith; bread for those favored by fate, stones for
those who hungered for truth!

"Nikolaus Osterroth, Clay Miner," in *The German Worker: Working Class Auto-biographies from the Age of Industrialization,* ed. and trans Alfred Kelly (Berkeley and Los Angeles: University of California Press, 1987), 185. †Arnold Dodel (1843–1908), German botanist and popularizer of Darwinism.

Secondary Literature: Eric Dorn Brose, *Christian Labor and the Politics of Frustration in Imperial Germany* (Washington, DC: Catholic University of America Press, 1985).

WILHELM OSTWALD (1853–1932)
German Physical Chemist

This property of adaptation facilitates and assures nourishment. If we take the fundamental idea developed by Darwin, that that predominates in the world which by virtue of its properties endures the longest time, then it is evident that a body which teleologically preserves and elaborates its nourishment will live longer than a similar body without this property. Moreover, by the general process of adaptation, these "teleological" properties come to be more greatly developed and more readily exercised in the body that lives longer, so that its long life gives it another advantage over its rival. Thus we can understand how this property of adaptation, which at first is to be conceived of as a purely physico-chemical quality is found developed in all organisms.

Wilhelm Ostwald, *Natural Philosophy,* trans. Thomas Seltzer (New York: Henry Holt, 1910), 173.

A very important point in this general idea is the transmission of memory from parents to offspring. The great riddle of heredity, which caused Darwin so much thinking without a corresponding result, may be brought somewhat nearer to a solution by the aid of this same concept of memory. A general view of the facts of generation and propagation shows us that the life of the offspring is nothing more nor less than the continuation of the life of the parents.

Wilhelm Ostwald, *Individual and Immortality* (Boston: Houghton Mifflin, 1906).

Secondary Literature: Hans-Günther Körber, "Carl Wilhelm Wolfgang Ostwald," in *Dictionary of Scientific Biography,* ed. Charles Coulston Gillispie, 16 vols. (New York: Scribner's, 1970–80), 10:251–52; Milic Capek, "Wilhelm Ostwald," in *Encyclopedia of Philosophy,* ed. Paul Edwards, 8 vols. (New York: Macmillan, 1967), 6:5–7.

RICHARD OWEN (1804–1892)
ENGLISH ANATOMIST

[December 1859]

You made a remark in our conversation something to the effect that my book could not probably be true as it attempted to explain so much. I can only answer that this might be objected to any view embracing two or three classes of facts. Yet I assure you that its truth has often and often weighed heavily on me; and I have thought that perhaps my book might be a case like [William Sharp] Macleay's* quinary system [of classification]. So strongly did I feel this that I resolved to give it all up, as far as I could, if I did not convince at least two or three competent judges. You smiled at me for sticking myself up as a martyr; but I assure you if you had heard the unmerciful and, I think, unjust things said of my book and to me in a letter by an old and very distinguished friend you would not wonder at me being sensitive, perhaps ridiculously sensitive. Forgive these remarks. I should be a dolt not to value your scientific opinion very highly. If my views are *in the main* correct, whatever value they may possess in pushing on science will now depend very little on me, but on the verdict pronounced by men eminent in science.

Charles Darwin to Owen, 13 December 1859, in *The Life of Richard Owen,* by Rev. Richard Owen, 2 vols. (London: John Murray, 1895), 2:90–91.

The octavo volume [. . .] which made its appearance towards the end of last year, has been received and perused with avidity, not only by the professed naturalist, but by that far wider intellectual class which now takes interest in the higher generalizations of all the sciences. The same pleasing style which marked Mr. Darwin's earliest work, and a certain disposition and sequence of his principal arguments, have more closely recalled the attention of thinking men to the hypothesis of the inconstancy and transmutation of species, than had been done by the writings of previous advocates of similar views. Thus, several, and perhaps the majority, of our younger naturalists have been seduced into the acceptance of the homeopathic form of the transmutative hypothesis now repented to them by Mr. Darwin, under the phrase of "natural selection."

Richard Owen, "Darwin on the Origin of Species," *Edinburgh Review* 11 (1860): 487–532, reprinted in *Darwin and his Critics,* by David Hull (Cambridge, MA: Harvard University Press, 1973), 175–213, on 175–76.

Secondary Literature: Nicholas Rupke, *Richard Owen: Victorian Naturalist* (New Haven, CT: Yale University Press, 1994).

JAMES PAGET (1814–1899)
English Surgeon and Pathologist

Darwin had the rare power of taking the common things that other men waste, and out of them making the grandest material of scientific work. So that it is vain to say, in any branch of practice, "I have no opportunity for scientific enquiry; I cannot investigate this; I can contribute nothing to that which I see the scientific members of the profession are doing." It requires merely the opportunity of a practice in the country, and the mind and resolution of Darwin, to bring great pathological conclusions out of the most ordinary facts of daily life in general practice.

James Paget, address to the British Medical Association, Metropolitan Counties branch, 17 January 1883, in *Memoirs and Letters of Sir James Paget,* ed. Stephen Paget (London: Longmans, Green, 1908), 331.

All these variations in diseases should be studied as Darwin studied the variations of species. Let me be clear in saying, as Darwin studied; for in the pursuit of new knowledge he may be a model to all, as he has been to me so far as I could imitate him. He, I know, would have studied these things, not by deduction, as from a law exactly formulated and from which he could trace the course of every change, but by a most careful collection of facts; facts to be seen in specimens and read in full records and stored in museums, and by a study as complete for every case as if no law of evolution had ever been discovered.

James Paget, "On Some Rare and New Diseases," Bradshawe Lecture, College of Surgeons, London, 13 December 1882, in *Memoirs and Letters of Sir James Paget,* ed. Stephen Paget (London: Longmans, Green, 1908), 328.

A letter from Professor Charcot[†] served us as an introduction to his famous friend Sir James Paget. The latter invited George Hugo and me to lunch at his classically Londonesque house. [. . .] As we sat down, he said to us, "These seats have often been occupied by Darwin and Huxley*." The authors of *The Origin of Species* and *The Crayfish* were intimate friends of Paget.

Memoirs of Léon Daudet, ed. Arthur Kingsland Griggs (New York: Dial, 1925), 128. †Jean-Martin Charcot (1825–93), French neurologist famous for his studies of hysteria and the use of hypnosis in diagnosing and treating it.

[Paget's] house from morning to night was in a whirl of excitement, but it never lost its feeling of home. The incessant hospitality, the confusion of tongues, the coming and going of all the masters of medicine and surgery with their disciples, the meeting of H.R.H. the Prince of Wales and H.I.H. the Crown Prince of Germany with Darwin, Pasteur, Virchow,† Huxley*, [John] Tyndall*, and other great personages—all these festivities were still "at home"; he could not easily imagine hospitality anywhere else: and the house, somehow, got through the work.

Memoirs and Letters of Sir James Paget, ed. Stephen Paget (London: Longmans, Green, 1908), 318. †Rudolph Virchow (1821–1902), German pathologist, anti-Darwinian.

Secondary Literature: M. Jeanne Peterson, *Family, Love, and Work in the Lives of Victorian Gentlewomen* (Bloomington: Indiana University Press, 1989).

FRANCIS TURNER PALGRAVE (1824–1897)
English Poet and Critic

July 21 1871. We have spent four agreeable days at the Palace at Exeter: I had one long walk with the Bishop [Frederick Temple],† and a really good discussion on Darwin and cognate topics. He was at his best on such points: large and wise and liberal.

Francis Turner Palgrave: His Journals and Memories of Life, ed. Gwenllian F. Palgrave (London: Longmans, Green, 1899), 134. †See Alfred Newton entry.

[Alfred Lord] Tennyson*, ever watchful of natural detail, was pleased if he felt that he had put successfully into verse some little noticed phenomenon. Yet the pleasure (which C. Darwin also must have often known) of going closely true to real fact, the sense almost of absolute contact with Nature, was the predominant feeling.

Francis Turner Palgrave, *Landscape in Poetry from Homer to Tennyson* (London: Macmillan, 1897), 288.

Secondary Literature: Anne Ferry, *Tradition and the Individual Poem: An Inquiry into Anthologies* (Stanford, CA: Stanford University Press, 2001).

THEODORE PARKER (1810–1860)
AMERICAN CLERGYMAN

[February 1860]
Here in Rome I am out of the way of all books, except the Lives of the Saints, etc. But yet I learn of Mr. Darwin's work one of the most important works the British have lately contributed to science. He does not believe in Agassiz's* foolish notion of an interposition of God when a new form of lizard makes its appearance on the earth. Indeed, a God who only works by fits and starts is no God at all. Science wants a God that is a constant force and a constant intelligence, immanent in every particle of matter. The old theological idea of God is as worthless for science as it is for religion. I should like to live long enough to finish and print a course of sermons I preached in 1858, on "The Testimony of Matter and Mind to the Existence and Character of God." It certainly is the most important thing I have done in my life; but is left not fit for publication. If I don't do the work some one else will; a little later, but perhaps better.

Parker to Edward Desor, 24 February 1860, in *The World of Matter and the Spirit of Man,* by Theodore Parker, ed. George Willis Cooke (Boston: American Unitarian Association, 1907), 419–20.

Secondary Literature: Dean Grodzins, *American Heretic: Theodore Parker and Transcendentalism* (Chapel Hill: University of North Carolina Press, 2002).

THEOPHILUS PARSONS (1797–1882)
American Lawyer and Jurisprudent

[May 1860]

My position [. . .] is precisely this. It is always possible that offspring may be born, differing as much from their parents and kindred in the way of gain, of advantage, and of improvement, as we know that offspring have differed in the way of loss, of hindrance and of degradation; and therefore when I speak of extreme aberration I shall mean by it variation carried to this extent. [. . .]

Let this doctrine of the new creation of new species, by generative development through variation be accepted, and we have Darwin's theory of the origin of species by successive generation; and instead of opposing the theory of Agassiz*, it confirms it; because it adopts and reasserts the principle of new creations, and offers some explanation of the way in which they were made.

Theophilus Parsons, review of *On the Origin of Species,* in *American Journal of Science and Arts,* 2nd ser., 30 (July 1860): 1–13, on 1–4.

Professor Parsons has published in the same "Silliman" [i.e., *American Journal of Science and Arts*][†] a speculative paper correcting my notions, worth nothing.

Darwin to Charles Lyell, August 1860, in *The Life and Letters of Charles Darwin,* ed. Francis Darwin, 2 vols. (New York: Appleton, 1887), 2:124. [†]The *American Journal of Science* was popularly referred to by the name of its founder and editor, Benjamin Silliman (1779–1864), a chemist and professor of science at Yale.

Secondary Literature: Paula Blanchard, *Sarah Orne Jewett* (New York: Da Capo, 2002).

WALTER PATER (1839–1894)
English Critic and Essayist

Apr. 1859. Pater, [J. W.] Hoole, [Robert Henry] Wood, and Moorhouse,[†] and one other stayed up in Oxford together. Being the only men left in college, they dined in each other's rooms, and had much animated discussion over Darwin and other rising writers.

Thomas Wright, *Life of Walter Pater,* 2 vols. (New York: G. P. Putnam, 1907), 1:174.
†Rev. M. B. Moorhouse was active in the British and Foreign Bible Society and published sermons and books of stories in verse.

The divergence between Pater and me [. . .] steadily increased till one evening I painfully remember when he and another who had been reading Darwin's *Origin of Species* proved with remorseless logic that if God was everywhere—*we* could have no existence at all! I tried to rebut the fallacy, but in vain, and that night, for the first, and last, time in my life, I threw myself on my bed without prayer. I awoke the next morning as from a hideous nightmare, and cast myself on my knees in the deepest gratitude that revelation was as much superior to logic as light to darkness. I told them that we must argue no more on these subjects, and if I remember aright we confined our discussions to secular matters, and there was always a certain constraint from that time forward.

M. B. Moorhouse to Thomas Wright, 21 December 1903, in *Life of Walter Pater,* by Thomas Wright, 2 vols. (New York: G. P. Putnam, 1907), 1:203.

And nature, together with the true pedigree and evolution of man also, his gradual issue from it, was still all to learn. The delightful tangle of things! it would be the delightful task of man's thoughts to disentangle that. Already Bruno† had measured the space which Bacon would fill, with room perhaps for Darwin also. That Deity is everywhere, like all such abstract propositions, is a two-edged force, depending for its practical effect on the mind which admits it, on the peculiar perspective of that mind. To Dutch Spinosa [*sic*], in the next century, faint, consumptive, with a hold on external things naturally faint, the theorem that God was in all things whatever, annihilating, their differences suggested a somewhat chilly withdrawal from the contact of all alike. In Bruno, eager and impassioned, an Italian of the Italians, it awoke a constant, inextinguishable appetite for every form of experience—a fear, as of the one sin possible, of limiting, for oneself or another, that great stream flowing for thirsty souls, that wide pasture set ready for the hungry heart.

Walter Pater, "Giordano Bruno," E-texts for Victorianists, 239, http://www.victori anprose.org/texts/Pater/Works/giordano_bruno.pdf. †Giordano Bruno (1548–1600), Italian cosmologist and proponent of heliocentrism, burned at the stake as a heretic.

Secondary Literature: Denis Donoghue, *Walter Pater: Lover of Strange Souls* (New York: Knopf, 1995).

CHARLES HENRY PEARSON (1830–1894)
ENGLISH-AUSTRALIAN HISTORIAN

The upper ten thousand, who learn their Darwin from stray passages in novels and the talk of dinner-tables, are admirably fitted to be a receptive medium of knowledge; and it would be absurd to expect that the mass of men can ever acquire it for themselves. To say, however, that it is these people that have the firmest grip on their knowledge, simply because they have got at results, and do not trouble themselves about processes, is to invert the scientific point of view, which regards the method—being generally applicable—as more important than the individual fact. Neither is there much more conviction in Mr. [Samuel] Butler's* argument that though the possession of knowledge is not incompatible with beauty, its pursuit seems to be so since a great many scientific and literary men look ugly and disagreeable.

Mr. Butler holds that our most solid notions are memories from past existences which we bring with us into the world, and that whatever we learn here is acquired in a self-conscious and bewildering manner. [. . .] The present generation can add nothing to them except by transmitting unconscious processes of reasoning to its own descendants. Certainly, if this is the true theory of man, Mr. Butler's cynicism is justified. Naturally enough his ultimate theory of society is that the professions will assuredly one day become hereditary. In this way the transmitted instincts of a hundred generations of painters and lawyers, instead of vitalizing and diversifying the whole race, will be concentrated in castes of unapproachable excellence!

Charles Henry Pearson, *Reviews and Critical Essays*, ed. H. A. Strong (London: Methuen, 1896), 72, 75.

Secondary Literature: John Tregenza, *Professor of Democracy: The Life of Charles Henry Pearson, 1830–1894, Oxford Don and Australian Radical* (New York: Cambridge University Press, 1968).

KARL PEARSON (1857–1936)
ENGLISH STATISTICIAN AND EUGENICIST

Perhaps some of the modern critics of Darwin will be less ready to consider adaptations as "not explicable" by natural selection, but due to the "precise chemical nature of protoplasmic metabolism," or to "an internal fate, expressible in terms of dominant chemical constitution," if they once grasp that physics and chemistry in their turn render nothing "explicable," but merely, like natural selection itself, are shorthand descriptions of changes in our sense-impressions.

Karl Pearson, *The Grammar of Science*, 2nd ed. (London: Adam & Charles Black, 1900), 356.

Secondary Literature: Theodore M. Porter, *Karl Pearson: The Scientific Life in a Statistical Age* (Princeton, NJ: Princeton University Press, 2004).

DOM PEDRO II (1825–1891)
BRAZILIAN EMPEROR

Refute Darwinism! Towards what end? One refutes a shaky theory, or a dangerous doctrine, or an absurd system. But one does not refute fairy tales. [...] Now when we have arrived at the heart of this work, certain scruples come to mind. I ask myself whether, in taking this fantastic book of Darwin's seriously, might I not have been duped by some of the mystifications prepared by Darwin himself?

This would not be the first instance of an author loosening up and in some way putting the fibers of his brain in abeyance in order to write a humorous book. Didn't Homer write the *Battles of the Frogs and the Rats,* Virgil *The Gnat,* Erasmus, *In Praise of Folly,* Montesquieu, *The Persian Letters?* Might not Darwin have done the same with *The Origin of Species?*

Constantin James, *L'hypnotisme expliqué dans sa nature et dans ses actes: Mes entretiens avec S. M. L'Empereur Dom Pedro sur le Darwinisme* (Paris: Société Bibliographique, 1888), 67–89, on 72–73, trans. TFG.

Many times I have recommended to my young aristocrats the benefits of reading the works of Darwin, because I am a champion of the truth, and the more I read the more I am convinced that there is only one truth: that all the sciences converge at one point—truth.

Dom Pedro II, quoted in *História de Dom Pedro II, 1825–1891: Fastígio, 1870–1880*, by Heitor Lyra (São Paulo: USP, 1977), 226, trans. TFG.

I got home [...] to find a telegram from "my friend" the Emperor of Brazil, saying that he would be out at 6 this morning; & sure enough here he has been, with his heart—set on seeing you—I pleaded your illness &c &c. but it was no good. First he suggested that I should write to [Richard] Owen* & offer himself you & me to *dejeuner!!!* Of course I would have none of that; then he wanted me to ask himself & you to *dejeuner* at Kew with me, to which I could only express myself proud & pleased if you were well enough to come—an alternative which he met by suggesting that if so He & I, should go to Down & *dejeuner* with you! He made me promise to write at once, to you to this effect. Pray my dear fellow do not suppose that I in any way ever countenanced all this. I can wriggle out of a bad hypothesis, but not out of such a pertinacious old bird as this Emperor. What is to be done?

Joseph Dalton Hooker to Darwin, 14 June 1877, Darwin Correspondence Project, letter 11000, text not yet available online.

Secondary Literature: Lila Moritz Schwarcz, *The Emperor's Beard: Dom Pedro II and the Tropical Monarchy of Brazil*, trans. John Gledson (New York: Hill & Wang, 2004).

BENJAMIN PEIRCE (1809–1880)
AMERICAN MATHEMATICIAN

[30 June 1860]
There has been a vast [amount] of discussion in the other section [of the British Association for the Advancement of Science meeting at Oxford] about Darwin's book, which is occupying all the attention of England.

I heard a very sharp pass between [Richard] Owen* and Huxley*, and a long and earnest one between the Bishop of Oxford [Samuel Wilberforce*] and Huxley, at which I should think that there must have been a thousand persons present. The Bishop is one of the most eloquent men I ever heard, he's known here as Soapy Sam, and the slippery character of the divine was apparent in all his argument. His power of language was wonderful and the revulsion? with which he seized upon his opponents' views and exposed them to the torture was a model of logical display.

Excerpt from Peirce's journal of his 1860 European trip, Benjamin Peirce Papers, Harvard University Archives, in *Science in Nineteenth-Century America: A Documentary History*, ed. Nathan Reingold (New York: Hill & Wang, 1964), 197–98.

Secondary Literature: Edward R. Hogan, *Of the Human Heart: A Biography of Benjamin Peirce* (Bethlehem, PA: Lehigh University Press, 2008).

CHARLES SANDERS PEIRCE (1839–1914)
AMERICAN PHILOSOPHER

I was away surveying in the wilds of Louisiana when Darwin's great work appeared, and although I learned by letters of the immense sensation it had created, I did not return until early in the following summer when I found [Chauncey] Wright* all enthusiasm for Darwin, whose doctrines appeared to him as a sort of supplement to those of [John Stuart] Mill*. I remember well that I then made a remark to him which, although he did not assent to it, evidently impressed him enough to perplex him. The remark was that these ideas of development had more vitality by far than any of his other favorite conceptions and that though they might at that moment be in his mind like a little vine clinging to the tree of Associationalism, yet after a time that vine would inevitably kill the tree. He asked me why I said that and I replied that the reason was that Mill's doctrine was nothing but a metaphysical point of view to which Darwin's, which was nourished by positive observation, must be deadly.

Charles S. Peirce, "On Phenomenology," in *The Essential Peirce: Selected Philosophical Writings*, 2 vols. (Bloomington: Indiana University Press, 1998), 2:158.

Secondary Literature: Carl B. Hausman, *Charles S. Peirce's Evolutionary Philosophy* (Cambridge: Cambridge University Press, 1993); Louis Menand, *The Metaphysical Club* (New York: Farrar, Straus, & Giroux, 2001).

BENITO PÉREZ GALDÓS (1843–1920)
Spanish Novelist

"Tell me, Señor Don José, what do you think of Darwinism?"

The young man could not help smiling at such ill-timed pedantry, and would willingly have led the youth on, and drawn out a full display of his infantile vanity; but thinking it more prudent not to be too intimate with either uncle or nephew, he simply answered:

"I really cannot say I think anything of the theories of Darwin, for I hardly know anything about them. The work in my profession does not allow me much time for such studies."

"Now," said the priest, laughing, "everything is reduced to the idea that we are descended from monkeys. If it were only said of certain persons, it would be quite right."

"The theory of natural selection," added Jacinto, "has, I am told, many partisans in Germany."

"I do not doubt that," said the priest, "in Germany one might feel that this theory might be true—as far as regards [Otto von] Bismarck*."

B. Pérez Galdós, *Doña Perfecta: A Tale of Modern Spain* (London: Samuel Tinsley, 1880), 64–65.

Secondary Literature: T. E. Bell, *Galdós and Darwin* (London: Tamesis Books, 2006).

DMITRY PISAREV (1840–1868)
Russian Social Critic

The starting-point, the very appearance of organic life, is still an unsolved problem, for as yet no naturalist has succeeded in preparing in his laboratory out of inorganic or organic substances a single living organism even of the simplest kind; but the process of development and degeneration of organic forms has been explained to a considerable extent by the English naturalist Charles Darwin, who in 1859 published his famous work *The Origin of Species*. This brilliant thinker, whose knowledge is enormous, took in all the life of nature with such a broad view and entered so deeply into all its scattered phenomena that he made a discovery which, perhaps, is unprecedented in the history of natural science. He discovered not an isolated fact, not a gland or a vein, not the function of some nerve, but a whole series of laws according to which all organic life on our planet is governed and varies. And he tells you so simply, proves them so irrefutably and bases his arguments on such obvious facts, that you, a common human, uninitiated in nature science, are in a state of continual astonishment only at not having thought out such conclusions yourself long ago. [. . .]

When the readers have acquainted themselves with Darwin's ideas, even through my weak and colorless sketch, I shall ask them whether we did rightly or wrongly in rejecting metaphysics ridiculing our fantasy and expressing complete scorn for our conventional aesthetics. Darwin, Lyell* and thinkers like them are the philosophers, the poets, the aestheticians of our time. [. . .]

It is worthwhile to note the honest, friendly relations existing between the best scientists of our time. Lyell and Hooker* constantly follow Darwin's work; Darwin consults them and they help him; over fifteen years Hooker continually informed him, sometimes of new facts, sometimes of his own critical observations. Wallace*, who came very near to Darwin's own conclusions, most trustingly sent him his memoir, and Darwin in turn spoke of it with great respect. Briefly, it is obvious that these men were interested in the success of their common cause, not in promoting their own personality or crushing dangerous rivals. The result of this was, first, their common cause prospered, and secondly, each of them achieved a degree of scientific fame that none of them would ever have acquired had they worked in isolation, jealously hiding from one another the facts they had discovered and not exchanging thoughts and remarks.

The broad intellectual development of these outstanding men makes them particularly capable of free association, and association in its turn gives them new strength and widens still more the horizon of their thought. So far voluntary and quite natural association has found application only in the higher spheres of scientific activity. Here there is no destructive war between competitors: here all honest people aim at the same goal and they rely on one another as friends: that is why we see that it is till only in the higher spheres of scientific activity that man can develop, maintain and ennoble all his truly human qualities and capacities; that is why we also see that science [. . .] is developing with extraordinary speed and leaving far behind all other branches of human activity.

Dmitry Pisarev, "Progress in the Animal and Vegetable Worlds," in *Selected Philosophical, Social and Political Essays* (Moscow: Foreign Language Publishing House, 1958), 303–9.

Secondary Literature: Adam Bruno Ulam, *Prophets and Conspirators in Prerevolutionary Russia* (New York: Transaction, 1998); Loren R. Graham, *Science in Russia and the Soviet Union: A Short History* (Cambridge: Cambridge University Press, 1993).

AUGUSTUS HENRY LANE FOX PITT RIVERS (1827–1900)
ENGLISH ARCHAEOLOGIST

Amongst the questions which anthropology has to deal with, that of the descent of man has been so elaborately treated, and at the same time popularized by Mr. Darwin, that it would be serving no useful purpose were I to allude to any of the arguments on which he has based his belief in the unbroken continuity of man's development from lower forms of life. Nor is it necessary for one to discuss the question of the *monogenesis* or *polygenesis* of man. On this subject Mr Darwin has shown how unlikely it is that races so closely resembling each other, both physically and mentally, and interbreeding as they invariably do, should on the theory of development have originated in different localities. Neither are we now, I think, in a position to doubt that civilization has been gradually and progressively developed, and that a very extended, though not by any means uniform, period of growth must have elapsed before we could arrive at the very high state of culture which we now enjoy.

Augustus Pitt Rivers, "Address to the Department of Anthropology," in *Report of the British Association for the Advancement of Science 1872* (London: John Murray, 1873), 158–59.

Secondary Literature: Marc Bowden, *Pitt Rivers: The Life and Archaeological Work of Lieutenant-General Augustus Henry Lane Fox Pitt Rivers* (Cambridge: Cambridge University Press, 1991).

PIUS IX (1792–1878)
Roman Catholic Pope

[Dr. Constantin James][†] refuted so well the aberrations of Darwinism. [. . .] A system that is repugnant at once to history, to the tradition of all peoples, to exact science, to observed facts, and even to reason itself, would seem to need no refutation. But the corruption of this age, the machinations of the perverse, the danger of the simple, demand that such fancies, altogether absurd though they are, should—since they borrow the mask of science—be refuted by true science.

Pius IX to Constantin James, 17 May 1877, in *L'hypnotisme expliqué dans sa nature et dans ses actes: Mes entretiens avec S. M. L'Empereur Dom Pedro sur le Darwinisme,* by Constantin James (Paris: Société Bibliographique, 1888), 67–89, on 84–85, trans. TFG. [†]Constantin James (1813–88), French physician and naturalist.

Secondary Literature: E. E. Y. Hayes, *Pio Nono: A Study of European Politics and Religion in the Nineteenth Century* (Garden City, NY: Doubleday, 1962).

FREDERICK POLLOCK (1845–1937)
English Jurist

The historical method is not the peculiar property of jurisprudence or any other branch of learning. It is the newest and most powerful instrument, not only of the moral and political sciences, but of a great part of the natural sciences, and its range is daily increasing. The doctrine of evolution is nothing else than the historical method applied to the facts of nature; the historical method is nothing else than the doctrine of evolution applied to human societies and institutions. When Charles Darwin created the philosophy of natural history (for no less title is due to the idea which transformed the knowledge of organic nature from a multitude of particulars into a continuous whole), he was working in the same spirit and towards the same ends as the great publicists who, heeding his field of labor as little as he heeded theirs, had laid in the patient study of historical fact the bases of a solid and rational philosophy of politics and law. Savigny,[†] whom we do not yet know or honor enough, and our own [Edmund] Burke, whom we know and honor, but cannot honor too much, were Darwinians before Darwin. In some measure the same may be said of the great Frenchman Montesquieu, whose unequal but illuminating genius was lost in a generation of formalists.

Frederick Pollock, "English Opportunities in Historical and Comparative Jurisprudence" (1883), in his *Oxford Lectures and Other Discourses* (London: Macmillan, 1890), 41–42. [†]Friedrich von Savigny (1779–1861), influential French jurist.

The separation of politics from ethics (including the ultimate metaphysical foundations of ethics), of which we spoke above, is not only not effected, but in many Continental schools would have been very lately regarded as impracticable; in some it would still be so. But it may be useful to point out why the English utilitarians always remained outside the main stream down to the time when the remnant of them was swept, some willing and others unwilling, into the floodtide of Evolutionism (not Darwinism: and this not merely because Herbert Spencer* claimed, and justly, to have been an evolutionist and made extensive applications of that way of thinking before Darwin declared himself, but because the corresponding movement in moral and political science was independent and earlier), they supposed themselves to be maintaining the teaching of experience against the whole brood of dogmatists.

Frederick Pollock, *History of the Science of Politics,* new rev. ed. (London: Macmillan, 1930), 118–19.

Secondary Literature: Neil Duxbury, *Frederick Pollock and the English Juristic Tradition* (London: Oxford University Press, 2004).

EDWARD BAGNALL POULTON (1856–1943)
ENGLISH BIOLOGIST

The reviewer in the *Athenaeum* for November 19, 1859, left the author "to the mercies of the Divinity Hall, the College, the Lecture Room, and the

Museum." Dr. [William] Whewell* for some years refused to allow a copy of the *Origin* to be placed in the library of Trinity College, Cambridge. My predecessor, Professor J. O. Westwood,† proposed to the last Oxford University Commission the permanent endowment of a Reader to combat the errors of Darwinism. Lyell* had difficulty in preventing [John William] Dawson* reviewing the *Origin* on hearsay, without having looked at it. No spirit of fairness can be expected from so biased a judge. [. . .]

It is remarkable to contrast the maturity, the balance, the judgment, with which Darwin put forward his views, with the rash and haphazard objections and rival suggestions advanced by his critics. It is doubtful whether so striking a contrast is to be found in the history of science—on the one side twenty years of thought and investigation pursued by the greatest of naturalists, on the other offhand impressions upon a most complex problem hastily studied and usually very imperfectly understood. It is not to be wondered at that Darwin found the early criticisms so entirely worthless.

Edward Bagnall Poulton, "Fifty Years of Darwinism," in *Fifty Years of Darwinism: Modern Aspects of Evolution* (New York: Henry Holt, 1909), 8–56, on 20–23. †John Obadiah Westwood (1805–93), English entomologist and archaeologist.

There is reason to believe that Professor [Alfred] Newton's* impressions of the result of the celebrated meeting of the British Association at Oxford in 1860 are more accurate than those of the eyewitness quoted in the *Life and Letters*. The latter has pictured a brilliant triumph for Huxley* in the renowned duel with the Bishop of Oxford. But I have been told by more than one of the audience that Huxley was really too angry to speak effectively, nor is this to be wondered at, considering the extreme provocation. Mr. William Sidgwick,† who was present and sympathized warmly with Huxley, has told me that this was his opinion. I have heard the same from the Rev. W. Tuckwell,‡ who also quoted a remark of the late Professor [George] Rolleston* tending in the same direction. Mr. Tuckwell said that it was clear that the audience as a whole was not carried away by Huxley's speech, but, on the contrary, was obviously shocked at it; and he contrasted that occasion with another at which he was also present, in the North, several years later, when Huxley replied to an opponent who, like the bishop, appealed to the theological prejudices of his hearers. But by that time the new teachings had been absorbed, and Huxley gained a signal triumph.

Edward B. Poulton, *Charles Darwin and the Theory of Natural Selection* (London: Cassell, 1896), 154–55. †William Sidgwick, headmaster of a grammar school in Skipton, father of the moral philosopher Henry Sidgwick. ‡William Tuckwell (1829–1919), politically radical schoolmaster.

Secondary Literature: Robinson M. Yost, "Edward Bagnall Poulton (1856–1943): Natural Selection of the Butterfly Mimicry," http://faculty.kirkwood.edu/ryost/poulton.htm (accessed 31 March 2009).

EZRA POUND (1885–1972)
American Poet

The mythological exterior lies on the moss in the forest
And questions him about Darwin.
And with a burning fire of phantasy
     He replies with "Deh! Nuvoletta [. . .]"

Ezra Pound, canto 29, in *The Cantos of Ezra Pound* (New York: New Directions, 1972), 144.

Secondary Literature: Ira Bruce Nadel, *Ezra Pound: A Literary Life* (New York: Palgrave Macmillan, 2004).

BADEN POWELL (1796–1860)
English Mathematician and Theologian

Just a similar skepticism [regarding the miraculous] *has been* evinced by nearly all the first physiologists of the day, who have joined in rejecting the development theories of Lamarck and the *Vestiges;* and while they have strenuously maintained successive creations, have denied and denounced the alleged production of organic life by Messrs. Crosse and Weekes,† and stoutly maintained the impossibility of spontaneous generation, in the alleged ground of contradiction to experience. Yet it is now acknowledged under the high sanction of the name of [Richard] Owen*, that "creation" is only another name for our ignorance of the mode of production; and it has been the unanswered and unanswerable argument of another reasoner that new species *must* have originated *either,* out of their inorganic elements, *or* out of previously organized forms; *either* development *or* spontaneous generation must be true: while a work has now appeared by a naturalist of the most acknowledged authority, Mr. Darwin's masterly volume on *The Origin of Species* by the law of "natural selection"—which now substantiates on undeniable grounds the very principle so long denounced by the first naturalists—*the origination of new species by natural causes:* a work which must soon bring about an entire revolution of opinion in favor of the grand principle of the self-evolving powers of nature.

Baden Powell, "On the Study of the Evidences of Christianity," in *Essays and Reviews: The 1860 Text and Its Reading,* ed. Victor Shea and William Whitla (Charlottesville: University of Virginia Press, 2000), 258. †Andrew Crosse (1784–1855), pioneer electricity experimenter who claimed to have materialized insects by "electrocrystallization," and William Henry Weekes (1790–1850), English surgeon who conducted electrical experiments in his laboratory.

Secondary Literature: Pietro Corsi, *Science and Religion: Baden Powell and the Anglican Debate, 1800–1860* (Cambridge: Cambridge University Press, 1988).

JOHN WESLEY POWELL (1834–1902)
AMERICAN GEOLOGIST AND EXPLORER

The discovery by Darwin and the masterly advocacy of evolution by Spencer*, through which the doctrine of the survival of the fittest was established, for a time gave a decided check to the theory. The blow struck by Spencer was especially efficient, for Spencer resolved all of the properties into force with a clearness which left no room to doubt his meaning.

A host of scientific men following Darwin and accepting the doctrine of the survival of the fittest have found it to be inadequate as a single theory of evolution. There are other laws, especially one expounded by Lamarck. I myself have set forth a new doctrine of evolution as that of culture [. . .] in which I shall attempt to prove that the fundamental law of evolution is the law of affinity by which bodies are incorporated, and hence that evolution is primarily telic.

John Wesley Powell, *Truth and Error or the Science of Intellection* (Chicago: Open Court, 1898), 95–96.

Secondary Literature: Donald Worcester, *A River Running West: The Life of John Wesley Powell* (New York: Oxford University Press, 2002).

JOHN PRICE (1803–1887)
Welsh Naturalist

[T]he formidable Llech, where, at the abrupt termination of a green zig-zag (distinctly visible from the sea), a very rugged and narrow path led the adventurous explorer past a little artificial dripping grotto, with a stone table in it, down to the very sea ledges. Here true Samphire *(Crithmum maritimum)* grows, which we used to get by shooting it down. *Asperugo Procumbens* grew here when we went down with Charles Darwin about 1824, but seems smothered by nettles long since. [. . .]

Reptiles—the writer, in company with Charles Darwin, caught a Viper on the Warren, about 1824, favored by Wellington boots and very strong gardening gloves. Holding him short by the neck, we let him bite at the glove, and emit a drop of clear fluid along the fang, which sank instantly into the leather.

When this had been done about five times, no more poison was left, and we killed, but did not eat him—a fact never satisfactorily explained.

John Price ("Old Price"), *Llandudno and How to Enjoy It* (Llandudno, Wales: Simpkin, Marshall, 1875).

MICHAEL PUPIN (1858–1935)
Serbo-American Electrical Engineer

A lively discussion was going on in those days [ca. 1875] between the biological sciences and theology. Huxley* and many other scientists championing the claims of Darwin's evolution theory and the theologians defending the claims of revealed religion, I was too young and too untutored to understand much of those learned discussions, but Bilharz [a fellow factory worker] followed them with feverish anxiety. His theological arguments did not appeal to me, and so far as I was concerned they lost even the little force they had when Bilharz turned them against what he called American mechanism and materialism, which he tried to make responsible for the alleged materialism of the evolution theory. His political and philosophical theories based upon blind prejudice created a gap between him and me which widened every day.

Michael Pupin, *From Immigrant to Inventor* (New York: Charles Scribner, 1925), 91.

Secondary Literature: Thomas Parke Hughes, *American Genesis: A Century of Invention and Technological Enthusiasm, 1870–1970* (Chicago: University of Chicago Press, 2004).

EDWARD BOUVERIE PUSEY (1800–1882)
English Theologian

The theory of Evolution seems to me one of the threatening clouds of the day. I fear that it will wreck the faith of many. It is very fascinating to a certain class of minds, and seems already to be a sort of gospel. A young man wrote to me on occasion of that sermon, that he believed in Evolution and in Genesis also, and supposed that they could be reconciled somehow; although he did not see how. I fear that, with most, Genesis would have to give way.

Darwin's *Descent of Man* was very distressing to me. Hitherto, Darwin had, in all his illustrations, kept himself to scientific facts, the variations or, if so be, fresh species of animals or plants of the same kind. In the *Descent of Man,* he claims to have done good service in "aiding to overthrow the doctrine of separate creations" (p. 61). He accepts (as you know) in principle all Haeckel's* genealogy of our ancestors, "still more simply organs than the lancelet or amphioxus" (p. 609).

To me, it would seem to stultify the whole of the Darwinian theory, to suppose a mere natural development up to man, including man's body, and then to suppose that this descendant from its ape-ancestors was, at once, endowed by God with all those magnificent gifts with which the Bible says He endowed us.

I can only hope that, in days which I shall not see, God may raise up some naturalists who may, in His hands, destroy the belief in our apedom.

I do not myself see the slightest difference between Darwin and Haeckel, except that Darwin assumes a First Cause, who, all those aeons ago, infused the breath of life into some primaeval forms, and has remained inactive (if, indeed, He is supposed to be a Personal Being) ever since.

Pusey to George Rolleston, Easter Tuesday 1879, in *Life of Edward Bouverie Pusey*, by Henry Parry Liddon, 4 vols. (London: Longmans, Green, 1897), 4:336.

Secondary Literature: C. Brad Faught, *The Oxford Movement* (University Park: Pennsylvania State University Press, 2004); Owen Chadwick, *The Victorian Church*, 2 vols. (New York: Oxford University Press, 1966–70).

JAMES JACKSON PUTNAM (1846–1918)
AMERICAN NEUROLOGIST

It was the cry of the church against Darwin, when he sought to "introspect" the history of life, and its echoes have drowned the voices of those who have sought to talk about the problems of sex, no matter with what earnestness. The cause of formal modesty and reticence has indeed had many noble martyrs. [. . .] But there is such a thing as paying too dear for this niceness, especially when, through the opposite course, we can have all that we should gain by this, and more besides. Strikingly enough, this outcry against one or another sort of investigation is never raised except with regard to our neighbors' efforts to find the truth; the purity of our own motives, the value of our own inquiries, provided they are genuine, rarely come in question.

James Jackson Putnam, *Addresses on Psycho-Analysis* (London: International Psycho-Analytic Press, 1921), 17.

Secondary Literature: George Prochnik, *Putnam Camp: Sigmund Freud, James Jackson Putnam and the Purpose of American Psychology* (New York: Other Press, 2006).

ARMAND DE QUATREFAGES (1810–1892)
FRENCH NATURALIST

I have had great pleasure, when occasion has offered, in defending the splendid researches made by Darwin in the natural sciences. For this reason, and at the risk of being considered narrow minded, enslaved to prejudices and unable to leave an old groove, etc. etc., I consider myself entitled to attack Darwinism, if I employ none but the weapons of science. [. . .]

Up to this point it is evident that I agree in all that Darwin has said on the struggle for existence and natural selection. I disagree with him when he attributes to them the power of modifying organized beings indefinitely in a given direction, so that the direct descendants of one *species* form *another species* distinct from the first.

Armand de Quatrefages, *The Human Species* (New York: Appleton, 1898), 93–95.

Secondary Literature: Jean-Christophe Sillard, "Quatrefages et le transformisme," *Revue de Synthèse,* 3rd ser., 95–96 (1979): 283–95.

R

SANTIAGO RAMÓN Y CAJAL (1852–1934)
Spanish Neurohistologist

[A]bout the year 1874 or 1875, I became acquainted with the fundamental works of Lamarck, Spencer*, and Darwin, and was able to taste the fruitful and elegant, though often unacceptable or exaggerated biogenetic hypotheses of Haeckel*, the spirited professor of Jena. Actually, the first refutation of Darwin's famous book, *The Origin of Species,* which came into my hands was written by Cánovas del Castillo!† It was sent to me from Madrid by one of the ardent admirers of the distinguished politician.

After more than a year had passed (1879), competitive examinations were announced for the vacant chair in Granada. Conscious of my defects, I had endeavored to overcome them so far as possible. I perfected myself in histological technique [. . .] ; I learned to translate scientific German; I acquired and studied conscientiously various German works on descriptive, general, and comparative anatomy; I posted myself in the modern theories of evolution, of which the standard-bearers at the time were the great Darwin, Haeckel, and Huxley*.

Santiago Ramón y Cajal, *Recollections of My Life,* trans. E. Horne Craigie (Cambridge, MA: MIT Press, 1989), 156–57, 255–56. †Antonio Cánovas del Castillo (1828–97), Spanish prime minister.

Secondary Literature: José María López Piñero, *Cajal* (Madrid: Debate, 2000); Richard L. Rapport, *Nerve Endings: The Discovery of the Synapse* (New York: Norton, 2005); *Cajal y la modernidad,* ed. Alejandro R. Díez Torre (Madrid: Ateneo de Madrid, 2008).

ANDREW CROMBIE RAMSAY (1814–1891)
Scottish Geologist

13 February [1848], Sunday [. . .] After lunch, [Edward] Forbes*, [Rich-ard] Owen*, Lyell*, and I had a walk in Sir John Lubbock's* park, and saw a number of things pleasant to look upon, in spite of a tendency to driz-zling. Nice cozy chat, too, before and after dinner. Darwin is an enviable man—a pleasant place, a nice wife, station neither too high nor too low, a good moderate fortune, and the command of his own time. After tea Mrs. Darwin and one of her sisters played some of Mendelssohn's duets, etc. etc., all very charming. I never enjoyed myself more. Forbes came to my room before going to bed, and gave me a sketch of his coming lecture on generic centers. Lyell is a much more amusing man than I gave him credit for. Mrs. Lyell is a charming person—pretty, lively, and full of faith in, and admiration of, her husband.

18th February.—Anniversary of Geol. Soc. Did not get down from my lecture till after the [Wollaston] medal had been given to, and acknowl-edged by, Dr. Buckland.[†] Sir Henry's[‡] address passed off very well. I sat mostly next Darwin.

For some years past Ramsay had been brooding upon what Darwin had so well enforced—the imperfection of the geological record. He was struck by the extraordinary gaps in the succession of organic remains, even where there was no marked physical interruption of the continuity of sedimentation. And he connected these gaps with geographical changes of which no other trace had survived. He had made a communication on this subject to the American Association at the Montreal meeting, which had attracted considerable attention among those present.

Sir Archibald Geikie, *Memoir of Sir Andrew Crombie Ramsay* (London: Macmil-lan, 1895), 123, 130, 276–77. [†]See Edward Forbes entry. [‡]Henry Thomas de la Beche. See Edward Forbes entry; the address referenced was his "anniversary address" as president of the Royal Geological Society.

Secondary Literature: Sandra Herbert, *Charles Darwin, Geologist* (Ithaca, NY: Cornell University Press, 2005).

WINWOOD READE (1838–1875)
English Explorer and Novelist

Thus finding that my outline of Universal History was almost complete, I determined in the last chapter to give a brief summary of the whole, filling up the parts omitted, and adding to it the materials of another work suggested several years ago by the *Origin of Species*. One of my reasons for revisiting Africa was to collect materials for this work, which I had intended to call "The Origin of Mind." However, Mr. Darwin's *Descent of Man* has left little for me to say respecting the birth and infancy of the faculties and affections. I, therefore, merely follow in his footsteps, not from blind veneration for a great master, but because I find that his conclusions are confirmed by the phenomena of savage life. On certain minor points I venture to dissent from Mr. Darwin's views, as I shall show in my personal narrative, and there is probably much in this work of which Mr. Darwin will disapprove. He must, therefore, not be made responsible for all the opinions of his disciple.

William Winwood Reade, *The Martyrdom of Man* (New York: Asa K. Butts, 1874), 7–8.

One day [in 1875] he came to me in trouble.[†] He had been reading the great work of Malthus—the *Essay on Population*—and said that it made him doubt the goodness of God. I replied with the usual common-place remarks; he listened to me attentively, then sighed, shook his head, and went away. A little while afterwards he read *The Origin of Species* which had just come out, and which proves that the Law of Population is the chief agent by which Evolution has been produced. From that time he began to show symptoms of insanity—which disease, it is thought, he inherited from one of his progenitors. He dressed always in black, and said that he was in mourning for mankind. The works of Malthus and Darwin, bound in somber covers, were placed on a table in his room; the first was lettered outside *The Book of Doubt,* and the second, *The Book of Despair.*

I kept myself awake, for I feared another dream; but the odors of the earth lingered in my nostrils, and its horrible cries still sounded in my ears. After all, I thought it was best to sleep if I could. Luckily, the last number of the *Quarterly Review* happened to be in the room, and I knew that Dr. Scott recommends this publication in cases of sleeplessness and nervous excitement. The article I selected was a perfect soporific—an essay on the Darwinian Theory, and before I had finished the preamble it had sent me to sleep. But on that fatal night even the *Quarterly Review* could not prevent me from dreaming; and, in fact, I dreamt of a review, for my third dream took me to a Demigod club where I found the following critique in a periodical lying on the table. I wrote it from memory as soon as I awoke.

The Review

[. . .] the development of matter to mind, of quadruped to man, of savage to civilized nations, is laudable enough as an idea; but how has it been carried out? As regards the first stage of the progress we have only to praise and admire; but how has progress been produced in the animated world? We are almost ashamed to explain a law which, in its recklessness of life and prodigality of pain, almost amounts to a crime. In cold forethought the Creator so disposed the forces of nature that more animated beings were born than could possibly obtain subsistence on the earth. This caused a struggle for existence, a desperate and universal war; the best and improved animals were alone able to survive, and so in time Evolution was produced. We shall not deny that there is a kind of perverted ingenuity in the composition of this law; but the waste of life is not less clumsy than it is cruel. By means of this same struggle for existence, man was raised from the bestial state and his early discoveries were made. Afterwards, ambition of fame, and later still more noble motives came into force, but that was towards the conclusion of the drama. At first, every step in the human progress was won by conflict, and every invention resulted from calamity. The most odious vices and crimes were at one time useful to humanity, while war, tyranny, and superstition assisted the development of man.

Winwood Reade, *The Outcast*, 3rd ed. (London: Chatto & Windus, 1875), 4, 32–33, 42–44. †Arthur Elliott, a parson who is losing his faith, comes to the author for advice.

Secondary Literature: Felix Driver, *Cultures of Exploration and Empire* (Oxford: Blackwell, 2001), chap. 5.

ERNEST RENAN (1823–1892)
FRENCH PHILOSOPHER AND HISTORIAN

Physiology and the natural sciences would have absorbed me, and I do not hesitate to express my belief—so great was the ardor which these vital sciences excited in me—that if I had cultivated them continuously I should have arrived at several of the results achieved by Darwin, and partially foreseen by myself.

Ernest Renan, *Recollections of my Youth,* trans. C. R. Pitman (London: Chapman & Hall, 1883), 228.

Secondary Literature: W. M. Simon, "Joseph Ernest Renan," in *Encyclopedia of Philosophy,* ed. Paul Edwards, 8 vols. (New York: Macmillan, 1967), 7:179–80; Alan Pitt, "The Cultural Impact of Science in France: Ernest Renan and the *Vie de Jésus,*" *Historical Journal* 43 (2000): 79–101.

HANS RICHTER (1843–1916)
AUSTRIAN CONDUCTOR

Richter was received in friendly fashion and after lunch he played to his host, mainly at the instructions of Mrs. Darwin who is very musical. A son is at present in Strasbourg and plays the bassoon very well. At half-past three Franke[†] and I took a beautiful walk back to the station. This day remains unforgettable to me. What kindness lay in the eyes of the great man! "I think you have several instruments in your pocket," he said after I had played a few pieces by Beethoven, [Richard] Wagner*, and Mozart.

Hans Richter's diary, 27 May 1881, quoted in *True Artist and True Friend: A Biography of Hans Richter,* by Christopher Fifield (Oxford: Clarendon, 1993), 158.
[†]Hermann Franke (1834–1919), German composer.

Secondary Literature: Christopher Fifield, *True Artist and True Friend: A Biography of Hans Richter* (Oxford: Clarendon, 1993).

CHARLES VALENTINE RILEY (1843–1895)
American Entomologist

I have had the pleasure on two occasions of visiting Darwin at his invitation. On the first occasion, in the summer of 1871, I was accompanied by Mr. J. Jenner Weir,[†] one of his life-long friends and admirers. From Mr. Weir I first learned that Darwin was, in one sense, virtually a confirmed invalid, and that his work had been done under physical difficulties which would have rendered most men of independent means vapid, self-indulgent, and useless members of society.

The Darwin residence is a plain, but spacious, old-fashioned house of the style so common in England, and which, with the surrounding well-kept grounds and conservatory, convey that impression of ease and comfort that belong to the average home of the English country gentleman. A noticeable feature is a bow window extending through three stories and covered with trellis and creepers. In Darwinian phrase, the environment was favorable for just such calm study and concentration as he found necessary to his health and his researches.

Upon introduction I was at once struck with his stature (which was much above the average, and I should say fully six feet,) his ponderous brow and long white beard—the moustache being cut on a line with the lips and slightly brown from the habit of snuff-taking. His deep-set eyes were light blue-gray. He made the impression of a powerful man reduced somewhat by sickness. The massive brow and forehead show in his later photographs, but not so conspicuously as in a life-sized head of him when younger, which hung in the parlor.

In the brief hours I then spent at Down the proverbial modesty and singular simplicity and sweetness of his character were apparent, while the delight he manifested in stating facts of interest was excelled only by the eagerness with which he sought them from others, whether while strolling through the greenhouse or sitting round the generously spread table.

Charles V. Riley, "Darwin's Work in Entomology," *Proceedings of the Biological Society of Washington DC* 1 (1882): 77, 78–79. [†]John Jenner Weir (1822–94), amateur entomologist, ornithologist, and informant of Darwin's.

Charles Robert Darwin was one of the original members of the London Entomological Society, of whom only six are yet living. He always took the keenest interest in the science of entomology, and drew largely from insects for illustrations in support of the theory with which his name will forever be associated. Indeed, I have the authority of my late associate editor of the *American Entomologist*, Benjamin Dann Walsh, who was a classmate of Darwin's at Cambridge, that the latter's love of natural history was chiefly manifested, while there, in a fine collection of insects; so that, as has been the case with so many noted naturalists, Darwin probably acquired from the study of insects that love of nature, which, first forever afterward, inspired him in his endeavors to win her secrets and interpret aright her ways!

[T]he changed aspect which natural history in general has assumed since the publication of the *Origin of Species* is perhaps more marked in entomology than in any other branch, for its author helped to replace ridicule by reason. During his voyage on the *Beagle* he collected a very large number of interesting species, especially in *Coleoptera*, and they formed the basis of many memoirs by Walker, Newman, and White,[†] and particularly by G. R. Waterhouse,[‡] who named *Odontoscelis Darwinii* after him.

In the light of Darwinism, insect structure and habit have come to possess a new significance and a deeper meaning. It has, in short, proved a new power to the working entomologist who, for all time, will hold in reverence the name of him who, more than any other man, helped to replace scholasticism by induction and who gave to the philosophic study of insects as great an impetus as did Linnaeus to their systematic study.

Charles V. Riley, "Darwin's Work in Entomology," *Proceedings of the Biological Society of Washington DC* I (1882): 70, 71, 77. [†]Francis Walker (1809–74), English entomologist, at the British Museum from 1837 to 1863; Edward Newman (1801–76), English entomologist, editor of the *Entomological Magazine*, the *Entomologist*, and the *Zoologist*; and Adam White (1817–79), English entomologist, in the Zoology Department of the British Museum. [‡]See Joseph Dalton Hooker entry.

The importance of insects as agents in cross-fertilization was scarcely appreciated, however, until the late Charles Darwin published the results of his researches on *Primula, Linum, Lythrum,* etc., and his elaborate work on the fertilization of orchids. The publication of these works gave to flowers a new significance and to their study almost as great an impulse as did his immortal *Origin of Species* to the general study of biology.

Charles Valentine Riley, "The Yucca Moth and Yucca Pollination," *Annual Report of the Missouri Botanical Garden (1892)* (St. Louis: Board of Trustees, 1892), 99–166, on 101.

Secondary Literature: Edward Oliver Essig, *A History of Entomology* (New York: Macmillan 1931); Christy Campbell, *The Botanist and the Vintner: How Wine Was Saved for the World* (Chapel Hill, NC: Algonquin Books, 2006).

ANNE THACKERAY RITCHIE (1837–1919)
English Author

We spent Sunday with Darwin to our great pride and delight.

Anne Thackeray Ritchie to W. W. F. Synge, 25 December 1881, in *Anne Thackeray Ritchie: Journals and Letters,* ed. Abigail Burnham Bloom and John Maynard (Columbus: Ohio State University Press, 1994), 231.

Secondary Literature: Philip J. Waller, *Writers, Readers, and Reputations: Literary Life in Britain, 1870–1918* (Oxford: Oxford University Press, 2006).

DAVID GEORGE RITCHIE (1853–1903)
Scottish Philosopher

Darwin disclaims the connection, which had been alleged in Germany, between the doctrine of natural selection and socialism. He sees clearly enough that his theory gives a *prima facie* support not to socialism, but to industrial competition. Yet he is amused at the idea of *The Origin of Species* having turned Sir Joseph Hooker* into "a jolly old Tory." "Primogeniture," he says, "is dreadfully opposed to selection: suppose the first-born bull was necessarily made by each farmer the begetter of his stock!" Still, he admits that English peers have an advantage in the selection of "beautiful and charming women out of the lower ranks," and thus get some benefit from the principle. In answering Mr. [Francis] Galton's* questions, Darwin describes his own politics as "Liberal or Radical" and this was in 1873, by which time Radicalism was no longer bound to out-and-out *laissez faire*.

Evolution, as applied to the whole of the universe, means a great deal more than the principle of natural selection. In the wider sense it is professedly applied to the guidance of life by [David Friedrich] Strauss* in his famous book, *The Old Faith and the New,* where military conquest, and social inequalities are expressly defended as right, because natural; and nothing but contempt is reserved for those who venture to hope for the abolition of war, who look beyond the limits of the nation or who dream of a better social order. It might be objected that in these passages we do not hear the voice of German science and philosophy, but of that reactionary military spirit which has infected the new German nation; and I think it could be shown that such sentiments are inconsistent with admissions that Strauss himself makes, although he and most German *savants* with him believe that they are a necessary consequence of the Evolutionist creed.

David G. Ritchie, *Darwinism and Politics,* 2nd ed. (London: Swan Sonnenschein, 1891), 7–9.

Secondary Literature: David Boucher and Andrew Vincent, *British Idealism and Political Theory* (Edinburgh: Edinburgh University Press, 2000).

WILLIAM BARTON ROGERS (1804–1882)
AMERICAN GEOLOGIST AND EDUCATOR

My dear William, . . . The only matter of any interest here is the appearance of Charles Darwin's book *On the Origin of Species by Means of Natural Selection.* It is a suggestive book, full of ingenious arguments in favor of the Lamarckian hypothesis. Huxley*, who bitterly criticized the *Vestiges,* has reviewed this work in terms of high commendation. When you read it you will often say, I think, that in his geology Darwin outdoes Lyell* himself in ignoring paroxysmal actions. That is its chief blemish with me.

Henry Darwin Rogers to William Barton Rogers, 23 December 1859, in *Life and Letters of William Barton Rogers,* 2 vols. (Boston: Houghton, Mifflin, 1896), 2:17–18.

The more I look into Darwin's argument the more I like it, save in the one particular of ignoring entirely violent and sudden physical changes. The calmness and truth-loving spirit of the book are truly remarkable. Much of it I know *you* will approve.

William Barton Rogers to Henry Rogers, 2 January 1860, in *Life and Letters of William Barton Rogers,* 2 vols. (Boston: Houghton, Mifflin, 1896), 2:19.

The next number of Silliman[†] will contain an elaborate review of Darwin's book by [Asa] Gray* who called some days ago to leave the proof-sheets with me. He has not affixed his name, and requested me to say nothing of it for the present. I hear that Mr. [George] Ticknor* has just heard from Lyell*, who speaks approvingly of Darwin's views. I anticipate many disciples for Darwin on this side of the Atlantic, in spite of the *"disastrous"* tendency of his views.

William Barton Rogers to Henry Rogers, February 1860, in *Life and Letters of William Barton Rogers,* 2 vols. (Boston: Houghton, Mifflin, 1896), 2:20. [†]See Theophilus Parsons entry.

I wrote to Huxley* awhile since in relation to Darwin's book and your liberal defense of it, and he replies much pleased. In a few years, opinion among the reflecting in Europe will be with Darwin, you may depend on it, as to the law of natural selection. As to the other point, the development of species from species, firmly as I believe in it, I think it will never be capable of a strictly scientific proof; no more can the opposite doctrine of supernatural creations, and therefore the main point to insist on now is toleration, and no dogmatizing.

Henry Rogers to William Barton Rogers, 24 February 1860, in *Life and Letters of William Barton Rogers,* 2 vols. (Boston: Houghton, Mifflin, 1896), 2:24.

Secondary Literature: A. J. Angulo, *William Barton Rogers and the Idea of MIT* (Baltimore: Johns Hopkins University Press, 2009).

GEORGE ROLLESTON (1829–1881)
English Anatomist and Physiologist

The interest in the early history of Man thus awakened by the problem of his relation to lower animals as raised by Darwin, came before Rolleston's mind a few months later in a more intense and personal form, in the Zoological Section of the British Association, at a memorable meeting, which Professor [Henry Wentworth] Acland,[†] who was present also, mentions twenty years after in his obituary notice of his friend. It so happened, that in 1860 a circumstance took place which tended materially to concentrate all the qualities of his nature on the highest biological questions, whether considered from the material or psychological point of view. The British Association met in Oxford, and the famous discussion on the hippocampus in the brain of man as compared with that of the higher apes took place between Professors [Richard] Owen[*] and Huxley[*]. Bishop [Samuel] Wilberforce[*] brought, as is well remembered by all scientific men, the forces of his ready wit and great reputation to bear against the sincere statements of the younger anatomist. Rolleston's indignation was fired, his sense of justice made him throw heart as well as head into the cause of what, at the moment, seemed the weaker man. It is not possible to say now to what extent that brief scene influenced the ardor and imagination of Rolleston. Be this as it may, all prejudice and even bias derived from the most refined Oxford culture was banished from his mind in dealing with the nature of man.

The controversy here referred to, one of importance in modern scientific history, lasted several years. It arose at the Oxford meeting just mentioned out of a botanical paper by Dr. Daubeny[‡] on the Sexuality of Plants, which went into criticism of Darwin's *Origin of Species*. Professor Owen, in the discussion which ensued, took up the question of differences between apes and man, asserting that the brain of the Gorilla presented more differences as compared with Man than with the lowest Quadrumana. This was met by Professor Huxley with a flat denial, he declaring that the brains of man and the highest monkeys differ less than the brains of the highest and lowest monkeys. Rolleston does not appear to have spoken at this time, nor on the occasion a day or two later when the Bishop of Oxford received a famous rebuke for the rhetorical device of perverting the Darwinian theory in order to make fun of it. But this problem of brain classification became an especial subject of Rolleston's study, and in January 1862 he delivered a lecture upon it at the Royal Institution.

Edward B. Tylor, "Life of Dr. Rolleston," in *Scientific Papers and Addresses,* by George Rolleston, ed. William Turner, 2 vols. (Oxford: Clarendon, 1884), xxxiii–xxxiv. †Henry Wentworth Acland (1815–1900), English physician, Regius professor of medicine at Oxford and founder of the Oxford University Museum. ‡Charles G. B. Daubeny (1795–1867), English chemist, botanist, and geologist.

Secondary Literature: Edward Bagnall Poulton, "George Rolleston," in *John Viriamu Jones and other Memories* (New York: Longmans, Green, 1911), 183–220.

GEORGE JOHN ROMANES (1848–1894)
English Biologist and Comparative Psychologist

We went over to pay a call upon Darwin [in July 1881]. He and his wife were at home, and as kind and glad to see us as possible. The servant gave our names wrongly to them, and they thought we were a very old couple whom they know, called Norman. So old Darwin came in with a huge canister of snuff under his arm—old Norman being very partial to this luxury—and looked very much astonished at finding us. He was as grand and good and bright as ever.

Romanes to Miss C. E. Romanes, 24 July 1881, in *The Life and Letters of George John Romanes,* ed. Ethel Duncan Romanes (New York: Longmans, Green, 1896), 129.

I have heard an eminent professor tell his class that the many instances of adaptation which Mr. Darwin discovered and described as occurring in orchids, seemed to him to tell more in favor of contrivance than in favor of natural causes; and another eminent professor once wrote to me that although he had read the *Origin of Species* with care, he could see in it no evidence of natural selection which might not equally well be adduced in favor of intelligent design. But here we meet with a radical misconception of the whole logical attitude of science. For, be it observed, the exception *in limine* to the evidence which we are about to consider, does not question that natural selection may not be able to do all that Mr. Darwin as-cribes to it: it merely objects to his interpretation of the facts, because it maintains that these facts might equally well be ascribed to intelligent design. And so undoubtedly they might, if we were all childish enough to rush into a supernatural explanation whenever a natural explanation is found sufficient to account for the facts.

George John Romanes, *The Scientific Evidences of Organic Evolution* (London: Macmillan, 1882), 7–8.

Secondary Literature: Donald Forsdyke, *The Origin of Species, Revisited: A Victorian Who Anticipated Modern Developments in Darwin's Theory* (Kingston, ON: McGill-Queen's University Press, 2001); T. A. Goudge, "George John Romanes," in *Encyclo-pedia of Philosophy,* ed. Paul Edwards, 8 vols. (New York: Macmillan, 1967), 7:205–6.

THEODORE ROOSEVELT (1858–1919)
AMERICAN PRESIDENT

His work as a naturalist now [ca. 1870] developed from a mere boys' pas-time into actual and more or less scientific study. In a musty little shop he discovered a former companion of Audubon's, a tall, clean shaven, white-haired old gentleman named Bell, who subsequently gave him lessons in taxidermy and spurred and directed his interest in collecting specimens for mounting and preservation. He had another friend at that time, Hil-borne West, a connection by marriage of his mother's, an intimate of many of the most noted scientists of his generation, who was neither an original thinker himself nor even a learned man, but who had the rare ability of explaining in words of one syllable the intricate theories and discoveries of his profounder brethren. It was through the clear interpretation of Hil-borne West that Theodore the younger made his first acquaintance with the ideas of Darwin and Huxley*, which were then shaking the founda-tions of science and religion.

Hermann Hagedorn, *The Boys' Life of Theodore Roosevelt* (New York: Harper, 1918), 38–39.

Indeed, I believe that already science has owed more than it suspects to the unconscious literary power of some of its representatives. Scientific writers of note had grasped the fact of evolution long before Darwin and Huxley*; and the theories advanced by these men to explain evolution were not much more unsatisfactory, as full explanations, than the theory of natural selection itself. Yet, where their predecessors had created hardly a ripple, Darwin and Huxley succeeded in effecting a complete revolution in the thought of the age, a revolution as great as that caused by the discovery of the truth about the solar system. I believe that the chief explanation of the difference was the very simple one that what Darwin and Huxley wrote was interesting to read. Every cultivated man soon had their volumes in his library, and they still keep their places on our book-shelves. But Lamarck and [Edward Drinker] Cope* are only to be found in the libraries of a few special students. If they had possessed a gift of expression akin to Darwin's, the doctrine of evolution would not in the popular mind have been confounded with the doctrine of natural selection and a juster estimate than at present would obtain as to the relative merits of the explanations of evolution championed by the different scientific schools.

Theodore Roosevelt, *History as Literature* (New York: Scribner's, 1913), 12.

A hundred years ago, even seventy or eighty years ago, before the age of steamboats and railroads, it was more difficult than at present to define the limits between this class and the next; and, moreover, in defining these limits I emphatically disclaim any intention of thereby attempting to establish a single standard of value for books of travel. Darwin's *Voyage of the Beagle* is to me the best book of the kind ever written; it is one of those classics which decline to go into artificial categories, and which stand by themselves; and yet Darwin, with his usual modesty, spoke of it as in effect a yachting voyage.

Theodore Roosevelt, *Through the Brazilian Wilderness* (New York: Charles Scribner's Sons, 1914), 345.

Secondary Literature: David H. Burton, "Theodore Roosevelt's Social Darwinism and Views on Imperialism," *Journal of the History of Ideas* 26 (1965): 103–18.

JOSIAH ROYCE (1855–1916)
American Philosopher

To begin, then, the exposition of what I take to be the spirit of pragmatism—thinking, judging, reasoning, believing—these are all of them essentially practical activities. One cannot sunder will and intellect. A man thinks about what interests him. He thinks because he feels a need to think. His thinking may or may not be closely linked to those more worldly activities which common sense loves to call practical. But the most remote speculations are, for the man who engages in them, modes of conduct. As contrasted with other men, the thinker, so far as his thoughts do not directly link themselves to the motor processes usually called practical, appears, when viewed from without, to be an inactive person. An anecdote records how a servant woman in Darwin's household ventured to suggest that the old gentleman must be so delicate as he was in health because, as she said, he lacked occupation, and only wandered about looking at his plants, or sat poring over his papers. Whether the anecdote is true or not, the thinker often seems to casual observers to be inhibited, held back from action, hopelessly ineffective. But this appearance we know to be but a seeming. The thinker plans motor processes, and in the end, or even constantly while he thinks, these processes get carried out. The thinker makes diagrams, arranges material objects in classes and in orderly series, constructs apparatus, adjusts with exquisite care his delicate instruments of precision; or he takes notes, builds up formulas, constructs systems of spoken or written words to express his thoughts; and ultimately, in expressing his thoughts, he may direct the conduct of vast numbers of other men, just as Darwin came to do.

Josiah Royce, "The Eternal and the Practical," *Philosophical Review* 13 (1904): 113–42, on 118–19.

Secondary Literature: John E. Smith, "Josiah Royce," in *Encyclopedia of Philosophy,* ed. Paul Edwards, 8 vols. (New York: Macmillan, 1967), 7:225–29; *Josiah Royce's Seminar, 1913–1914,* ed. Grover Smith (New Brunswick, NJ: Rutgers University Press, 1963).

CLÉMENCE ROYER (1830–1902)
French Philosopher and Translator

The doctrine of M. Darwin is the rational revelation of progress, pitting itself in its logical antagonism with the irrational revelation of the fall. These are two principles, two religions in struggle, a thesis and an antithesis of which I defy the German who is most proficient in logical developments to find a synthesis. It is a quite categorical yes and no between which it is necessary to choose, and whoever declares himself for one is against the other. [...] For myself, the choice is made: I believe in progress.

Clémence Royer, "Préface de la première édition," in her translation of Darwin, *De l'origine des espèces* (Paris: Guillaumin & V. Masson, 1866), quoted in "France," by Robert Stebbins, in *The Comparative Reception of Darwinism*, ed. Thomas F. Glick, 2nd ed. (Chicago: University of Chicago Press, 1988), 117–63, on 126.

I received, 2 or 3 days ago, a French translation of the *Origin* by a Madlle. Royer, who must be one of the cleverest and oddest women in Europe: is an ardent Deist, and hates Christianity, and declares that natural selection and the struggle for life will explain all morality, nature of man, politics, &c. &c.! She makes some very curious and good hits, and says she shall publish a book on these subjects.

Darwin to Asa Gray, June 1862, in *The Life and Letters of Charles Darwin*, ed. Francis Darwin, 2 vols. (New York: Appleton, 1887), 2:179.

[Jules] Gavarret[†] had a profound admiration for you, your character, and your remarkable work, as I am certain you have not forgotten. When [he] pronounced your name, everyone in the room stood up to cheer you. It is true that young people of great and generous aspirations and healthy, accurate judgments render homage to all that is good, beautiful, and true, even when it concerns a woman. That meeting I have never forgotten. It raised in my mind the same breath of enthusiasm. The next day I read the *Origin of Species* and learned of the immortal work of Darwin.

Dr. Madeleine Brès, addressing Royer, in a speech at a banquet honoring Royer, 10 March 1897, quoted in *"Almost a Man of Genius": Clémence Royer, Feminism, and Nineteenth-Century Science*, by Joy Harvey (New Brunswick, NJ: Rutgers University Press, 1997), 104–5. [†]Jules Gavarret (1809–90), French medical professor.

Secondary Literature: Joy Harvey, "Darwin in a French Dress: Translating, Publishing and Supporting Darwin in Nineteenth-Century France," in *The Reception of Charles Darwin in Europe,* ed. Eve-Marie Engels and Thomas F. Glick, 2 vols. (London: Continuum, 2008), 2:354–74.

JOHN RUSKIN (1819–1900)
English Essayist and Critic

While we were sitting over our wine after dinner, in came Dr. [Charles G. B.] Daubeny, one of the most celebrated geologists of the day—a curious little animal, looking through its spectacles with an air very *distinguee*—and Mr. Darwin, whom I had heard read a paper at the Geological Society. He and I got together, and talked all the evening.

John Ruskin to his father, 22 April 1837, in *The Life of John Ruskin,* by W. G. Collingwood (Boston: Houghton Mifflin, 1902), 61.

We were detained late in the country. On the 20th November [1868], Ruskin wrote: "I will come to-morrow and shall have very great pleasure in meeting Mr. Darwin." They had never before met, and each was interested to see the other. The contrast between them was complete, and each in his own way was unique and delightful. Ruskin's gracious courtesy was matched by Darwin's charming and genial simplicity. Ruskin was full of questions which interested the elder naturalist by the keenness of observation and the variety of scientific attainment which they indicated, and their animated talk afforded striking illustration of the many sympathies that underlay the divergence of their points of view, and of their methods of thought. The next morning Darwin rode over on horseback to say a pleasant word about Ruskin, and two days afterward Ruskin wrote, "Mr. Darwin was delightful."

Charles Eliot Norton, in *Letters of John Ruskin to Charles Eliot Norton,* ed. Norton, 2 vols. (Boston: Houghton Mifflin, 1905), 1:194–95.

All these materialisms, in their unclean stupidity, are essentially the work of human bats; men of semi-faculty or semi-education, who are more or less incapable of so much as seeing, much less thinking about, color;

among whom, for one-sided intensity, even Mr. Darwin must be often ranked, as in his vespertilian treatise on the ocelli of the Argus pheasant which he imagines to be artistically gradated, and perfectly imitative of a ball and socket. If I had him here in Oxford for a week, and could force him to try to copy a feather by Bewick,[†] or to draw for himself a boy's thumbed marble, his notions of feathers, and balls, would be changed for all the rest of his life. But his ignorance of good art is no excuse for the acutely illogical simplicity of the rest of his talk of color in the *Descent of Man.* Peacocks' tails, he thinks, are the result of the admiration of blue tails in the minds of well-bred peahens,—and similarly, mandrills' noses the result of the admiration of blue noses in well-bred baboons. But it never occurs to him to ask why the admiration of blue noses is healthy in baboons, so that it develops their race properly, while similar maidenly admiration either of blue noses or red noses in men would be improper, and develop the race improperly. The word itself "proper" being one of which he has never asked, or guessed, the meaning. And when he imagined the gradation of the cloudings in feathers to represent successive generations, it never occurred to him to look at the much finer cloudy gradations in the clouds of dawn themselves; and explain the modes of sexual preference and selective development which had brought them to their scarlet glory, before the cock could crow thrice. [. . .]

When therefore I said that Mr. Darwin, and his school, had no conception of the real meaning of the word "proper," I meant that they conceived the qualities of things only as their "properties," but not as their "becomingnesses"; and seeing that dirt is proper to a swine, malice to a monkey, poison to a nettle, and folly to a fool, they called a nettle but a nettle, and the faults of fools but folly; and never saw the difference between ugliness and beauty absolute, decency, and indecency absolute, glory or shame absolute, and folly or sense absolute.

John Ruskin, *Proserpina* (Philadelphia: Reuwee Watley & Walsh, 1891), 60–65.
[†]William Bewick (1795–1866), English painter and copyist.

[Among books omitted from Ruskin's list of the best hundred books:] Darwin.—Because it is every man's duty to know what he is, and not to think of the embryo he was, nor the skeleton that he shall be. Because, also, Darwin has a mortal fascination for all vainly curious and idly speculative persons, and has collected, in the train of him, every impudent imbecility in Europe, like a dim comet wagging its useless tail of phosphorescent nothing across the steadfast stars.

John Ruskin, *Arrows of the Chace,* in *The Works of John Ruskin,* ed. E. T. Cook and Alexander Wedderburn, vol. 34 (London: Longmans, Green, 1908), 586.

In the seventh volume of *Fors,* 91, [Ruskin] quotes with approval a letter of [Thomas] Carlyle's*, in which he says with contemptuous superiority, "A good sort of man is this Darwin, and well-meaning, but with very little intellect." Neither Carlyle, however, nor Ruskin seem [*sic*] to have ever rightly grasped the theory of Evolution. Ruskin says, for instance, "We might safely, even sufficiently, represent the general manner of conclusion in the Darwinian system by the statement that if you fasten a hair-brush to a mill-wheel, with the handle forward, so as to develop itself into a neck by moving always in the same direction, and within continual hearing of a steam-whistle, after a certain number of revolutions the hairbrush will fall in love with the whistle, they will marry, lay an egg, and the produce will be a nightingale." This is an amusing skit, but Mr. Darwin would, I need not say, have been much astonished to find himself credited with such a theory.

John Lubbock, *Essays and Addresses, 1900–1903* (Leipzig: Bernhard Tauchnitz, 1904), 50.

Secondary Literature: Raymond Williams, "John Ruskin," in *Encyclopedia of Philosophy,* ed. Paul Edwards, 8 vols. (New York: Macmillan, 1967), 7:234–35; Michael Wheeler, *Ruskin's God* (Cambridge: Cambridge University Press, 1999).

ARTHUR RUSSELL (1825–1892)
English Politician

We saw a great deal of the great Darwin at Holwood—he lives at Down not far off—a charming man, but with wretched health. Laura painted his picture—He is the greatest name in English science at this moment—the most universally known abroad.

Arthur Russell to Kate Russell, 9 September 1869, in *The Amberly Papers,* by Bertrand Russell and Patricia Russell, 2 vols. (New York: Norton, 1937), 2:450.

BERTRAND RUSSELL (1872–1970)
English Philosopher

Sometimes [in Russell's childhood] the conversation descended to more recent times, and I should be told how [Thomas] Carlyle* had called Herbert Spencer* a "perfect vacuum," or how Darwin had felt it a great honor to be visited by Mr [William] Gladstone*.

Bertrand Russell, *Autobiography* (London: Routledge, 1998), 15.

Evolutionism, in one form or another, is the prevailing creed of our time. It dominates our politics, our literature, and not least our philosophy. [Friedrich] Nietzsche*, pragmatism, [Henri] Bergson*, are phases in its philosophic development, and their popularity far beyond the circles of professional philosophers shows its consonance with the spirit of the age.

Bertrand Russell, *Our Knowledge of the External World as a Field for Scientific Method in Philosophy* (Chicago: Open Court, 1914), 11.

There is a further consequence of the theory of evolution, which is independent of the particular mechanism suggested by Darwin. If men and animals have a common ancestry, and if men developed by such slow stages that there were creatures which we should not know whether to classify as human or not, the question arises: at what stage in evolution did men, or their semi-human ancestors, begin to be all equal? Would *Pithecanthropus erectus,* if he had been properly educated, have done work as good as Newton's? Would the Piltdown Man have written Shakespeare's poetry if there had been anybody to convict him of poaching? A resolute egalitarian who answers these questions in the affirmative will find himself forced to regard apes as the equals of human beings. And why stop with apes? I do not see how he is to resist an argument in favor of Votes for Oysters. An adherent of evolution may maintain that not only the doctrine of the equality of all men, but also that of the rights of man, must be condemned as unbiological, since it makes too emphatic a distinction between men and other animals. There is, however, another aspect of liberalism which was greatly strengthened by the doctrine of evolution, namely the belief in progress. So long as the state of the world allowed optimism, evolution was welcomed by liberals, both on this ground and because it gave new arguments against orthodox theology. [Karl] Marx* himself, though his doctrines are in some respects pre-Darwinian, wished to dedicate his book to Darwin. The prestige of biology caused men whose thinking was influenced by science to apply biological rather than mechanistic categories to the world. Everything was supposed to be evolving, and it was easy to imagine an immanent goal. In spite of Darwin, many men considered that evolution justified a belief in cosmic purpose. The conception of organism came to be thought the key to both scientific and philosophical explanations of natural laws, and the atomic thinking of the eighteenth century came to be regarded as out of date. This point of view has at last influenced even theoretical physics. In politics it leads naturally to emphasis upon the community as opposed to the individual. This is in harmony with the growing power of the State; also with nationalism, which can appeal to the Darwinian doctrine of survival of the fittest applied, not to individuals, but to nations.

Bertrand Russell, *A History of Western Philosophy* (New York: Simon & Schuster, 1945), 724–26.

Secondary Literature: Caroline Moorhead, *Bertrand Russell: A Life* (London: Sinclair-Stevenson, 1992); Paul Edwards, William P. Alston, and A. N. Prior, "Bertrand Arthur Wilson Russell," in *Encyclopedia of Philosophy*, ed. Paul Edwards, 8 vols. (New York: Macmillan, 1967), 7:235–58.

S

GEORGE SANTAYANA (1863–1952)
AMERICAN PHILOSOPHER

In parts of biology which do not deal with man observers do not hesitate to refer in the same way to the pain, the desire, the intention, which they may occasionally read in an animal's aspect. Darwin, for instance, constantly uses psychical language: his birds love one another's plumage and their aesthetic charms are factors in natural selection. Such little fables do not detract from the scientific value of Darwin's observations, because we see at once what these fables mean. The description keeps close enough to the facts observed for the reader to stop at the latter, rather than at the language in which they are stated.

George Santayana, *The Life of Reason* (New York: C. Scribner's, 1905–6; Amherst, NY: Prometheus Books, 1998), 420–21.

[T]he professor of philosophy had to swim against rather a powerful current. Sometimes he succumbed to the reality; and if, for instance, he happened to mention Darwin, and felt a blank before him, he would add in a parenthesis, "Darwin, Charles, author of the *Origin of Species,* 1859; epoch-making work."

George Santayana, *Character and Opinion in the United States* (New York: Scribner's, 1921), 60.

That you should think Plato good but not true and at the same time follow Darwin with approval would seem to indicate that you instinctively think as I think. This, and your Latin [. . .] blood don't apparently suffice to make you feel at home in my Weltanschauung. What is the difficulty? You don't tell me or give me any hint of where it lies. Why is Plato good in spite of being wrong? I should say because his ethics and politics are right in principle. But his cosmology is mythical and made to fit his humanism miraculously. Having been planned on purpose to produce an ideal Athens and a perfect set of Athenians. Now, this is contrary to Darwin and must be abandoned. Although the Platonic myth may be excellent parables, illustrating the growth of human virtues, I therefore stick to Darwin (or in my case—rather to Lucretius and Spinoza) in my cosmology; but when I turn to the realm of Spirit (which has its perfectly natural place in animal life), I drop Darwin, Lucretius, and even Spinoza and stick to Plato, or rather to the *idea* of Christ. I have lately been writing a book on this last subject, which may show you what I mean, and I graft this Christian morality on the naturalistic stalk. Of course, if you hanker for a physically really *good* world, you will never find it, and it may seem to you discouraging spiritually that spirit should not rule the universe. That would seem to me to be a pity, and a lack of caution in not keeping truth and imagination in their respective places. Is that what makes you feel uncomfortable?

Santayana to Lieutenant Garcia, 26 February 1946, in *The Letters of George Santayana,* ed. William G. Holzberger, 8 vols. (Cambridge, MA: MIT Press, 2001–8), 7:221.

Secondary Literature: John McCormick, *George Santayana: A Biography* (New Brunswick, NJ: Transaction, 2003); Frederick A. Olafson, "George Santayana," in *Encyclopedia of Philosophy,* ed. Paul Edwards, 8 vols. (New York: Macmillan, 1967), 7:282–87.

FRANCISQUE SARCEY (1827–1899)
French Journalist and Drama Critic

The servant was assisting us to take off our overcoats when Darwin appeared. He had come into the hall to shake us by the hand. It is odd what ideas we get of people when we picture them to ourselves without having seen them. I expected to see a little stooping, wrinkled, suffering old man. I knew, indeed, that he was seventy-six, and we were told in our letter of introduction that he was not well. There approached me, however, a tall fine old man, as upright as a pin, and apparently as robust as an oak. The characteristic feature of the countenance is the prominence of the forehead and eyebrows, the latter being arched above eyes of inexpressible brightness and gentleness. I was reminded of the portraits of Goethe, whose eyes were sunk in the same way behind a projecting brow. Darwin received us very warmly. He smiled when he offered us his hand, and displayed thereby a set of white teeth. Everything about him denotes at once health and strength, and seems built to live till he is at least a hundred years old. He speaks freely of his age, and of the end which he foresees, and not without a shade of melancholy. "It is a pity to have to go," he said to Barbier,[†] "when one has still so many things to do. As I proceed in the study of nature, I discover vaster horizons, and I feel that I shall not have time to reach them." He limits his ambition to finishing two works, which he has already commenced. One is the life of his grandfather, who was a celebrated physician, and who, like himself, was overwhelmed with abuse by the contemptible and the bigoted of his time for having ventured to break a lance with certain of the prejudices which then prevailed. The other is a work upon the faculties of motion possessed by plants.

He introduced us into a spacious drawing room, the doors and windows of which opened upon a large garden. There were already a few persons in the room, and to these he presented us. He then sat down upon a very high armchair at the chimney-corner, where there burned (in the middle of summer!) a tolerably big coal fire. We were placed upon much lower seats, so that from the height of his arm chair, and with his tall figure, not an inch of which he lost, he towered above us and resembled one of those veterans whom Victor Hugo likes to draw In the *Legende des Siècles,* and whom be calls "Aieule." Darwin does not speak French, and he excused himself for it with great politeness. As he speaks slowly—and perhaps he had the kindness to moderate his utterance out of regard to the ignorance of his guest—I half understood the sense of what he had said, and Barbier explained to me the rest. He has a very harmonious voice, and I did not notice in him that guttural accent which always astonishes our French ears among his fellow-countrymen. He explained to us that, in his day, French was not taught to young pupils, and that, later, he had thought of learning it, but that an opportunity presented itself for his making a long voyage of exploration and study. This journey round the world did not last less than five years. On his return he was engaged in occupations so numerous and diverse that he no longer had leisure to think of the French language, which was not absolutely necessary to him. He carried already within him the germ of his book on *The Origin of Spe-*

cies. Will it be believed, he worked twenty years in the shadow of his retreat, isolating himself in his labor, without revealing the secret of his studies to the learned societies, without relaxation, in silence and patience, because he was sure of himself and of fame! All in the drawing-room, except Darwin, spoke French fluently. The company consisted of his son, his wife, two of his daughters-in-law, and another lady. We passed into the dining-room, and the ladies, naturally enough, asked me particulars about the Comédie Française and our Parisian theatres. The conversation was very lively. Darwin, who heard the ladies laugh, turned sometimes from the conversation which he was holding with Barbier in English, to ask what I had said. It was repeated to him, and he gave a good natured smile.

Darwin was afterwards good enough to take me into his study. I know not with what emotion I entered the room from which so many fine works have issued. I have passed my fiftieth year, yet my heart beat like that of a schoolboy about to be presented to Victor Hugo. It is very simple and severe, this study, enlivened by the sun shedding its light over it in waves. The walls, from top to bottom, are concealed with books. Two book-cases, also occupy the middle of the study. One is full of volumes, and, on the other are arranged all sorts of scientific instruments. After we had a good look round, Darwin took us back again into the drawing-room, where the rest of the company were awaiting us. The conversation was made very animated and pleasant by the good humor of the ladies. Barbier, however, watched the finger of the clock, and at the time which had been prescribed, he beckoned to me to take leave. Darwin rose from his high-chair, and, addressing himself to me, said gracefully and with such a clear articulation that I did not miss a word, "I have not been able to understand what you said to the ladies, and which amused them so much, but they will repeat it to me to-night, and I shall have much pleasure in hearing you in that way." I replied to this compliment by an inclination of the head, for I do not venture to speak a word of English. He gave us a last "shake hand," and we left, enchanted with our visit.

Francisque Sarcey, "Lettres de Londres, X," *Le XIX Siècle,* 19 June 1879. [†]Edmond Barbier (d. 1880), well-known French translator of Darwin's work.

Secondary Literature: Luigi de Anna, *Francisque Sarcey, professeur et journaliste* (Florence: R. Bemporad, 1919).

DOMINGO FAUSTINO SARMIENTO
(1811–1888)
PRESIDENT OF ARGENTINA

I have been familiar with Darwin's name for forty years when, embarked on the Beagle under Fitzroy's command, he visited the extreme south of the continent, for I knew the ship and its crew and, later on, [from] *The Voyage of the Beagle* which I had to cite not a few times with reference to [litigation over] the Straits [of Magellan]. You will recall that I was never very exercised about our southern territories, because I didn't think they were worth expending a single barrel of gunpowder in their defense, and I refrained from employing fantastic descriptions to excite the imagination of those people who still hoped to find the El Dorado for which our fathers sought in vain, in order to prevent a war over that Holy Sepulcher of traditionalist illusions. [...]

We must suppose that the Creator awoke in a really bad mood on the fifth day [of the Creation] and looked on with bemusement as he deposited in the Brazilian Amazon eighteen hundred different species of fish, so beautifully disciplined that even today they preserve the habitats assigned to each species. Darwin simplified the task [of classification] by explaining the variability of organic forms according to their needs and location. It is a well-known fact that what induced him to suspect it was a little bird of Chilean origin, which he encountered in the Galapagos archipelago which, without ceasing to be the same [species] had modified its beak—short, large, thick, or thin—according to whether in its habitat it found insects, seeds, grains or hard nuts to eat, in the same way that the eucalyptus, almost the only tree in Australia, which is so familiar to us, has acquired one hundred forms, according to whether the terrain is swampy or dry, low-lying or mountainous.

Domingo Faustino Sarmiento, "Discurso en honor de Darwin" (30 May 1882), in his *Discursos populares* (Buenos Aires: Imprenta Europea, 1883), 408–29, on 410–13, trans. TFG.

Secondary Literature: Aline Helg, "Race in Argentina and Cuba, 1880–1930: Theory, Policies, and Popular Reaction," in *The Idea of Race in Latin America, 1870–1940,* ed. Richard Graham (Austin: University of Texas Press, 1990), 37–70.

MINOT JUDSON SAVAGE (1841–1918)
American Unitarian Minister

And once more to-day, in the person of Darwin, the true Creator is scouted and ridiculed because he did not make the world and man after the fashion that an unknown, barbaric Hebrew laid down for him.

Minot Judson Savage, *Belief in God* (Boston: George H. Ellis, 1888), 147.

Well, what but that is the history of orthodoxy all through? It fights every thing new as long as it can. Then it re-interprets the Bible, and finds it all there, and benevolently takes it under the wing of revelation. It won't be ten years before a fast and firm alliance will be patched up between even Darwin and Moses. Moses will be made out the original Darwinian. Just so they treated Newton: they cursed his gravitation as long as they could; and now for two hundred years have been using the great law to glorify the Jewish conception of a God who taught a flat world "founded on the seas and established on the floods."

Minot Judson Savage, *Bluffton: A Story of To-Day* (Boston: Lee & Shepard, 1878), 76.

There is not a single question of the age, that for present, practical, pressing importance, begins to approach the one that Spencer* and Darwin and Haeckel* have raised. You might as well say that because the sun is ninety-two millions of miles away, its influence, and the laws of its life and shining, are of no practical importance to Boston.

Minot Judson Savage, *The Religion of Evolution* (Boston: Lockwood, Brooks, 1877).

DARWIN

O God, thy "Holy Church infallible"
Did place thee on the "Index," in the name
Of thy son Kepler, who with single aim
Sought out thy starry steps, and dared to tell
Thy secret, that the world had failed to spell
For ages. And now, once again, the shame

Of thy true prophet, banned with evil fame,
The chorus of the Church's curse doth swell.
But, as did Kepler, so hath Darwin done!
With childlike seeking, he found out the way
Where God's mysterious feet had trod before,
And humbly followed. Planet thus and sun
Hold one's high fame in keeping; and for aye
Men's loving lips will tell the other o'er.

Minot Judson Savage, *Poems* (Boston: George H. Ellis, 1882), 154.

Secondary Literature: James R. Moore, *The Post-Darwinian Controversies* (Cambridge: Cambridge University Press, 1979).

AUGUST SCHLEICHER (1821–1868)
German Linguist

Open Letter to Ernst Haeckel*, 1863:

You would leave me no peace until I began reading Bronn's† translation of the much discussed work by Darwin, *On the Origin of Species* [...]. I have complied with your request; I have waded through the whole of the book, in spite of its being rather clumsily arranged, and heavily written in a curious kind of German, and the greater part of the work I was tempted to read again and again. [...] In supposing that Darwin's *Origin of Species* would please me, you were thinking, no doubt [...] of my amateur gardening and botanizing. I confess that our gardening represents many [...] an opportunity of observing for example that "struggle for life" which we are wont to decide in favor of our chosen pets, and which, in the language of ordinary life, goes by the name of "weeding." [...]

Yet, my dear friend, you were not altogether on the right track. [...] Darwin's views and theory struck me in a much higher degree, when I applied them to the science of language.

What Darwin lays down of the animal creation in general, can equally be said of the organisms of speech—nay it is quite accidentally that I pronounced an opinion coinciding in a remarkable degree with Darwin's views on "the struggle for life" on the extinction of ancient forms, on the widely-spread varieties of individual species in the field of speech as far back as the year 1860. [. . .] Can you wonder now that the book has made so strong an impression on me?

August Schleicher, *Darwinism Tested by the Science of Language,* trans. Alexander V. M. Bikkers, in *Linguistics and Evolutionary Theory: Three Essays by August Schleicher, Ernst Haeckel, and Wilhelm Bleek,* ed. Konrad Koerner (Amsterdam: John Benjamins, 1983), 13–16, originally published in German as *Darwinsche Theorie und die Sprachwissenschaft* (Weimar: Hermann Böhlau, 1863). †Heinrich Georg Bronn (1800–1862), German paleontologist, translated the *Origin of Species* into German.

Secondary Literature: Guy Deutscher, *The Unfolding of Language: An Evolutionary Tour of Mankind's Greatest Invention* (New York: Metropolitan Books, 2005).

ARTHUR SCHOPENHAUER (1788–1860)
German Philosopher

It is significant to find that Schopenhauer, the brilliant thinker, regarded the *Origin of Species* as one of the empirical soapsud or barber books produced by exact investigation, which he thoroughly despised from his metaphysical point of view.

Wilhelm Bölsche, *Haeckel, His Life and Work,* trans. Joseph McCabe (London: T. Fisher Unwin, 1906), 132.

Secondary Literature: Patrick Gardiner, "Arthur Schopenhauer," in *Encyclopedia of Philosophy,* ed. Paul Edwards, 8 vols. (New York: Macmillan, 1967), 7:325–32; Barbara Hannan, *The Riddle of World: A Reconsideration of Schopenhauer's Philosophy* (New York: Oxford University Press, 2009).

JOSEPH A. SCHUMPETER (1883–1950)
Austrian-American Economist

The reader is urgently advised to peruse [*Origin of Species*] carefully. It is one of the most important pieces of scientific history ever written, and presents a case study about one of the objects of our interest—the ways of the human mind and the mechanisms of scientific advance. In addition, it elucidates a concept that plays some role in our own story, the concept of Inadequate Acknowledgment of Priorities. Darwin illustrates the meaning of this concept by presenting an ideal instance of what is Adequate Acknowledgment. In everything he did, that man was a living and walking compliment to himself and also to the economic and cultural system that produced him—a point recommended to the reader whenever he feels like ruminating on the civilization of capitalism (and, incidentally, about more modern forms of organization of research).

In the first place, the *Origin of Species* and the *Descent of Man* make one of the biggest patches of color in our picture of that period's *Zeitgeist*. Their secular importance for mankind's cosmic conceptions is comparable with that of the heliocentric system. They were widely read by the general public, passionately discussed, and effective in refurnishing the bourgeoisie's mental house, though it seems that, in most cases, this new furniture did not oust metaphysical furniture that still existed but only occupied empty space. Our fundamental beliefs and attitudes are beyond the power of any book to make or shake; in particular, I do not think that any cultivated person will find his faith destroyed through reading Darwin, provided that person has any faith to destroy.

Joseph Schumpeter, *History of Economic Analysis* (New York: Oxford University Press, 1954), 444n19, 445.

Secondary Literature: Thomas K. McCraw, *Prophet of Innovation: Joseph Schumpeter and Creative Destruction* (Cambridge, MA: Harvard University Press, 2007).

ALBERT SCHWEITZER (1875–1965)
Alsatian Physician, Musician, and Philosopher

About Schmiedeberg and his friend Schwalbe, the anatomist,[†] the following delightful story circulated at the university [of Strasbourg]. Schwalbe was due to give a lecture on anthropology to the Adult Education Center of an Alsatian town and would of course have to mention the Darwinian theory. When he told Schmiedeberg of his fear that he might give offense, the latter replied: "Don't spare them! Tell them all about Darwinism, only take care not to use the word 'monkey,' and they'll be quite satisfied both with Darwin and with you."

Albert Schweitzer, *Out of My Life and Thought* (Baltimore: Johns Hopkins University Press, 1989), 105. †Oswald Schmiedeberg (1838–1921), German pharmacologist, and Gustav Schwalbe (1844–1916), German anatomist and physical anthropologist.

Secondary Literature: James Brabazon, *Albert Schweitzer: A Biography* (Syracuse, NY: Syracuse University Press, 2000).

ADAM SEDGWICK (1785–1873)
English Geologist

[November 1859]

This visit was rendered notable by my being taken by Grote† to luncheon at Trinity Lodge with Dr. [William] Whewell*. The main incident was that, during luncheon, Adam Sedgwick, the old geologist, came in in a state of great excitement, and addressed Whewell to this effect: "Well, Master, what do you think I've been doing all the morning? Reading Darwin's new book on the *Origin of Species* that has just come into my hands." He, thereupon, indulged in a vehement diatribe against Darwin—in which Whewell concurred—for setting aside the Creator in accounting for the Universe. Most curious and remarkable was his defiance of Darwin's evolution to bring about the races of animals and man as we find them—remarking with vehemence, "I'll give you the Bank of Eternity to draw upon." He was, of course, unaware at that time of the limits put by physical authorities upon the age of the solar system. Sedgwick had made himself conspicuous by showing up the well-known *Vestiges* in the *Quarterly Review,* and he now felt much in the same mood with Darwin.

Alexander Bain, *Autobiography* (London: Longmans, Green, 1904), 257–58. †John Grote (1813–66), English clergyman and moral philosopher.

[24 December 1859]

My dear Darwin, I write to thank you for your work *On the Origin of Species.* [. . .] If I did not think you a good-tempered, and truth-loving man, I should not tell you that [. . .] I have read your book with more pain than pleasure. Parts of it I admired greatly, parts I laughed at till my sides were almost sore; other parts I read with absolute sorrow, because I think them utterly false and grievously mischievous. You have deserted—after a start in that tram-road of all solid physical truth—the true method of induction, and started off in machinery as wild, I think, as Bishop Wilkins's locomotive that was to sail with us to the moon. Many of your wise conclusions are based upon assumptions which can neither be proved nor disproved. Why then express them in the language and arrangements of philosophical induction? As to your grand principle—natural selection— what is it but a secondary consequence of supposed, or known, primary facts? Development is a better word, because more close to the cause of the fact. For you do not deny causation. I call (in the abstract) causation the will of God; and I can prove that He acts for the good of His creatures. He also acts by laws which we can study and comprehend. Acting by law, and under what is called final cause, comprehends, I think, your whole principle. You write of "natural selection" as if it were done consciously by the selecting agent. 'Tis but a consequence of the presupposed development, and the subsequent battle for life. This view of nature you have stated admirably, though admitted by all naturalists and denied by no one of common sense. We all admit development as a fact of history; but how came it about? Here, in language, and still more in logic, we are point-blank at issue. There is a moral or metaphysical part of nature as well as a physical. A man who denies this is deep in the mire of folly. [...]

Lastly, then, I greatly dislike the concluding chapter—not as a summary, for in that light it appears good—but I dislike it from the tone of triumphant confidence in which you appeal to the rising generation [...] and prophesy of things not yet in the womb of time, nor (if we are to trust the accumulated experience of human sense and the inferences of its logic) ever likely to be found anywhere but in the fertile womb of man's imagination.

The Life and Letters of the Reverend Adam Sedgwick, ed. John Willis Clark and Thomas McKenny Hughes, 2 vols. (Cambridge: Cambridge University Press, 1890), 2:356–58.

Secondary Literature: Sandra Herbert, *Charles Darwin, Geologist* (Ithaca, NY: Cornell University Press, 2005).

KARL GOTTFRIED SEMPER (1832–1893)
GERMAN ZOOLOGIST

No one at the present day disputes the fact that the Darwinian theory has exerted an extensive influence not only on the development of the natural sciences, but in other branches of study; it would be superfluous here to bring forward any proofs of this. It is equally recognized that it is to this influence that modern zoology owes its most essential pretensions to be regarded as of equal estimation with other sciences. But it may be advisable to pause for a moment at the question, In what way is it that this influence has affected zoology? since in this book we have to deal exclusively with this science.

Darwin showed the possibility of discovering the path which nature struck out in order to produce her endless variety of animal forms, and of detecting the means she has employed in her task. Hence first arose those efforts, so natural in the zoologist, to acquire some comprehension of the succession in time of the different types in the animal kingdom, since all who recognize Darwin's teaching must regard it not as an arbitrary and lawless assemblage of independent species, but, on the contrary, as a great family of organisms of which the individual members, whether living or extinct, are united by a real, and not merely fanciful, bond of close affinity. The search for the natural genealogy of these families of organisms is one of the grandest of the problems propounded to modern zoology by the great English philosopher.

Karl Semper, *The Natural Conditions of Existence as They Affect Animal Life* (London: Kegan Paul, Trench, 1883), 1–2.

Secondary Literature: Ernst Mayr, "Carl Gottfried Semper," in *Dictionary of Scientific Biography,* ed. Charles Coulston Gillispie, 16 vols. (New York: Scribner's, 1970–80), 12:299.

GIUSEPPE SERGI (1841–1936)
ITALIAN ANTHROPOLOGIST

There is no need of recapitulating the facts which relate to variations in the human cranium, nor of seeking their causes, since investigations of

Darwin, Wallace and others concerning variability of organisms, well known to all students of biology. I would simply state that the various phenomena of variation are repeated in man, and, for the case in point, the human cranium.

We know that the so-called "species" of the animal kingdom have forms derived from some variations of characteristics, and that they are such because the variations from the mother-species are permanent and become transmitted by heredity. These forms may be called "varieties" of the "species," or races, according to some, or subspecies, according to others. We will call them "varieties," because the name indicates their immediate origin. According to Darwin, a variety is a species in the process of formation, because it still bears many characteristics of the species from which it is derived, and cannot become an independent form, like the species itself, until it acquires still more diverging characteristics.

If we apply this principle to the human cranium, we should first learn if man comprises a single species, as many anthropologists believe, or has many species. In the first case, the typical variations of the cranium would certainly be varieties; if, however, there are several human species, the problem becomes more complicated. In that case the varieties might be of one species, and a primitive type be found to which it is allied. But if of such primitive types there were several, these would form several species which should be grouped under one genus.

I cannot venture the solution of the general question regarding the unity or plurality of the human species, considering the actual state of my personal observations, limited to Southern Europe, especially the Mediterranean, to Oriental Europe, and to the Kurgans of Russia. I should examine Asia, Africa, Oceanica, America, Central and Northern Europe, before being able to give an opinion on such a problem. I will call therefore varieties only, human varieties, the typical forms of the cranium which are clearly distinguished from each other by their own and diverging characteristics, while I will suppose that such varieties may converge in different species, of which I cannot now give the type nor characteristics. [...]

Calling the typical forms of the cranium "varieties," we have the advantage of finding the differences or individual variations of the same type, and also certain differences which cannot be reduced to individual variations, but which are equally repeated as diverging characteristics of the same variety: these constitute subordinate groups or "subvarieties."

The "subvariety" therefore diverges from the "variety" by a new characteristic which retains in a persistent manner. We have an easy means of recognizing varieties and subvarieties, and of distinguishing them from individual variations. The latter are not repeated, or if there is repetition it is accidental; varieties are repeated by groups more or less large, which, in addition, have individual variations, the subvarieties also repeat in lesser groups that characteristic or those characteristics of the variety from which they are derived.

Giuseppe Sergi, *The Varieties of the Human Species: Principles of Classification* (Washington, DC: Smithsonian Institution, 1894), 19–20.

Secondary Literature: Peter D'Agostino, "Craniums, Criminals, and the 'Cursed Race': Italian Anthropology in American Racial Thought, 1861–1924," *Comparative Studies in Society and History* 44 (2002): 319–43.

NATHANIEL SOUTHGATE SHALER (1841–1906)
AMERICAN GEOLOGIST AND PALEONTOLOGIST

WILLIAM STIMPSON (1832–1872)
AMERICAN ZOOLOGIST

William Stimpson was a naturalist of no mean capacity. If he had not been turned to species-describing, a task akin to "gerund grinding," he would have come to largeness. As it was, his keen interest in animals of all kinds, his real love for them, made him something much better than his printed work. It was his affection for creatures as well as his general wit that quickly brought us together. Stimpson was much my senior, probably by ten years or more. He had rather cut loose from Agassiz*, for he had a fierce independence of spirit which did not allow him to profit by mastery. Yet he now and then worked in the laboratory, at that time on molluscs. We used to debate the Darwinian hypothesis privately, for to be caught at it was as if it is for the faithful to be detected in a careful study of a heresy. We had both read the *Vestiges of the Natural History of Creation,* Lamarck's *Philosophie Zoologique,* and first the Darwin-Wallace* papers and then the newly published *Origin of Species.* Agassiz had given a large part of his lectures in one term to denouncing these works, and to the assertion that species were absolute creations. He never even suggested how the special creation came about, and when, at the end of a lecture, I pressed him for some conception of how a species first appeared, he stated that it was a "thought of God," thereby showing the curious mysticism which lay at the foundation of his nature. The logic of these views bothered Stimpson less than it did me, because he was a man of facts and not fancies. He was puzzled by the transitional varieties between many of the species of molluscs he was studying, especially those occurring among the freshwater gasteropods. On one occasion I saw him throw one of these vexatious shapes on the floor, after he had studied it for a long time, put his heel upon it and grind it to powder, remarking, "That's the proper way to serve a damned transitional form."

The Autobiography of Nathaniel Southgate Shaler (Boston: Houghton Mifflin, 1909), 128–29.

Secondary Literature: David Livingstone, *Nathaniel Southgate Shaler and the Culture of American Science* (Tuscaloosa: University of Alabama Press, 1987); Alfred Goldsborough Mayer, *Biographical Memoir of William Stimpson* (Washington, DC: National Academy of Sciences, 1918).

GEORGE BERNARD SHAW (1856–1950)
ENGLISH DRAMATIST

One day the richest and consequently most dogmatic of my uncles came into a restaurant where I was dining, and found himself, much against his will, in conversation with the most questionable of his nephews. By way of making myself agreeable, I spoke of modern thought and Darwin. He said, "Oh, that's the fellow who wants to make out that we all have tails like monkeys." I tried to explain that what Darwin had insisted on in this connection was that some monkeys have no tails. But my uncle was as impervious to what Darwin really said as any Neo-Darwinian nowadays.

Bernard Shaw, *Back to Methuselah: A Metabiological Pentateuch* (New York and London: Oxford University Press, 1947), vii.

Secondary Literature: Julie Sparks, "The Evolution of Human Virtue: Precedents for Shaw's 'World Betterer' in the Utopias of Bellamy, Morris, and Bulwer-Lytton," in *Shaw and Other Matters,* ed. Susan Rusinko (Selinsgrove, PA: Susquehanna University Press, 1998), 63–82.

HENRYK SIENKIEWICZ (1846–1916)
Polish Novelist

[1893]

I noticed that our conservatives crowded round Stawowski, not so much out of curiosity to hear what he said as rather with a certain watchful coquetry. Here, and maybe in other countries, this party has little courage. They looked at the speaker with insinuating smiles, as if they would say: "Although conservatives, nevertheless—" Ah! That "nevertheless" was like an act of contrition, a kind of submission. This was so evident that I who am a sceptic as to all party spirit, began to contradict Stawowski, not as a representative of any party, but simply as a man who is of a different opinion. My audacity excited some astonishment. The matter in question was the position of the working-men. Stawowski spoke of their hopeless condition, their weakness and incapacity for defending themselves; the audience which listened to his words grew every minute larger, when I interrupted:

"Do you believe in Darwin's theory, the survival of the fittest?"

Stawowski, who is a naturalist by profession, took up the challenge at once.

"Of course I do," he said.

"Then allow me to point out to you that you are inconsequent. If I, as a Christian, care for the weak and defenceless, I do so by the doctrine of Christ; but you, from a standpoint of a struggle-for-life existence, ought to see it in a different light: they are weak, they are foolish, consequently bound to succumb; it is a capital law of nature—let the weaker go to perdition. Why is it you do not take it this way? Please explain the contradiction."

Whether Stawowski was taken aback by the unexpected opposition, or whether he really had never put the two things together, the fact was that he was at a loss for a ready answer, grew confused, and did not even venture upon the expression "altruism," which, after all, says very little.

Henryk Sienkiewicz, *Without Dogma*, trans. Iza Young (Boston: Little, Brown, 1893), 45–46.

Secondary Literature: Waclaw Lednicki, *Henryk Sienkiewicz: A Retrospective Synthesis* ('s-Gravenhage, Netherlands: Mouton, 1960).

GEORGE GAYLORD SIMPSON (1902–1984)
AMERICAN PALEONTOLOGIST

Darwin was no philosopher. He seems to have been quite unaware that his objective approach to the facts of life had implicit philosophical premises. He had no world view, no cosmic ontology, or rather he took the one he had for granted without introspective identification. His method, never formally learned, but acquired by a sort of mental osmosis, was first to postulate some relationship among available facts, and then painstakingly to seek any additional facts that might check and more fully delineate the possible relationships. It was at this level that he learned that the basic relationships among all the groups of organisms are those of phylogenetic affinity or organic evolution. Simultaneously (for these two aspects were never clearly distinguished in Darwin's work) he sought some proximate explanation of those relationships, of the resemblances and differences and of the degrees of affinity involved in them. His explanation at this second level was natural selection, and that explanation, now enriched, is still seen as valid. Darwin knew there were other factors beyond this that up to this point had to be taken for granted. His only extensive effort concerning these factors was unfortunately oriented by a tradition that had reached him through Lamarck and many others, that of the inheritance of acquired characters. This one avenue of research in which Darwin was (albeit rather unconsciously) carrying on from Lamarck ended in a fiasco, a hypothesis on inheritance that Darwin's admirers would prefer to forget. [. . .] Even here, although here without success, he was following his method of postulation checked and amplified by induction, and not the philosophical deductive method of Lamarck.

George Gaylord Simpson, *This View of Life* (New York: Harcourt Brace, 1964), 50–51.

In 1955, one hundred twenty-two years after Darwin and sixty-eight after Ameghino,[†] I also sat and mused at Monte Hermoso. [. . .] Here I had a feeling of awe, or even a sense of piety, as it is a place made holy, in a proper sense, by the fact that those two truly great men had stopped and worked and mused here long before me.

George Gaylord Simpson, *Discoverers of the Lost World* (New Haven, CT: Yale University Press, 1984), 26. [†]Florentino Ameghino (1854–1911), Argentine paleontologist.

Secondary Literature: Leo F. Laporte, *George Gaylord Simpson: Paleontologist and Evolutionist* (New York: Columbia University Press, 2000).

RICHARD SIMPSON (1820–1876)
English Catholic Author and Editor

[February 1860]
I have made no alteration in [my] Darwin [review][†] for I cannot satisfy myself that the introduction is theological—except in the sense that it is a controversy about a "simplicity" theory which is as applicable to metaphysics or politics as to theology, and as destructive wherever it is applied as it is to theology.

Simpson to Lord Acton, 15 February 1860, in *The Correspondence of the Lord Action and Richard Simpson,* ed. Josef L. Altholz et al. (Cambridge: Cambridge University Press, 1973), 44. [†]Simpson's review of *Origin* in *Rambler,* n.s., 2 (1860): 361–76.

Secondary Literature: Joseph Altholz, *The Liberal Catholic Movement in England: The "Rambler" and its Contributors, 1848–1864* (London: Burns & Oates, 1960).

ISAAC BASHEVIS SINGER (1902–1991)
Polish-American Yiddish Novelist

My brother had brought home a brochure about Darwin which contained a chapter about Malthus. Making sure my father shouldn't see me, I read the book in a single day. Malthus proved in a way that couldn't be clearer that countless creatures were born to die, for otherwise the world would fill up with so many creatures that everyone would starve to death or simply become crushed. Wars, plagues, and famines sustained life on this earth. Darwin went even further and maintained that the continuous struggle for food or sex is the origin of all species. The Cossacks who massacred the Jews, the Russians, the Tartars, all the tribes who kept on killing each other, actually implemented the plans of Creation.

Isaac Bashevis Singer, *Love and Exile* (Garden City, NY: Doubleday, 1984), 18.

Patricia resembled the young man, a big, good-natured young woman, one of those female giants that came from the West endowed with the inherited strength of generations of gentile pioneers. When she smiled, Grein was struck by her teeth, which were as sharp and pointed as an animal's. These teeth proved more clearly than any arguments Darwin's theory of evolution.

"I beg you, Shloymele, don't talk about such things. As it is, people don't live long. So what's the sense of killing one another? Doesn't the Angel of Death take care of enough? I'm telling you, Shloymele, the wicked are mad."

"They rule the world."

"What kind of world? What do all these wars achieve? Why can't they straighten things out once and for all? I've just been through the newspapers. All you read about is robbery, thievery, violence."

"That's how the world works. If you knew Darwin, you'd know that the struggle for survival is the force that created the world."

"I've read; I've heard. It's nonsense, nonsense . . ."

"Then why did God arrange things so that when two stags encounter the same hind, they lock horns with each other until one of them falls? That's not a result of free will. Deer aren't wicked. They do what's in their nature to do."

"Today's Jew wants to outdo Esau in insolence, in debauchery, even in bloodthirstiness. That's the bitter truth."

"Yes, it is bitter. But what can we do? Forty years ago, when I first read Darwin's *Origin of Species*, I knew that the earth with all its green mountains and fertile valleys was nothing but a slaughterhouse."

Isaac Bashevis Singer, *Shadows on the Hudson,* trans. Joseph Sherman (New York: Farrar, Straus & Giroux, 1988), 191, 395–96, 447.

He agreed with Darwin that man was descended from the ape, and he contended that humans were the only intelligent beings in the cosmos.

Ezriel smiled to himself. "How difficult it is to rid oneself of superstition. It possessed an atavistic power; though, according to [August] Weismann*, acquired characteristics could not be inherited. Well, that hadn't been definitely established. And whey couldn't they both be correct, Lamarck and Darwin?

"[A]m I a self-murderer, as Olga calls me?" He bit his lips. Perhaps among Jews he was an exception. But among Christians there were millions like him, after all. They studied Darwin and attended church.

Isaac Bashevis Singer, *The Manor and the Estate,* trans. Joseph Singer (Madison, WI: Terrace Books, 2004), 222, 416, 795.

Secondary Literature: Edward Alexander, *Isaac Bashevis Singer* (Boston: Twayne, 1980).

OSBERT SITWELL (1892–1969)
English Essayist

Consider the Epochs of Effort. The Victorian Era was the most consciously devoted to this curious ideal; the chief effect being, in reality, to sell something. The singular fact emerges that Charles Darwin, albeit a great man of whom most of the Victorians intensely disapproved, nevertheless through the medium of his theory of "The Survival of the Fittest," did much, however unconsciously, to give them support. He imparted to their often iniquitous proclivities an ethical foundation; because, for each rival merchant knocked into the workhouse, for each businessman assassinated, for every native murdered or enslaved in order that his land might be appropriated, the persons responsible for these results could, when occasionally their consciences stung them, always comfort themselves with the reflection: "It can't be helped. . . . Survival of the Fittest, and all that. . . ." The theory was applied to everything in the universe; the very laws of the universe itself were interpreted in terms of effort and Will.

Osbert Sitwell, *Pound Wise* (London: Hutchinson, 1963), 101.

My father was fond of animals, and I have often seen him go up to the cage of a monkey, and talk to the sad-eyed but *insouciant* inmate. [. . .] His special interest in the simian tribes was no doubt a tribute to the part they played in the evolutionary theories of his great hero Charles Darwin.

Osbert Sitwell, *Tales My Father Taught Me* (London: Hutchinson, 1963), 60.

Secondary Literature: Philip Ziegler, *Osbert Sitwell* (New York: Knopf, 1999).

GEORGE WASHBURN SMALLEY (1833–1916)
ENGLISH JOURNALIST

The hour of the funeral was noon. Holders of chapterhouse tickets were expected to arrive not later than half-past eleven. At that hour a company of perhaps 200 people had assembled. [. . .] One's first look at the group in the chapter-house showed how brilliant—if one may use such a word— was the company gathered to pay the last tribute to Darwin. Some of the greatest names in England belonged to the men who clustered at the top of the steps. The ten nearest are the pall-bearers; life-long friends and disciples of Darwin some of them: Mr. Huxley*, whose resolute face is softened by the sense of a double bereavement, by the loss of his leader and his comrade; Sir Joseph Hooker*, Mr. [William] Spottiswoode,[†] President of the Royal Society; Sir John Lubbock*, the Duke of Devonshire, the Duke of Argyll*, the American Minister [James Russell Lowell*], Lord Derby, Mr. A. R. Wallace*, and Canon [Frederic William] Farrar*. The last name gives rise to the same reflections which come to everybody who thinks for a moment of the significance of the burial of Darwin in West-minster Abbey. What has Darwin to do with the Abbey or the Abbey with him?

What place has a dignitary of the Church of England by the coffin of the foremost man of science of his age? Dean Stanley[‡] used to say the Abbey was something more than a church, that it was the fitting tomb of the heroes of England—their resting-place and monument; and so it is. England, or so much of England as is modern and liberal, would have cried out at the exclusion of Darwin from the national shrine. And I sup-pose we may take Canon Farrar's presence and his share in the ceremonial as an act of personal respect and of ecclesiastical compensation. It is not twenty years since divines of the Church of England anathematized Dar-win as a heretic—to use no harsher term. Her advocates said then what a Roman Catholic advocate has said since Darwin's death—that a man ca-pable of inventing a theory which led straight to atheism must be knave or fool or both. The relative intelligence of the devotees of the two Churches—of Rome and of England—may be measured by the breadth of their divergence to-day on this point.

George Washburn Smalley, *London Letters and Some Others,* 2 vols. (New York: Harper & Brothers, 1891), 1:70–71. [†]See Thomas Carlyle entry. [‡]Arthur Penrhyn Stanley (1815–81), liberal theologian, dean of Westminster.

Secondary Literature: Joseph James Matthews, *George W. Smalley: Forty Years a Foreign Correspondent* (Chapel Hill: University of North Carolina Press, 1973).

MARY SOMERVILLE (1780–1872)
Scottish Popular Astronomer and Physical Geographer

Mr. [John] Murray[†] has kindly sent me a copy of Darwin's recent work on the *Descent of Man.* Mr. Darwin maintains his theory with great talent and with profound research. His knowledge of the characters and habits of animals of all kinds is very great, and his kindly feelings charming. It is chiefly by the feathered race that he has established his law of selection relative to sex. The males of many birds are among the most beautiful objects in nature; but that the beauty of nature is altogether irrelative to man's admiration or appreciation, is strikingly proved by the admirable sculpture on Diatoms and Foraminifera, beings whose very existence was unknown prior to the invention of the microscope. The Duke of Argyll* has illustrated this in the *Reign of Law*, by the variety, graceful forms and beautiful coloring of the humming birds in forests which man has never entered. In Mr. Darwin's book it is amusing to see how conscious the male birds are of their beauty; they have reason to be so, but we scorn the vanity of the savage who decks himself in their spoils. Many women without remorse allow the life of a pretty bird to be extinguished in order that they may deck themselves with its corpse. In fact, humming birds and other foreign birds have become an article of commerce. Our kingfishers and many of our other birds are on the eve of extinction on account of a cruel fashion.

I have just received from Frances Power Cobbe* an essay, in which she controverts Darwin's theory, so far as the origin of the moral sense is concerned. It is written with all the energy of her vigorous intellect as a moral philosopher, yet with a kindly tribute to Mr. Darwin's genius. I repeat no one admires Frances Cobbe more than I do. I have ever found her a brilliant, charming companion, and a warm, affectionate friend. She is one of the few with whom I keep up a correspondence. To Mr. Murray I am indebted for a copy of [Edward Burnett] Tylor's* *Researches on the Early History of Mankind, and the Development of Civilization*—a very remarkable work for extent of research, original views, and happy illustrations. The gradual progress of the pre-historic races of mankind has laid a foundation from which Mr. Tylor proves that after the lapse of ages the barbarous races now existing are decidedly in a state of progress towards civilization. Yet one cannot conceive human beings in a more degraded state than some of them are still; their women are treated worse than their dogs. Sad to say, no savages are more gross than the lowest ranks in England, or treat their wives with more cruelty.

Mary Somerville, *Personal Recollections, from Early Life to Old Age* (Boston: Roberts Brothers, 1874), 357–59. †John Murray III (1808–92), Darwin's publisher.

Secondary Literature: Kathryn A. Neeley, *Mary Somerville: Science, Illumination and the Female Mind* (Cambridge: Cambridge University Press, 2001); Bernard Lightman, *Victorian Popularizers of Science: Designing Science for New Audiences* (Chicago: University of Chicago Press, 2007).

HERBERT SPENCER (1820–1903)
English Political and Social Theorist

I am quite well and getting on satisfactorily with my next number. Only yesterday I arrived at a point of view from which Darwin's doctrine of "Natural Selection" is seen to be absorbed into the general theory of Evolution as I am interpreting it.

Herbert Spencer, letter of 9 June 1863, in *Autobiography*, 2 vols. (London: Williams & Norgate, 1904), 2:99–100.

A good deal of this chapter ["Human Population in the Future"] retains its original form; and the above paragraph is reprinted verbatim from the *Westminster Review* for April, 1852, in which the views developed in the foregoing hundred pages were first sketched out. This paragraph shows how near one may be to a great generalization without seeing it. Though the process of natural selection is recognized; and though to it is ascribed a share in the evolution of a higher type; yet the conception must not be confounded with that which Mr. Darwin has worked out with such wonderful skill, and supported by such vast stores of knowledge. In the first place, natural selection is here described only as furthering direct adaptation—only as aiding progress by the preservation of individuals in whom functionally-produced modifications have gone on most favorably. In the second place, there is no trace of the idea that natural selection may, by co-operation with the cause assigned, or with other causes, produce *divergences* of structure; and of course, in the absence of this idea, there is no implication, even, that natural selection has anything to do with the origin of species. And in the third place, the all-important factor of variation—"spontaneous," or incidental as we may otherwise call it—is wholly ignored. Though use and disuse are, I think, much more potent pauses of organic modification than Mr. Darwin supposes—though, while pursuing the inquiry in detail, I have been led to believe that direct equilibration has played a more active part even than I had myself at one time thought; yet I hold Mr. Darwin to have shown beyond question, that a great part of the facts—perhaps the greater part—are explicable only as resulting from the survival of individuals which have deviated in some indirectly-caused way from the ancestral type. Thus, the above paragraph contains merely a passing recognition of the selective process; and indicates no suspicion of the enormous range of its effects, or of the conditions under which a large part of its effects are produced.

Herbert Spencer, *The Principles of Biology*, 2 vols. (New York: Appleton, 1867), 2:501n.

Since the publication of Darwin's *Descent of Man* there has been a great sensation about the theory of the development of mind—essays in the magazines on Darwinism and Religion, Darwinism and Morals, Philosophy and Darwinism, all having reference to the question of mental evolution, and all proceeding on the supposition that it is Darwin's hypothesis. As no one says a word in rectification, and as Darwin himself has not indicated the fact that the *Principles of Psychology* was published five years before the *Origin of Species*, I am obliged to gently indicate it myself.

Spencer to Edward Livingston Youmans, 3 June 1871, in *Edward Livingston Youmans: Interpreter of Science for the People*, by John Fiske (New York: Appleton, 1894), 267.

Organic evolution being a part of Evolution at large [...] had to be explained in physical terms: the changes produced by functional adaptation [...] and the changes produced by "natural selection" had both to be exhibited as resulting from the redistribution of matter and motion everywhere and always going on. Natural selection as ordinarily described, is not comprehended in this universal redistribution. It seems to stand apart as an unrelated process. The search for congruity led first of all to perception of the fact that what Mr. Darwin called "natural selection," might more literally be called survival of the fittest. But what is survival of the fittest, considered as an outcome of physical actions? The answer presently reached was this: the changes constituting evolution tend ever towards a state of equilibrium. On the way to absolute equilibrium or rest, there is in many cases established for a time, a moving equilibrium—a system of mutually-dependent parts severally performing actions subserving maintenance of the combination. Every living organism exhibits such a moving equilibrium—a balanced set of functions constituting its life;

and the overthrow of this balanced set of functions or moving equilibrium is what we call death. Some individuals in a species are so constituted that their moving equilibria are less easily overthrown than those of other individuals; and these are the fittest which survive, or, in Mr. Darwin's language, they are the select which nature preserves.

Herbert Spencer, *Autobiography,* 2 vols. (London: Williams & Norgate, 1904), 2:100–101.

[Darwin on Spencer]
Herbert Spencer's* conversation seemed to me very interesting, but I did not like him particularly, and did not feel that I could easily have become intimate with him. I think that he was extremely egotistical. After reading any of his books, I generally feel enthusiastic admiration for his transcendent talents, and have often wondered whether in the distant future he would rank with such great men as Descartes, Leibnitz, etc., about whom, however, I know very little. Nevertheless I am not conscious of having profited in my own work by Spencer's writings. His deductive manner of treating every subject is wholly opposed to my frame of mind. His conclusions never convince me: and over and over again I have said to myself, after reading one of his discussions—"Here would be a fine subject for half-a-dozen years' work." His fundamental generalizations (which have been compared in importance by some persons with Newton's laws!)—which I daresay may be very valuable under a philosophical point of view, are of such a nature that they do not seem to me to be of any strictly scientific use. They partake more of the nature of definitions than of laws of nature. They do not aid one in predicting what will happen in any particular case. Anyhow they have not been of any use to me.

The Autobiography of Charles Darwin, 1809–1882, ed. Nora Barlow (London: Collins, 1958), 108–9.

Secondary Literature: Jack Kaminsky, "Herbert Spencer," in *Encyclopedia of Philosophy,* ed. Paul Edwards, 8 vols. (New York: Macmillan, 1967), 7:523–27; Valerie A. Haines, "Spencer, Darwin, and the Question of Reciprocal Influences," *Journal of the History of Biology* 24 (1991): 409–31; Barry Werth, *Banquet at Delmonico's: Great Minds, the Gilded Age, and the Triumph of Evolution in America* (New York: Random House, 2009).

FRANKLIN MONROE SPRAGUE (1841–1926)
AMERICAN CLERGYMAN

The war of Protoplasms. Dr. Robert Tuttle Morris[†] in an address on "Warfare as Natural History," described man as a group of protoplasmic cells that were in constant warfare. The war was between two strong types of varietal hybrids, each trying to get control of the other. In this struggle, he says, a "strong, haughty, nouveau-riche protoplasm is in conflict with an equally strong, old patrician protoplasm." Dr. Morris, says the *New York Times,* diagnosed the European conflict as a "free-for-all show-down between Mr. Darwin of England and Mr. Treitschke[‡] of Germany." It is thus a battle of protoplasms.

Darwin, however, had another theory beside that of the survival of the fittest; this was the "mutual dependence of protoplasms." This means a recognition by the fighting protoplasmic cells of each other's rights, and an arrangement brought about by the Hague congress, whereby the warring protoplasms will submit to an "international mind." Dr. Morris believes this theory will prevail. The German Treitschke flouts this idea, declares war to be a biological necessity and demands that German protoplasm should ruthlessly crush all other national protoplasms.

Franklin Monroe Sprague, *Made in Germany* (Boston: Pilgrim, 1915), 207. [†]Robert Tuttle Morris (1857–1945), American surgeon. [‡]Heinrich von Treitschke (1834–96), German historian and Social Darwinist.

Secondary Literature: Howard H. Quint, *The Forging of American Socialism: Origins of the Modern Movement,* 2nd ed. (Indianapolis: Bobbs-Merrill, 1964).

JOSEPH STALIN (1878–1953)
RUSSIAN DICTATOR

Evolution prepares for revolution and creates the ground for it; revolution consummates the process of evolution and facilitates its further activity.

Similar processes take place in nature. The history of science shows that the dialectical method is a truly scientific method: from astronomy to sociology, in every field we find confirmation of the idea that nothing is eternal in the universe, everything changes, everything develops. Consequently, everything in nature must be regarded from the point of view of movement, development. And this means that the spirit of dialectics permeates the whole of present-day science.

As regards the forms of movement, as regards the fact that according to dialectics, minor, quantitative changes sooner or later lead to major, qualitative changes—this law applies with equal force to the history of nature. Mendeleev's "periodic system of elements" clearly shows how very important in the history of nature is the emergence of *qualitative* changes out of *quantitative* changes. The same thing is shown in biology by the theory of neo-Lamarckism, to which neo-Darwinism is yielding place.

Pascal and Leibnitz were not revolutionaries, but the mathematical method they discovered is recognized today as a scientific method. Mayer[†] and [Hermann von] Helmholtz[*] were not revolutionaries, but their discoveries in the field of physics became the basis of science. Nor were Lamarck and Darwin revolutionaries, but their evolutionary method put biological science on its feet. [...] Why, then, should the fact not be admitted that, in spite of his conservatism, Hegel succeeded in working out a scientific method which is called the dialectical method?

Joseph Stalin, "Anarchism or Socialism?" *Akhali Tskhovreba (New Life),* 21, 24, 28 June and 9 July 1906, reprinted in Stalin, *Works,* 13 vols. (Moscow: Foreign Languages Publishing House, 1954), 1:304–6. [†]Julius Robert von Mayer (1814–78), German physicist, pioneer of thermodynamics.

The Party cannot be neutral towards religion, and it conducts antireligious propaganda against all religious prejudices because it stands for science, whereas religious prejudices run counter to science, because all religion is the antithesis of science. Cases such as occur in America, where Darwinists were prosecuted recently, cannot occur here because the Party pursues a policy of defending science in every way. (In 1925 [...] a trial took place in the state of Tennessee, U.S.A., which attracted world-wide attention. A college teacher named John Scopes was tried for teaching Darwin's theory of evolution. The American reactionary obscurantists found him guilty of violating the laws of the state and fined him.)

Joseph Stalin, "Interview with the First American Labor Delegation," *Pravda,* 15 September 1927, reprinted in Stalin, *Works,* 13 vols. (Moscow: Foreign Languages Publishing House, 1954), 10:97–158, on 138.

Secondary Literature: Nikolai Krementzov, *Stalinist Science* (Princeton, NJ: Princeton University Press, 1997).

THOMAS STEBBING (1835–1926)
ENGLISH CLERGYMAN AND NATURALIST

The genus *Geryon*, Kroyer, 1837, may claim a passing notice as one of those instances in which systematic arrangement finds itself at fault. It is sometimes placed among the Cyclometopa and sometimes among the Catometopa. Mr. Miers[†] says that it is very nearly allied to *Pseudorhombila* and *Pilumnoplax* in the latter, and to *Galene* in the former group. That, on the theory of the evolution of different groups from a common stem, such inosculant forms are almost sure to occur, has long been recognized. Darwin himself humorously admits that while as a theorist he delighted in coming across them, as a naturalist engaged in classification he found them an unmitigated nuisance.

Thomas R. R. Stebbing, *A History of Crustacea: Recent Malocostraca* (New York: Appleton, 1893), 93. [†]John Miers (1789–1879), English botanist.

Such being the character of Mr. Darwin's own Work, the handful of Essays and Letters contained in the present volume, supporting the same views by almost the same arguments, may seem a superfluous contribution to the literature of the question. And so it would be if all who condemn and ridicule Darwinism would be at pains to study Mr. Darwin's Work. But opinions passed upon it and allusions made to it in common conversation and in popular lectures often testify to nothing except supreme ignorance of its general merits. To judge by such hearsay, one might believe that Mr. Darwin had lived all his life shut up in a dove-cote, and never seen or examined any other living creature than a pigeon. Another estimate will dismiss the whole subject, scathed with indignant laughter, by simply explaining, that, according to this fatuous theory, man is descended from a monkey. Naturally no well-minded persons will consent to be *pithecoid* in origin, whether they know what *pithecoid* means or not; still less can a theory be accepted as moral and good, according to which, as some will tell you, the giraffe lengthened its neck by a series of stretchings, and the elephant acquired a trunk by continually pulling its own nose. A

disinterested advocate will perhaps be allowed to deprecate these burlesque and ignorant representations, and to strip from what is merely vulgar prejudice the guise of magnanimity and fine feeling. [...] The whole
subject is a great one, and worthy of attention, claiming earnest thought
and varied learning to decide upon it in all its bearings; it cannot be disposed of by caricaturing; it cannot be settled in deference to any religious
prepossession; it must be examined with open eyes, and with the full candor of mind which great subjects demand, and which great subjects nobly
repay. [...]

It is a [...] mistake to suppose that the truth of the development-
theory in any way hinges upon the possibility of constructing an effigy of
the first vertebrate either as it actually was, or to suit an anti-Darwinian's
notions of what it ought to have been. According to the development-
theory, it must have been the product of innumerable antecedent factors,
itself the heir of many far-descended and often modified characters; and
yet, for all that, it will probably have been a far simpler organism than the
simplest modern vertebrate.

Thomas R. R. Stebbing, *Essays on Darwinism* (London: Longmans, Green, 1871),
iv–v, 158.

Secondary Literature: Eric Mills, "Amphipods and Equipoise: A Study of T. R. R.
Stebbing," *Connecticut Academy of Arts and Sciences Transactions* 44 (1972): 237–56.

JOHN STEINBECK (1902–1968)
AMERICAN AUTHOR

Our collecting ends were different from those ordinarily entertained. In
most cases at the present time, collecting is done by men who specialize in
one or more groups. Thus, one man interested in hydroids will move out
on a reef, and if his interest is sharp enough, he will not even see other
forms around him. For him, the sponge is something in the way of his
hydroids.

Collecting large numbers of animals presents an entirely different aspect and makes one see an entirely different picture. Being more interested in distribution than in individuals, we saw dominant species and
changing sizes, groups which thrive and those which recede under varying

conditions. In a way, ours is the older method, somewhat like that of Darwin on the *Beagle*. He was called a "naturalist." He wanted to see everything, rocks and flora and fauna; marine and terrestrial. We came to envy this Darwin on his sailing ship. He had so much room and so much time. He could capture his animals and keep them alive and watch them. He had years instead of weeks, and he saw so many things. Often we envied the inadequate transportation of his time—the *Beagle* couldn't get about rapidly. She moved slowly along under sail. And we can imagine that young Darwin, probably in a bos'n's chair hung over the side, with a dip-net in his hands, scooping up jellyfish. When he went inland, he rode a horse or walked. This is the proper pace for a naturalist. Faced with all things he cannot hurry. We must have time to think and to look and to consider. And the modern process—that of looking quickly at the whole field and then diving down to a particular—was reversed by Darwin. Out of long consideration of the parts he emerged with a sense of the whole. Where we wished for a month at a collecting season and took two days, Darwin stayed three months. Of course he could see and tabulate. It was the pace that made the difference. And in the writing of Darwin, as in his thinking, there is the slow heave of a sailing ship and patience of waiting for a tide. The results are bound up with the pace.

John Steinbeck, *The Log from the Sea of Cortes* (New York: Penguin, 1977), 50–51.

Secondary Literature: Roy Simmonds, *A Biographical and Critical Introduction of John Steinbeck* (Lewiston, NY: E. Mellen, 2000).

LESLIE STEPHEN (1832–1904)
English Author and Critic

Since I wrote to you last I have read Mr. Chauncey Wright's* book, or nearly all, and—to say the truth—found it a tolerably tough morsel. [. . .] Perhaps I am a little spoilt by article-writing and inclined to value smartness of style too highly. The only point which struck me unpleasantly in the substance of the book was his rather over-contemptuous tone about Spencer* and [George Henry] Lewes*. I don't doubt that his criticisms of Spencer are tolerably correct, though I can't see that Spencer really means to concede so much to the enemy as C. W. supposes; but I confess that

Lewes seems to me to be a remarkably acute metaphysician, and one who will make his mark. . . . Anyhow, Wright must be a great loss. Nobody can mistake the soundness and toughness of his intellect, and his thorough honesty of purpose. I had the pleasure the other day of showing the book to the great Darwin, who had already received a copy from you. He was in town for a few days, and most kindly called upon me. You may believe that I was proud to welcome him, for of all eminent men that I have ever seen he is beyond comparison the most attractive to me. There is something almost pathetic in his simplicity and friendliness. I heard a story the other day about a young German admirer, whom [John] Lubbock* took to see him. He could not summon up courage to speak to the great man; but, when they came away, burst into tears. That is not my way; but I can sympathize to some extent with the enthusiastic Dutchman.

Stephen to Charles Eliot Norton, 5 May 1877, in *The Life and Letters of Leslie Stephen,* ed. Frederic William Maitland (New York: G. P. Putnam, 1906), 300–301.

To add something more cheerful, I may say that I saw Darwin the other day, who seemed to be well for him, and as cheerful and pleasant as usual. There are few people whom I admire more, and I could envy more, if anything were to be got out of envying. I think that I wrote to you about that wonderful meeting of "liberal thinkers," which was got up by [Moncure D.] Conway* in the summer. It seems to be coming to life again—rather to my disgust, to say the truth. Huxley* and [John] Tyndall* are going to take it up, and I shall have to join, if it is launched. [. . .] Oh, Lord, what bosh will be poured forth if we get the freethinkers together for a palaver!

Stephen to Charles Eliot Norton, 5 December 1878, in *The Life and Letters of Leslie Stephen,* ed. Frederic William Maitland (New York: G. P. Putnam, 1906), 330.

To appeal to "experience" [the Utilitarians] have to make the whole universe incoherent, while to get general laws they have to treat variable units as absolutely constant. "External circumstances" must account for all variation, though it is difficult to see how everything can be "external." The difficulty has now appeared in history proper, and the attempt to base a sociology upon a purely individualist assumption. This may help to explain the great influence of the Darwinian theories. They marked the point at which a doctrine of evolution could be allied with an appeal to experience. Darwin appealed to no mystical bond, but simply to verifiable experience. He postulated the continuance of processes known by observation, and aimed at showing that they would sufficiently explain the present as continuous with the past. There was nothing mystical to alarm empiricists, and their consequent adoption of Darwinism implied a radical change in their methods and assumptions. The crude empiricism was transformed into evolutionism. The change marked an approximation to the conceptions of the opposite school when duly modified, and therefore in some degree a reconciliation. "Intuitions" no longer looked formidable when they could be regarded as developed by the race instead of mysteriously implanted in the individual mind. The organic correlations were admissible when they were taken to imply growth instead of supernatural interference, and it was no longer possible to regard "natural kinds" as mere aggregates of arbitrarily connected properties. I need not ask which side really gained by the change, whether Darwinism inevitably leads to some more subtle form of atomism, or whether the acceptance of any evolution does not lead to idealism—to a belief in a higher teleology than Paley's—and the admission that mind or "spirit" must be the ultimate reality. Such problems may be treated by the philosopher of the future.

Leslie Stephen, *The English Utilitarians*, 3 vols. (London: Duckworth, 1900), 3:374–75.

Secondary Literature: J. B. Schneewind, "Leslie Stephen," in *Encyclopedia of Philosophy*, ed. Paul Edwards, 8 vols. (New York: Macmillan, 1967), 8:14–15.

ROBERT LOUIS STEVENSON (1850–1894)
Scottish Author

I have arrived, as you see, without accident; but I never had a more wretched journey in my life. I could not settle to read anything; I bought Darwin's last book in despair, for I knew I could generally read Darwin, but it was a failure. However, the book served me in good stead; for when a couple of children got in at Newcastle, I struck up a great friendship with them on the strength of the illustrations.

Stevenson to Mrs. Sitwell, 1 September 1873, in *The Letters of Robert Louis Stevenson*, ed. Sidney Colvin, 2 vols. (New York: Charles Scribner's Sons, 1911), 1:67.

Work? Work is now arrested, but I have written, I should think, about thirty chapters of the South Sea book; they will all want rehandling, I dare say. Gracious, what a strain is a long book! The time it took me to design this volume, before I could dream of putting pen to paper, was excessive; and then think of writing a book of travels on the spot, when I am continually extending my information, revising my opinions, and seeing the most finely finished portions of my work come part by part in pieces. Very soon I shall have no opinions left. And without an opinion, how to string artistically vast accumulations of fact?! Darwin said no one could observe without a theory; I suppose he was right: it is a fine point of metaphysic; but I will take my oath, no man can write without one—at least the way he would like to—and my theories melt, melt, melt, and as they melt the thaw-waters wash down my writing, and leave unideal tracts—wastes instead of cultivated farms.

Stevenson to Henry James, 29 December 1890, in *Letters and Miscellanies of Robert Louis Stevenson*, vol. 2, *Letters to His Family and Friends* (New York: Charles Scribner's Sons, 1905), 255–56.

Secondary Literature: Claire Harman, *Myself and the Other Fellow: A Biography of Robert Louis Stevenson* (New York: HarperCollins, 2005).

WILLIAM STIMPSON
See NATHANIEL SOUTHGATE SHALER
and WILLIAM STIMPSON

JAMES HUTCHISON STIRLING (1820–1909)
ENGLISH PHILOSOPHER

Mr. Darwin is described as a tall man, six feet in height, broad-shouldered but not noticeably so, with a spare body and thin legs. His hair was brown, and his complexion, as I am tempted to interpret his own "rather sallow" and his son's "ruddy rather than sallow," a rustic reddish fair. From the circumference of it, "22¼ inches," his hat would be, as the manufacturers have it, at least a 7; which medium size was, as I take it, that in Kant's case also. The first of the portraits, of which a photograph, dated 1854, seems to have been the original, may allowably, from its place and otherwise, be assumed to be, generally, the most characteristic. It represents Mr. Darwin as, at the age of forty-five, he was just in his prime. With checked vest, checked neckcloth, and a certain honest, matter-of-fact look, it is an English squire-like face we see there. The head, bald, rises and rounds finely. The eyes, overhung by unusually projecting shaggy brows, look out honestly. They seem as if they had been made both for and by observation. The ridge above them is so steep that one might almost think a cleaver had struck across the line beneath it. The nose is quite what we might expect from Fitz-Roy's dislike to it as inexpressive whether of energy or quickness. It is shortish, smallish, turned-upish, dumpyish, common; it has an insignificant, and withal an innocent look. The mouth in this portrait is a very remarkable feature; and it is well seen, the face being beardless as yet, and framed only by a plain, close, gentlemanly side-whisker. It is the expression of it that is remarkable. In the other portraits the beard so far hides the mouth; but we might almost fancy the expression in question to have disappeared from these. Especially in the last of them, all is serene, composed, and assured now (it is taken within a year of his death); there is a reflective look in it—almost a look, indeed, of rather sad reflection. On the other hand, it is the eyebrows that are the prominent feature in the remaining portrait. Each is shaggier now, with a terrier-like look; and the face itself seems smaller somehow, more set-like—is it as still in battle that it is set?

Returning now to the squire-like portrait of the checked waistcoat, the checked neckerchief, and the shaven face between the gentlemanly side-whiskers, with the fine bald forehead rounding down to the rugged ridge over the honest eyes, succeeded by the insignificant nose, the peculiar mouth, and the broadish chin, it is the mouth was specially remarked on. One fancies there is an expression on it as though hiding simple gratification at the compliment (of the sitting), but returning to a usual surface, as it were, of habitual unpretending plainness and unreflected, yet considerate, sincerity. And yet again that mouth seems almost to be saying, you are looking at me, and I fear you do not see much in me—I am not quite sure that you do not see an ignoramus in me, which perhaps I am and perhaps—with a twist of the chin!—I am not. But, with whatever shade of contradicting defiance, there is at the same time an expression, amiable and good, half of admitted, half of denied slowness.

James Hutchison Stirling, *Darwinianism: Workmen and Work* (Edinburgh: T. & T. Clark, 1894), 145–46.

Mr. Huxley* was to Mr. Darwin "his good and kind agent for the propagation of the Gospel"—Natural Selection; and surely—with all these Speeches, Addresses, Lectures, Essays, Reviews, Articles, and other relative efforts on the part of Mr. Huxley before him—not without reason. In these respects we cannot venture to attempt to render an accurate list of all Mr. Huxley's labors; but we shall at least not be wrong in saying that— There were his Speeches at Oxford in answer to [Richard] Owen* and to the Bishop [Samuel Wilberforce*]—there were his Addresses or Lectures at the Royal Institution and the School of Mines—there were his printed Essays—there were his Reviews in the *Westminster,* the *Contemporary,* and elsewhere—and there were his articles in *Nature,* the *Encyclopedia Britannica,* and the *Times.* The speeches at Oxford and the article in the *Times* may be specially signalized. In the former reference, Mr. Darwin wrote on July 20th, 1860, to Mr. Huxley himself: "From all that I hear from several quarters, it seems that Oxford did the subject great good—it is of enormous importance, the showing the world that a few first-rate men are not afraid of expressing their opinion." And in the latter reference, Mr. Francis Darwin (ii. 254) avows: "There can be no doubt that this powerful essay, appearing as it did in the leading daily journal, must have had a strong influence on the reading public." As for the lecture at the Royal Institution, too, while the prestige of the position is not to be forgotten, it is to be acknowledged that no words whatever could have been more admirably calculated to move the audience. As for Mr. Huxley it was even in Mr. Huxley himself to move. He was the king of the amphitheatre, and not a man of the day was a greater favorite with the public; inasmuch as he, perhaps, was the very ablest writer of information which, while instructive and expressly scientific, was, at the same time, also interesting, entertaining, and in the highest degree lucid.

The *Origin of Species* was published on the 24th November 1859. [Huxley's] *Times* article appeared on 26th December 1859. The lecture at the Royal Institution was delivered on the 10th February 1860. And of the speeches at Oxford the dates were 28th and 30th June 1860. Even from as much as this, then, it was impossible but that the subject must have been in most mouths in England in the course of a few months. As we all know, all in England is done by parties, and everything that appears in England is of no use whatever until it is made an affair of party. It was not different with the origin of species. Creation *or* Evolution became the party-question of the day; and it was debated at a temperature that was perfectly suffocating. Lecture-rooms rang with the subject, and not a periodical in the kingdom but glowed red-hot with it. I say Creation or Evolution: for, as usual, any nicety of distinction was not to be understood; and, whatever might be peculiar and specific in natural selection, it itself must mean, and could mean—"to the general"—only evolution. We would just suggest for the moment that this at once was a wandering from the question; which, as a question, was not of evolution as evolution, but of Mr. Darwin's special, proper, and particular theory of evolution. [. . .] The success of the book depended on the belief of the public that it was the product of work at first hand, and not of compilation at second—work at first hand and of the greatest naturalist in existence.

James Hutchison Stirling, *Darwinianism: Workmen and Work* (Edinburgh: T. & T. Clark, 1894), 173–79.

Secondary Literature: Amelia Hutchinson Stirling, *James Hutchinson Stirling: His Life and Work* (London: T. Fisher Unwin, 1912).

LYTTON STRACHEY (1880–1932)
English Author

Lytton Strachey had been reading Darwin in 1907. He wrote to Dorothy Bussy[†] about it [in May 1907]: "Did you realize that [Darwin] was one of the great stylists? Huxley[*], whom they all praise, is a sign-board by comparison!" Their cousin, Walter Raleigh's[‡] book on Shakespeare had just appeared: "My belief is that it is quite impossible to write decently about Shakespeare unless you're a master of style. (If only Darwin had done it!)."

Walter C. Lubenow, "Authority, Honor and the Strachey Family, 1817–1974," *Historical Research* 76 (2003): 512–34, on 524. †Dorothy Bussey, née Strachey (1865–1960), English novelist and translator. ‡Walter Alexander Raleigh (1861–1922), Scottish professor of English.

Secondary Literature: John Keith Johnstone, *The Bloomsbury Group: A Study of E. M. Forster, Lytton Strachey, Virginia Woolf, and Their Circle* (London: Secker & Warburg, 1954).

DAVID FRIEDRICH STRAUSS (1808–1874)
German Theologian

Natural Science has long endeavored to substitute the evolutionary theory in place of the conception of creation, so alien to her spirit; but it was Charles Darwin who made the first truly scientific attempt to deal seriously with this conception, and trace it throughout the organic world.

Nothing is easier than to ridicule the Darwinian theory, nothing cheaper than those sarcastic invectives against the descent of man from the ape, in which even the better class of reviews and newspapers are still so fond of indulging. But a theory whose very peculiarity is the interpolation of intermediate members, thus linking the seemingly remote in an unbroken chain of development, and indicating the levers by means of which Nature achieved the progressive ascension in this process of evolution—this theory surely no one can suppose himself to have refuted by bringing two formations of such utterly different caliber as ape and man in their present condition into immediate contact with each other, and utterly ignoring those intermediate gradations which the theory partly proves, partly assumes.

That the orthodox, the believers in Revelation and in miracles, should brandish their repugnance and its accompanying weapon, ridicule, against Darwin's theory, is perfectly intelligible. They know what they are about, and have good reason, too, in combating to the uttermost a principle so inimical to them. But those sarcastic newspaper writers, on the other hand—do they, then, belong to the faithful? Certainly not, as regards the vast majority; they swim with the stream of the times, and have nothing to say to miracles, or to the intervention of a Creator in the course of

Nature. Very well; how, then, do they explain the origin of man, the evolution of the organic from the inorganic, if they find Darwin's explanation so ludicrous? Do they intend evolving primeval man a human organism, however rude and unformed he be, immediately from the inorganic: the sea, the mud of the Nile, etc.? They are hardly so daring; but do they realize that the choice only lies between the miracle—the divine artificer—and Darwin? [...]

The theory is unquestionably still very imperfect; it leaves an infinity of things unexplained, and moreover, not only details, but leading and cardinal questions; it rather indicates possible future solutions than gives them already itself. But be this as it may, it contains something which exerts an irresistible attraction over spirits athirst for truth and freedom. It resembles a railway whose track is just marked out. What abysses will still require to be filled in or bridged over, what mountains to be tunneled, how many a year will elapse ere the train full of eager travelers will swiftly and comfortably be borne along and onwards! Nevertheless, we can see the direction it will take; thither it shall and must go, where the flags are fluttering joyfully in the breeze. Yes, joyfully, in the purest, most exalted, spiritual delight. Vainly did we philosophers and critical theologians over and over again decree the extermination of miracles; our ineffectual sentence died away, because we could neither dispense with miraculous agency, nor point to any natural force able to supply it, where it had hitherto seemed most indispensable. Darwin has demonstrated this force, this process of Nature; he has opened the door by which a happier coming race will finally cast out miracles. Every one who knows what miracles imply will praise him as one of the greatest benefactors of the human race.

David Friedrich Strauss, *The Old Faith and the New: A Confession*, ed. Mathilde Blind (New York: Henry Holt, 1873), 202–5.

Secondary Literature: Harris Horton, *David Friedrich Strauss and His Theology* (Cambridge: Cambridge University Press, 1973); Hayden V. White, "David Friedrich Strauss," in *Encyclopedia of Philosophy*, ed. Paul Edwards, 8 vols. (New York: Macmillan, 1967), 8:25–26.

GEORGE TEMPLETON STRONG (1820–1875)
AMERICAN LAWYER AND DIARIST

[18 February 1860]
Spent this evening diligently cutting the leaves of Darwin's much discussed book on *The Origin of Species* and making acquaintance with its general scope and aim. It's a laborious, intelligent, and weighty book. First obvious criticism on it seems this: that Darwin has got hold of *a* truth which he wants to make out *the* one generative law of organic life. Because he shews that the fauna and flora of a group of islands near a certain continent are so like those of that continent, though differing specifically therefrom, and so unlike those of other regions more remote, as to make it probable that they are the offspring of the continental species modified by the altered conditions of their new habitat, he considers himself entitled to affirm that *all* beasts, birds, and creeping things, from mammal to medusa, are developments from one stock, and that man is the descendant of some archaic fish, with swimming bladder improved into lungs, that flying fish have by successive minute steps of progress through countless ages become albatrosses, and flying squirrels bats. But I suspect that He who created and upholds this great marvelous system of various harmonies life is not obliged to conform with any *one* Law of Creation and preservation that Darwin's or any other finite intellect can discover.

Darwin asks rather large concessions. You must begin by giving him thousands of millions of millions of years [. . .] for the operation of his law of progress, and admit that the silence of the stratified record of those ages as to its operation and existence may be explained away. [. . .] The area covered by scientific research and by experiments in breeding for the last century is equivalent (in considering Darwin's theory) to scores of thousands of years of recorded observation in a single district. But however this may be, man's experience for, we will say, only four thousand years furnishes no instance of the development of new functions or new organs by any animal or vegetable organism.

March 6. Looking further into Darwin's *Origin of Species* this evening. Though people who don't like its conclusions generally speak of it as profound and as a formidable attack on received notions, I timidly incline to think it a shallow book, though laboriously and honestly written. [. . .] To him, as to the physicists of the last one hundred years, the notion of a supernatural creative power is repugnant and offensive.

The Diary of George Templeton Strong: The Civil War, 1860–1865, ed. Allan Nevins and Milton Halsey Thomas (New York: Macmillan, 1952), 10–11, 13.

Secondary Literature: Vera Brodsky Lawrence, *Strong on Music: The New York Music Scene in the Days of George Templeton Strong,* 3 vols. (New York: Oxford University Press, 1988–99).

ARTHUR SULLIVAN (1842–1900)
ENGLISH COMPOSER

Darwin, in his *Descent of Man,* says: "Neither the enjoyment nor the capacity of producing musical notes are faculties of the least direct use to man in reference to his ordinary habits of life." Physiologically he is probably correct, but as soon as merely rudimentary actions are left, as soon as existence becomes life, his statement is completely false. Indeed, music is, as the same philosopher elsewhere says, bound up in daily life, and a necessity of existence.

Arthur Lawrence, *Sir Arthur Sullivan: Life Story, Letters and Reminiscences* (Chicago: Herbert S. Stone, 1900), 275–76.

Secondary Literature: Percy M. Young, *Sir Arthur Sullivan* (London: J. M. Dent, 1971).

BILLY SUNDAY (1862–1935)
AMERICAN EVANGELIST

Billy Sunday ridiculed Darwin's theory of evolution, scored all unbelievers in the truth of the Bible, and cited Nature's laws in its support and illustration of points in his sermon at the tabernacle last night, when he preached on the subject, "Nuts for skeptics to Crack." [...] Darwin's writings were nature fakes, in the evangelist's opinion, and the scientists who believed this world was composed by "some chemical force, or the fortuitous concurrence of atoms, were in the school for feeble-minded."

[...] Sunday declared that if one accepted Darwin's theory he could not believe in the Bible.

"Sunday Hits Darwin for 'Nature Faking,'" *New York Times,* 25 May 1917.

Secondary Literature: Robert Francis Martin, *Hero of the Heartland: Billy Sunday and the Transformation of American Society, 1862–1935* (Bloomington: Indiana University Press, 2002).

ALGERNON CHARLES SWINBURNE (1837–1909)
English Poet

The evidence that the late Mr. Darwin was the real author of the poems attributed to Lord Tennyson does not need the corroboration of any cryptogram; but if it did, Miss Lesbia Hume, of Earlswood, has authorized me to say that she would be prepared to supply any amount of evidence to that effect. The first book which brought Mr. Darwin's name before the public was a record of a voyage on board the Beagle. In a comparatively recent poem, written under the assumed name of Tennyson, he referred to the singular manner in which a sleeping dog of that species "plies his function of the woodland." In an earlier poem, "The Princess," the evidence derivable from allusions to proper names—that of the real author and that of the pretender—is no less obvious. [...] The princess asks if the prince has nothing to occupy his time—"quoit, *tennis,* ball—no games?" The prince hears a voice crying to him "Follow, follow, thou shalt *win.*" Here we find half the name of Darwin—the latter half—and two thirds of the name of Tennyson—the first and second third—at once associated, contrasted, and harmonized for those who can read the simplest of cryptograms.

"Dethroning Tennyson," *Nineteenth Century* 23 (January 1888): 127–30, reprinted in *Littell's Living Age,* 5th ser., 61 (1888): 443–44.

Secondary Literature: Gowan Dawson, "Charles Darwin, Algernon Charles Swinburne and Sexualized Responses to Evolution," in *Darwinism, Literature and Victorian Respectability* (Cambridge: Cambridge University Press, 2007), 26–81.

J. M. SYNGE (1871–1909)
Irish Playwright

When I was about fourteen [ca. 1886] I obtained a book of Darwin's. It opened in my hands at a passage where he asks how can we explain the similarity between a man's hand and a bird's or bat's wings except by evolution. I flung the book aside and rushed out into the open air—it was summer and we were in the country—the sky seemed to have lost its blue and the grass its green. I lay down and writhed in an agony of doubt. My studies showed me the force of what I read, [and] the more I put it from me the more it rushed back with new instances and power. I was appalled. Till then I had never doubted and never conceived that a sane and wise man or boy could doubt. I had of course heard of atheists but as vague monsters that I was unable to realize. Now it seemed I was degraded, a playfellow of Judas, and incest and parricide were [bound to follow] the blasphemous ideas that possessed me. My memory does not record how I returned home nor how long my misery lasted. I know only that I got the book out of the house as soon as possible and kept it out of sight, saying to myself logically enough that I was not yet sufficiently advanced in science to weigh his arguments, I would do better to reserve his work for future study. Soon afterwards I turned my attention to works of Christian evidence, reading them at first with pleasure, soon with doubt, and at last in some cases with derision.

My study of insects had given me a scientific attitude—probably a crude one—which did not and could not interpret nature as I heard it interpreted from the pulpit. By the time I was sixteen or seventeen I had renounced Christianity after a good deal of wobbling, although I do not think I avowed my decision quite soon. I felt a sort of shame in being thought an infidel, a term which I have always used with reproach.

The Autobiography of J. M. Synge, ed. Alan Price (Chester Springs, PA: Dufour, 1966), 22–23.

Secondary Literature: Weldon Thornton, *J. M. Synge and the Western Mind* (New York: Barnes & Noble, 1979).

HIPPOLYTE TAINE (1828–1893)
French Critic and Historian

What if Darwin be right? What if organic matter arrives at evolution through the mere fact of adaptation and selection? What if the phenomena you quote only simulate the preconception of a definite design, just as flame seems to incorporate the desire to ascend, and water the reverse? [. . .] I much fear that this may be but metaphorical language, a convenient solution like Newton's attraction, and possibly you may take it thus yourself. You yourself offer the strongest argument against it by proving that this *nisus* [Latin, "impulse"] is not universal, by stating that conditions are often more than hostile to it, by quoting the abortion of four millions of cod's eggs for every one that is hatched, by stating that the great pachydermals of Siberia were killed off by the glacial period. In short, I should like to see you analyze coldly and in detail the theory of causation. Whether Darwin's solution be true or not true matters little; another similar procedure might have produced living organization; but to take the question generally, there results from his hypothesis that the accumulated effects of a working cause (adaptation to environment, survival of the fittest) give to the observer an illusion as of an abstract cause. In this—thanks to the naturalists—metaphysics has advanced during the last twenty years.

Taine to Ernest Renan, 3 June 1876, in *Life and Letters of H. Taine,* trans. E. Sparvel-Bayly (London: Archibald Constable, 1908), 163–64.

Secondary Literature: Milič Čapek, "Hippolyte-Adolphe Taine," in *Encyclopedia of Philosophy,* ed. Paul Edwards, 8 vols. (New York: Macmillan, 1967), 8:76–77; Pascale Seys, *Hippolyte Taine et l'avènement du naturalisme: Un intellectual sous le Second Empire* (Paris: Harmattan, 1999).

JAMES MONROE TAYLOR (1848–1916)
AMERICAN EDUCATOR

I am afraid you and I will have to assume that there is some higher power necessary to account for an evolution which is after all only a method of work and not in any sense a kind of origin. By the way, let me recur to Wallace*, who of course lacks balance in some ways, we know, but who in the final pages of his book on Darwinism makes some very true remarks from the point of view of a scientist solely, on the different method which Darwin pursues in these chapters on The Mental and Moral Life [*Descent of Man*, chap. 5] from that which characterizes his other work. The fact is that if he was going to account for the higher qualities by natural selection, he had to do something different. The questions that I want you to answer me are, first, whence came the egg prior to all life on the earth; and second, is the plain life that the egg discloses identical with thinking, aspiration and the appreciation of goodness?

Taylor to Aaron Treadwell, 17 October 1902, in *The Life and Letters of James Monroe Taylor: The Biography of an Educator,* ed. Elizabeth Hazelton Haight (New York: Dutton, 1919), 182.

Secondary Literature: The Life and Letters of James Monroe Taylor: The Biography of an Educator, ed. Elizabeth Hazelton Haight (New York: Dutton, 1919).

WILLIAM BERNHARDT TEGETMEIER (1816–1912)
ENGLISH NATURALIST AND PIGEON BREEDER

My friendship with Yarrell[†] ended only with his life, and to him I owe my personal introduction to Darwin. [. . .] At one of these exhibitions [of pigeon fanciers] I heard a voice which said, "Oh, here's Tegetmeier; he will tell you all about these birds better than I can." I turned round, and saw Yarrell with a stranger, whom he introduced as Mr. Darwin. [. . .] Darwin was at that time accumulating evidence for his large work on the *Variation of Animals*. [. . .] How eagerly he embraced the opportunity of adding to his materials may be inferred from the fact that the next morning brought him to my little country cottage at Wood Green. When he went away he took with him a box of skulls and other specimens, many of which have been engraved in his well-known volumes on variation, and this led to a friendship which lasted until his death.

Tegetmeir, quoted in *A Veteran Naturalist, Being the Life and Work of W. B. Tegetmeier,* by E. W. Richardson (London: Witherby, 1916), 101–2. †William Yarrell (1784–1856), English naturalist.

Variations, however, of a much more striking character, not unfrequently occur in single cases of wild birds; but when they take place in a state of nature, they are not very likely to be propagated, inasmuch as a bird with any variation of plumage or form will almost of necessity mate with one of the ordinary character, the offspring again do the same, so that in a very few generations all trace of any singular variation is apt to be lost. In a state of domesticity, however, any singular variation would be noticed, and, by careful selection of breeding stock, would be perpetuated, and even increased. In this manner all the different breeds have been produced. [...] The recurved feathers of the Jacobin and other breeds, the long beak of the Carrier, the length of plumage and limb in the Pouter, &c., &c., all owe their origin to natural variations which have been perpetuated and intensified by the careful selection exercized by the breeders through many successive generations. We know that this view is widely opposed to the general ideas of persons who have not very carefully studied the subject, and would therefore call attention to the following passage from *The Origin of Species,* by Mr. Charles Darwin, in which the facts bearing on this question are very fully stated:—"Great as the differences are between the breeds of pigeons, I am fully convinced that the common opinion of naturalists is correct, namely, that all have descended from the Kock Pigeon *(Columba livia),* including, under this term, several geographical races or sub-species, which differ from each other in the most trifling respects."

W. B. Tegetmeier, *Pigeons: Their Structure, Varieties, Habits, and Management* (London: George Routledge, 1873), 27–28.

Secondary Literature: Edmund W. Richardson, *A Veteran Naturalist: Being the Life and Work of W. B. Tegetmeier* (London: Withby, 1916).

PIERRE TEILHARD DE CHARDIN (1881–1955)
French Paleontologist

A century ago evolution (so-called) could still be regarded as a mere local hypothesis, framed to meet the problem of the origin of species (and, more particularly, of human origins). Since that time, however, we cannot avoid recognizing that it has included and now dominates the whole of our experience. "Darwinism" and "transformism" are words that already have only an historical interest. From the lowest and least stable nuclear elements up to the highest living beings, we now realize, nothing exists, nothing in nature can be an object of scientific thought except as a function of a vast and single combined process of "corpuscularization" and "complexification," in the course of which can be distinguished the phases of a gradual and irreversible "interiorization" (development of consciousness) of what we call (without knowing it) matter. [. . .] [E]volution has in a few years invaded the whole field of our experience; but, what is more, since we can feel ourselves swept up and sucked up in its convergent flood, this evolution is giving new value, as material for our action, to the whole domain of existence: precisely in as much as the appearance of a peak of unification at the higher term of cosmic ferment is now objectively providing human aspirations (for the first time in the course of history) with an absolute direction and an absolute end.

Pierre Teilhard de Chardin, *Christianity and Evolution* (1969; reprint, New York: Harcourt Harvest, 2002), 238–39.

If we look around us (and more particularly in the conservative sections of the religious world) we are amazed to note how obstinately the truly infantile idea persists that the word "evolution" does no more than disguise a mere "local" dispute between biologists who are divided on the question of the origin of living species. We still hear the word "Darwinism" used as a synonym for "evolutionism." As though what has happened in a mere half century goes for nothing: the integration of the various trends recorded, more or less independently, by all groups of scientific researchers without exception, which is every day making it more evident that the ontogenesis of the microcosm (which each one of us represents) has no physically possible significance or context unless it is restored to its correct place not only in the phylogenesis of some zoological branch but in the very cosmogenesis of an entire universe; and that it is in the perception of this fundamental dynamic unity that we find the essence of the great modern advance represented by the idea of evolution.

Pierre Teilhard de Chardin, *Activation of Energy,* trans. René Hague (New York: Harcourt Brace Jovanovich, 1971), 256.

Secondary Literature: David Lane, *The Phenomenon of Teilhard* (Macon, GA: Mercer University Press, 1996).

ALFRED LORD TENNYSON (1809–1892)
English Poet

[19 November 1859]
A. Much interested by Darwin's book which Mr. [Francis Turner] Palgrave* is so good as to send him, not knowing we have already a copy.

19 April 1861: No reading, we talk of Darwin.

Lady Tennyson's Journal, ed. James O. Hoge (Charlottesville: University Press of Virginia, 1981), 140, 156.

[28 March 1871]
I told [Tennyson] what Sir R. Martin† had said about the French being ugly, deformed dwarves. He said when he traveled in France he did not notice it. The women, he had heard, had skulls shaped like Simian apes. Asked me if I had read Darwin's book [*Descent of Man*]. Rather believed theory. Had written "In Memoriam" without having seen *Vestiges of Creation,* was quite excited when he read it. Others accused him of having copied from it. There is nothing degrading in the theory.

[29 March 1871]
Talked of Darwin's new book, for which he had taken a great fancy.

Tennyson at Aldworth: The Diary of James Henry Mangles, ed. Earl A. Knies (Athens: Ohio University Press, 1984), 63, 67. †Richard Biddulph Martin (1838–1916), English banker and Liberal Party leader.

Secondary Literature: Christopher Ricks, *Tennyson* (London: Palgrave Macmillan, 1989); Philip J. Waller, *Writers, Readers, and Reputations: Literary Life in London, 1870–1918* (Oxford: Oxford University Press, 2006).

WILLIAM ROSCOE THAYER (1859–1923)
American Historian

We rejoice to find Darwin worthy of being the prophet of a new dispensation—Darwin, the strong, quiet, modest man, harassed hourly by a depressing ailment, but patient under suffering, and preferring truth to the triumph of his own opinions or to any other reward.

William Roscoe Thayer, "Biography, I. General Introduction," in *Lectures on the Harvard Classics* (New York: P. F. Collier, 1914), 163–80, on 173.

But why should we seek farther for evidence of the danger of trying to fit history to any theory when we, and the whole world, have been struggling to break loose from the coils of a misinterpreted phrase? I do not believe that the atrocious war into which the Germans plunged Europe in August, 1914, and which has subsequently involved all lands and all peoples, would ever have been fought, or at least would have attained its actual gigantic proportions, had the Germans not been made mad by the theory of the survival of the fittest. The Germans are the most amazing doctrinaires the world has ever seen; they are also the greatest pedants. Whatever subject attracts their attention, obsesses them; and to be obsessed means to lose contact with the normal measures and perspectives of life.

So the phrase, "the survival of the fittest," obsessed them. Studying only the animal kingdom, they concluded that fitness was won by and depended upon brute force. The species possessing the greatest amount of force was, therefore, the fittest. Any of us, though we be not naturalists, can see how untrue this conclusion is, even when applied to the animal world. Frail creatures survive in spite of all the efforts of the strong creatures which prey upon them; and some of the frail have a far longer geologic ancestry than has the lion or the elephant. Insect tribes which flit hither and thither at the will of a passing breeze, date back aeons on aeons to conditions when no mammal trod the earth. If brute force alone were the test of fitness to survive, how could this be?

But we see, of course, that the vital consideration is, what do you mean by fitness? The fishes have a certain fitness which enables them to swim and to live under water; snakes have another by which they glide; insects and birds are fitted to fly; animals and man to walk and run. If you examine all these creatures, on the physical side alone, you find that something besides strength, physical force, has accounted for their being able to adjust themselves to their environment. Now, when we discover that at a certain point in mankind's evolution *moral* considerations come in, we see that as the race develops morals play a more and more important part in determining fitness to survive. The higher races, like the higher individual types, cease to regard the possession of power—brute power, enabling them to kill or enslave their neighbors—as their final aim. In a family the brothers who are physically stronger do not beat their weaker sisters; in society, we do not allow the brawny man of six feet two, merely because he is big, to persecute or destroy the little man of five feet. Civilization lives by ideals, by standards with which the girth of a man's chest or the thrust of his thighs has nothing to do.

The Germans, however, in their obsession, left all this out. Hindenburg, colossal in form and brutish in nature, could knock down, trample, and destroy Goethe, shall we say that he thereby could prove that he was fitter than Goethe to survive? At any rate, in the imaginary conflict, he survived, and Goethe didn't.

William Roscoe Thayer, "Vagaries of Historians," *American Historical Review* 24 (1919): 183–95.

Secondary Literature: The Letters of William Roscoe Thayer, ed. Charles Downer Hazen (Boston: Houghton Mifflin, 1926).

D'ARCY WENTWORTH THOMPSON (1860–1948)
English Mathematical Biologist

Darwin found no difficulty believing that "natural selection will tend in the long run to reduce *any part* of the organization, as soon as, through changed habits, it becomes superfluous, without by any means causing some other part to be largely developed in a corresponding degree. And conversely, that natural selection may perfectly well succeed in largely developing an organ without requiring as a necessary compensation the reduction of some adjoining part (OS 6th ed., 118)." This view has been developed into a doctrine of the independence of single characters [...] especially by the paleontologists. Thus [Henry Fairfield] Osborn* asserts a "principle of hereditary correlation, combined with a principle of *hereditary separability*, whereby the body is a colony, a mosaic, of single individual and separable characters." I cannot think that there is more than a very small element of truth in this doctrine. [...] We may look upon the coordinated parts, now as related and fitted *to the end or function of* the whole, and now as related to or resulting *from the physical causes* inherent in the entire system of forces to which the whole has been exposed, and under whose influence it has come into being.

A "principle of discontinuity" [...] is inherent in all our classifications, whether mathematical, physical or biological; and the infinitude of possible forms, always limited, may be further reduced and discontinuity further revealed by imposing conditions—as, for example, that our parameters must be whole numbers, or proceed by *quanta*, as the physicists say. The lines of the spectrum, the six families of crystals, [John] Dalton's atomic law, the chemical elements themselves, all illustrate this principle of discontinuity. In short, nature proceeds *from one type to another* among organic as well as inorganic forms; and these types vary according to their own parameters and are defined by physico-mathematical conditions of possibility. In natural history Cuvier's "types" may not be perfectly chosen nor numerous enough, but *types* they are; and to seek for stepping stones across the gaps between is to seek in vain forever.

This is no argument against the theory of evolutionary descent. It merely states that formal resemblance, which we depend on as our trusty guide to the affinities of animals within certain bounds or grades of kinship and propinquity, ceases in certain other cases to serve us, because under certain conditions it ceases to exist. Our geometrical analogies weigh heavily against Darwin's conception of endless small continuous variations; they help to show that discontinuous variations are a natural thing, that "mutations"—or sudden changes, greater or less—are bound to have taken place, and new "types" to have arisen now and then. Our argument indicates, if it does not prove, that such mutations, occurring on a comparatively few definite lines, or plain alternatives, of physico-mathematical possibility, are likely to repeat themselves: that the "higher" protozoa, for instance, may not spring from or through one another, but severally from the simpler forms; or that the worm-type, to take another example, may have come into being again and again.

D'Arcy Wentworth Thompson, *On Growth and Form,* 2nd ed., 2 vols. (Cambridge: Cambridge University Press, 1952), 2:1019–20, 1094–95.

Secondary Literature: Ruth Darcy Thompson, *D'Arcy Wentworth Thompson, the Scholar-Naturalist, 1860–1948* (London: Oxford University Press, 1948).

HENRY DAVID THOREAU (1817–1862)
American Naturalist and Philosopher

As the woodchuck dines chiefly on crickets, he will not be at much expense in seats for his winter quarters. Since the anatomical discovery, that the *thyroid* gland, whose use in man is *nihil,* is for the purpose of promoting digestion during the hibernating jollifications of the woodchuck, we sympathize less at his retreat. Darwin, who hibernates in science, cannot yet have heard of this use of the above gland, or he would have derived the human race from the Mus Montana, our woodchuck, instead of landing him flat on the *Simiadae,* or monkey.

Thoreau, quoted in *Thoreau, the Poet-Naturalist,* by William Ellery Channing, new ed. (Boston: Charles E. Goodspeed, 1902), 221.

Secondary Literature: Alfred I. Tauber, *Henry David Thoreau and the Moral Agency of Knowing* (Berkeley and Los Angeles: University of California Press, 2001); David Spooner, *Thoreau's Vision of Insects and the Origins of American Entomology* (Philadelphia: Xlibris, 2002).

GEORGE TICKNOR (1791–1871)
American Literary Historian

While the narrowness of Puritan Protestantism was thus slowly yielding, before the advances of social civilization, it was not yet strenuously attacked, either by the influx of a foreign population bringing with it its own foreign creed, or by the cold skepticism of what is called modern thought. For many years after this there was but one Roman Catholic church in Boston. At the same time the means of intellectual training were infinitely less than they are now. Books were scarce, and there were no large libraries rich with the spoils of learning. But a taste for reading and a love of knowledge were generally diffused, and there were few homes of those in comfortable circumstances where there was not at least a closetful of good books. These were carefully, almost reverently, read; and such reading was productive of sound intellectual growth. Johnson was the favorite author in prose, and Pope in verse. [James] Hervey's *Meditations* and [Johann Georg] Zimmermann *On Solitude* were popular books, and the glittering monotony of Darwin found admirers and imitators.

Life, Letters and Journals of George Ticknor, ed. George S. Hillard, 7th ed., 2 vols. (Boston: James R. Osgood, 1877), 1:18–19.

You have read Darwin on the *Origin of Species,* and know what he means by natural selection. If it were possible so to frame the institutions of a country, that the population should be chiefly kept up by breeding from those individuals, male and female, which are the best morally and intellectually; this selection would tend, according to Darwinian principles, to improve the race, and if an opposite system were persevered in an opposite result might be expected. Perhaps the best experiment hitherto made is that which we owe to the Inquisitors, and the result in regard to the deterioration of the race is perhaps as satisfactory as Darwin could desire.

Charles Lyell to Ticknor, 8 February 1867, in *Life, Letters, and Journals of Sir Charles Lyell, Bart,* 2 vols. (London: John Murray, 1881), 2:411–12.

Secondary Literature: David B. Tyack, *George Ticknor and the Boston Brahmins* (Cambridge, MA: Harvard University Press, 1967).

KLIMENT TIMIRIAZEV (1843–1920)
RUSSIAN BIOLOGIST

[1877]
Knowing that Darwin was something like a church elder and much loved by the people of Downe I boldly addressed the first man I met with the

question, how to get to Mr. Darwin, to which he replied, somewhat reproachfully, "You mean, Doctor Darwin? This is his garden, only you have
to approach the house from the other end." Since then I have had many
occasions to note that the English, even the commoners, set great store by
academic titles. For example, the people at Brentwood, where [John]
Ruskin* lived, invariably referred to him as "Professor." [...]

The door was answered by an old butler, most probably the one of
whom Francis Darwin said in his reminiscences, "We had come to see
him as a member of the family." He looked at me with a mixture of surprise and admonition: surprise, because I had come on foot and admonition because, as everyone else in the family he was afraid of intruders. [...] It
was a usual English parlor with a mantle-piece, a veritable "family hearth"
around which were the seats usually occupied by its inhabitants, with
Darwin's comfortable chair and a smaller one, a writing desk, apparently
the favorite place of Mrs. Darwin. Along the walls and in the corners
there were a few *établissements* and on the wall facing the fireplace two
windows with a door in between. Near the left window there was a small
writing desk on curved legs with all sorts of bric-à-brac which obviously
belonged to Mrs. Darwin. Everywhere there was the simplicity and cozi-
ness of an English home. The door opened into the garden without a
single step or even a threshold—*de plain pied*, as the French say—right
onto a space covered, as in most European gardens by fine gravel, very
inconvenient for flimsy shoes but a good protection against mud and slush
so common on our roads. There was a light gallery running the length of
the parlor forming what was locally called a verandah, and under it were
flowerpots and easy-chairs, including Darwin's chair with a high back
known from numerous photographs. [...]

A few minutes later, and quite unexpectedly, Darwin entered the room.
I have already had occasion to describe my first impression of him. It must
be said that the familiar portrait of him with a long grey beard was not yet
known at the time. The only known portrait of him was one in the German edition of his *Origin of Species* (and in my book *Charles Darwin*).
That portrait, dating back to the 1850s shows him as a man of about forty,
well-shaven and with trimmed side-whiskers and because the portrait
showed him from the waist up one could not help seeing him in one's
mind's eye as a shortish plump man looking rather like a businessman or
perhaps a sportsman but certainly not a profound and great thinker. And

now I was confronted with an impressive old man with a large grey beard, deep-sunken eyes, whose calm and gentle look made you forget about the scientist and think about the man. I couldn't help comparing him to an ancient sage or an Old Testament patriarch, a comparison which has often been quoted since. [. . .]

But when our conversation drifted to serious scientific subjects it assumed a purely English character. Learning that I was studying plant physiology he immediately confounded me with a question: "You must have felt it odd to find yourself in a country where there is not a single botanist-physiologist?" [. . .] From this question and the conversation that followed, I guessed that I had come to Downe at a very opportune moment (though I only learnt it with certainty many years later). It is known that after publishing his *Origin of Species* and other works which look at particular sides of his theory Darwin concentrated on botany, more specifically experimental and physiological botany. All this special research was designed to prove the usefulness of his theory as a "working hypothesis." At the time of my visit he was already engaged together with Francis on a piece of research that provided the content of a volume called *The Power of Movement in Plants.* [. . .]

Then he asked me what else, apart from Kew interested me in Britain from the botanical point of view. I said that I was planning to go to Rothamsted (a well-known agronomy experimental station, the first in Europe) and said that in terms of the teaching about "the struggle for survival" the current experiments of the changing composition of the meadow flora due to the use of fertilizers was of some interest. While I was talking he made signs to his son and when I finished he said reproachfully, "You see, the man has come from across the world and he is going to Rothamsted tomorrow and we still have not been there." And again, it was only many years later when the first volume of his letters appeared, that I learnt that Darwin was at the time planning a large series of experiments with artificial cultures as a means of changing forms and had entered into correspondence on this matter with Gilbert,[†] a well-known chemist from Rothamsted. At about the same time he conceived with amazing penetration his experiments in obtaining artificial plant growth (in nuts, etc.) and methods of experimental study of the laws of evolution. In the 30 years since that time no progress has been made on that matter. [. . .]

From botany we went on to discuss science in general. Darwin was particularly pleased that he had found among young Russian scientists some ardent supporters of his theory, referring most frequently to the name of Kovalevsky. When I asked him whom of the two brothers he had in mind—probably Alexander [O. Kovalevski]*, the zoologist—he replied, "No I believe that Vladimir's work on paleontology is even more important." [. . .] In the midst of this conversation Darwin startled me with a question: "Tell me, why do these German scientists quarrel so much among themselves?" "You are in a better position to judge," I replied. "How is that? I've never been to Germany." "Yes, but this must be another proof of your theory: there are probably too many of them. It's another example of the struggle for survival." He was taken aback for a second and then burst out laughing heartily. [. . .]

At that time Darwin was working on an answer to the accusation that he had not proved how insects benefited from the animal food and that this process was not eating but decay under the influence of bacteria. I saw a series of pans with *Drosera* turf; each was divided into two halves by a tin plate; the leaves of one were fed with meat and the leaves of the other were without any meat food and it was obvious that the former were much larger than the latter.

Showing me his nurslings Darwin spoke in a very pacific tone as if defending and justifying himself that he "was probably right," and that the results of the experiment spoke in his favor. Meanwhile we know now from the memoirs of his son that no objection vexed him as much as this one.

When we returned home coffee had arrived, and the talk assumed a more general character. It is known that Darwin was obliged to rest in the afternoons and that during that time his wife read aloud to him. For the most part these were novels of somewhat inferior quality but with a happy end. But sometimes by way of an exception he had something serious read to him. On that occasion the book on his desk was McKenzie-Walles' well-known book about Russia. I must say that in spite of the 15-odd years that had passed since the abolition of serfdom in Russia many people in Europe still remembered that peaceful revolution that had liberated some 20 million peasants (with land) in particular in the light of the bloody war for the liberation of Negroes in America. During his round-the-world voyage Darwin had come to hate slavery, and this led him (and not him alone) to see the future of the Russian people in the rosiest of colors. Another question that interested him was free thought that was beginning to manifest itself in Russia. "A society in which such books as [Henry Thomas] Buckle's* History of Civilization (a fact probably borrowed from McKenzie-Walles) are wide-spread and where they freely read books by Lyell* and Darwin's The Origin of Man," he said, "cannot revert to traditional views on the basic questions of science and life." [. . .]

When [Darwin] left, Mr. Francis offered to show me his study. It is by now well known through photographs—a small room with an ordinary fireplace, a simple writing desk in the middle and a small couch on which the tireless worker lay down when his illness got the better of him. What struck me about the study was the complete absence of what we call a "library." Darwin was known for his rather original attitude to books. If anyone could sincerely despise him it was the book lovers or rather the book maniacs who value the book as an object and would not allow themselves to cut an old publication in order not to depreciate its value to the antiquarian, or provide cheap trash with precious bindings. Darwin valued the book only for what he needed in it and so he tore out the pages that he needed and thus avoided cluttering up his desk and room. An even more modest room was shown to me on the upper storey. It was apparently occupied by Francis himself and also housed a kind of make-shift laboratory for experiments involved in Darwin's last major work *The Power of Movement in Plants* which he had already begun.

Kliment Timiriazev, "A Visit to Darwin at Down," http://www.archipelago.org/vol9/timiriazev.htm. This article, first published in the newspaper *Rosskiye Vedomosti*, is taken from Timiriazev's book *Nauka i Demokratiya: Sbornik statei 1904–1919* [Science and democracy: Collected articles, 1904–1919] (Moscow: Izd-vo, 1920). †Joseph Henry Gilbert (1817–1901), English chemist.

Secondary Literature: Alexander Vucinich, *Darwin in Russian Thought* (Berkeley and Los Angeles: University of California Press, 1988).

EDWARD BRADFORD TITCHENER (1867–1927)
American Psychologist

I do not think that Darwin ever had a profound interest in poetry; the scientific temperament was too strong in him. The historical plays in which as a schoolboy he took "intense delight" probably interested him in the main as stories. The poets whom he read during his plastic period probably attracted him, in large measure, by their felicity of language; the cult of words and phrases is characteristic of adolescence, and is curiously different from a real appreciation of style—into which it may or may not develop, according to the temperament of the reader. On the other side, I think that Darwin's poetic leanings were much more pronounced and much more persistent than those of the average man of science. By his own unconscious confession, and by the evidence of his written works, his mind was leavened with poetic feeling; all through his mature life he is ready with quotation when the occasion calls; and the very poignancy of his regret for the loss of poetry witnesses to his poetic endowment. If and in so far as he did lose his poetic interests, the loss was due, not specifically to his occupation with science, but generally to the combination of a stupendous life-work with continued ill-health.

Edward Bradford Titchener, "Poetry and Science: The Case of Charles Darwin," *Popular Science Monthly* 74 (1909): 43–47, on 47.

Secondary Literature: Donald A. Dewsbury, "Edward Bradford Titchener: Comparative Psychologist?" *American Journal of Psychology* 110 (1997): 449–56.

LEO TOLSTOY (1828–1910)
Russian Novelist

The prominent writers on aesthetics in England during the current century have been Charles Darwin (to some extent), Herbert Spencer*, Grant Allen*, Ker, and Knight.[†]

According to Charles Darwin (1809–1882—*Descent of Man,* 1871), beauty is a feeling natural not only to man, but also to animals, and consequently to the ancestors of man. Birds adorn their nests and esteem beauty in their mates. Beauty has an influence on marriages. Beauty includes a variety of diverse conceptions. The origin of the art of music is the call of the males to the females.

Leo Tolstoy, *What Is Art?* trans. Aylmer Maude (New York: Thomas Y. Crowell, 1899), 30. [†]William Paton Ker (1855–1923), Scottish literary scholar, and William Angus Knight (1836–1916), Scottish philosopher.

I hold with Darwin that violent struggle is a law of nature which overrules all other laws.

E. M. de Vogüé, quoted in Leo Tolstoy, *The Kingdom of God Is Within You,* trans. Constance Garnett, 2 vols. (London: William Heinemann, 1894), 1:230.

All discussion of the possibility of establishing peace instead of everlasting war, is the pernicious sentimentality of phrasemongers. There is a law of evolution by which it follows that I must live and act in an evil way; what is to be done? I am an educated man, I know the law of evolution, and therefore I will act in an evil way. "Entrons au palais de la guerre." There is the law of evolution, and therefore there is neither good nor evil, and one must live for the sake of one's personal existence, leaving the rest to the action of the law of evolution.

Leo Tolstoy, *The Kingdom of God Is Within You,* trans. Constance Garnett, 2 vols. (London: William Heinemann, 1894), 1:232.

[N]ow he comes at once upon a literature in which the old creeds do not even furnish matter for discussion, but it is stated baldly that there is nothing else—evolution, natural selection, struggle for existence—and that's all.

Leo Tolstoy, *Anna Karenina* (Plain Label Books, 1959), 1084.

Secondary Literature: Daniel P. Todes, *Darwin without Malthus: The Struggle for Existence of Russian Evolutionary Thought* (New York: Oxford University Press, 1989).

PAUL TOPINARD (1830–1911)
French Anthropologist

The struggle for existence is not so general nor merciless as some extreme disciples of Darwin would maintain. There are frequent lulls. Many species do not have antagonistic wants; the animal is not always possessed of blind hunger; he does not always covet the place of his neighbor; his motives for quarreling are sometimes extremely slight. The Carnivora are the born enemies of the species that constitute their food, but the Herbivora have only a desire for plants, fruits, roots, barks, etc. Both the one and the other have their moments of necessary repose. Rest is as imperious a want as activity.

Paul Topinard, *Science and Faith or Man as an Animal, and Man as a Member of Society* (Chicago: Open Court, 1899), 94.

Secondary Literature: Jennifer Michael Hecht, *The End of the Soul: Scientific Modernity, Atheism, and Anthropology in France* (New York: Columbia University Press, 2003).

MARY TREAT (1830–1923)
AMERICAN BOTANIST

Nothing yet in the history of carnivorous plants comes so near to the animal as [bladderwort]. I was forced to the conclusion that these little bladders are in truth like so many stomachs, digesting and assimilating animal food. What it is that attracts this particular larva into the bladders is left for further investigation. But here is the fact that animals are found there, and in large numbers, and who can deny that the plant feeds directly upon them? The why and wherefore is no more inexplicable than many another fact in nature. And it only goes to show that the two great kingdoms of nature are more intimately blended than we had heretofore supposed, and, with Dr. Hooker*, we may be compelled to say, "our brother organisms—plants." About the 1st of December, after I had made most of my observations, I wrote to Dr. Asa Gray* and to Mr. Darwin, both on the same day, telling them of my discovery. Dr. Gray then informed me that Mr. Darwin had been engaged in the same work on Utricularia, and also sent me a note from him, bearing date Aug. 5. From this note it would appear that at that date he had not worked the matter up as far as I had—at least had not found so many imprisoned animals; but with his superior facilities he may have far outstripped me.

Mary Treat, "Plants that Eat Animals," *American Naturalist* 9 (1875): 658–62, on 662.

Secondary Literature: Dawn Sanders, "Mary Treat and Her Traps," *Nature First,* no. 49 (Autumn 2007): 21–22.

HENRY BAKER TRISTRAM (1822–1906)
English Ornithologist and Clergyman

[October 1859]

Although the historic meeting at the Linnean Society appeared to produce but little effect, one distinguished naturalist publicly accepted the theory of natural selection before the publication of *The Origin of Species*, and therefore as the direct result of Darwin's and Wallace's* joint paper. This great distinction belongs to Canon Tristram, as Professor [Alfred] Newton* has pointed out in his Presidential Address to the Biological Section of the British Association at Manchester in 1887 (*Reports*, 727), at the same time expressing the hope "that thereby the study of Ornithology may be said to have been lifted above its fellows."

Canon Tristram's paper, "On the Ornithology of Northern Africa" [. . .] was published in *The Ibis*, vol. i., October, 1859. The important conclusions alluded to above are contained at the end of the section upon the species of desert larks: "Writing with a series of about 100 larks of various species from the Sahara before me, I cannot help feeling convinced of the truth of the views set forth by Messrs. Darwin and Wallace in their communications to the Linnean Society, to which my friend Mr. A. Newton last year directed my attention. [. . .] It is hardly possible, I should think, to illustrate this theory better than by the larks and chats of North Africa." In all these birds we trace gradual modifications of coloration and of anatomical structure, deflecting by very gentle gradations from the ordinary type; but when we take the extremes, presenting most marked differences. [. . . These differences] "have a very direct bearing on the ease or difficulty with which the animal contrives to maintain its existence."

Edward B. Poulton, *Charles Darwin and the Theory of Natural Selection* (London: Cassell, 1896), 92–93.

Tristram had been the first zoologist of any note who, at the instance of [Alfred] Newton*, publicly accepted the Darwinian views by his paper in the *Ibis* of October, 1859; his re-conversion at Oxford to the old faith, perhaps inspired by a feeling of loyalty to the Bishop, was a source of disappointment to Newton, who sought (unavailingly) to show him the error of his ways.

A. F. R. Wollaston, *Life of Alfred Newton* (New York: Dutton, 1921), 120.

Secondary Literature: Patrick Armstrong, *The English Parson-Naturalist: A Companionship Between Science and Religion* (Leominister, Herfordshire, UK: Gracewing, 2000).

ANTHONY TROLLOPE (1815–1882)
English Novelist

I am afraid of the subject of Darwin. I am myself so ignorant on it, that I should fear to be in the position of editing a paper on the subject.

Trollope to J. E. Taylor, 25 September 1868, in *The Letters of Anthony Trollope*, ed. N. John Hall, 2 vols. (Stanford, CA: Stanford University Press, 1983), 1:447.

Trollope in one of his novels gives as a maxim of constant use by a brickmaker—"It is dogged as does it"—and I have often and often thought that this is the motto for every scientific worker.

Darwin to G. R. Romanes, 1877, in *More Letters of Charles Darwin*, ed. Francis Darwin, 2 vols. (New York: Appleton, 1903), 1:370–71.

Secondary Literature: Philip J. Waller, *Writers, Readers, and Reputations: Literary Life in Britain, 1870–1918* (Oxford: Oxford University Press, 2006).

LEON TROTSKY (1879–1940)
Russian Politician

Darwin's theory of the origin of species embraces the development of the vegetable and animal kingdoms in its entirety. The struggle for existence and natural and genetic selection proceeds unceasingly and without interruption. But if an observer were to have sufficient time at his disposal for making observations—say a millennium as the smallest unit of measurement—he would doubtless establish before his eyes that there are long epochs of a relative equilibrium of life when the workings of the laws of selection are almost imperceptible and the species preserve their relative stability and appear to be an embodiment of platonic ideal types. But there are also epochs when the equilibrium between the animal, vegetable and geographical factors is upset, epochs of geo-biological crises when the laws of natural selection assert themselves with all their fierceness and lead the development over the corpses of vegetable and animal species. In this gigantic perspective Darwin's theory stands before us above all as the theory of critical epochs in the development of the vegetable and animal world.

Leon Trotsky, "Political Profiles: Karl Kautsky," http://www.marxists.org/archive/trotsky/profiles/kautsky.htm, originally published in *War and Revolution*, vol. 1.

Secondary Literature: Geoff Swain, *Trotsky* (New York: Longman, 2006).

MARK TWAIN (1835–1910)
AMERICAN AUTHOR

I often think that the highest compliment ever paid to my poor efforts was paid by Darwin through President Eliot,[†] of Harvard College. At least, Eliot said it was a compliment, and I always take the opinion of great men like college presidents on all such subjects as that.

I went out to Cambridge one day a few years ago and called on President Eliot. In the course of the conversation he said that he had just returned from England, and that he was very much touched by what he considered the high compliment Darwin was paying to my books, and he went on to tell me something like this:

"Do you know that there is one room in Darwin's house, his bedroom, where the housemaid is never allowed to touch two things? One is a plant he is growing and studying while it grows" (it was one of those insect-devouring plants which consumed bugs and beetles and things for the particular delectation of Mr. Darwin) "and the other some books that lie on the night table at the head of his bed. They are your books, Mr. Clemens, and Mr. Darwin reads them every night to lull him to sleep."

My friends, I thoroughly appreciated that compliment, and considered it the highest one that was ever paid to me. To be the means of soothing to sleep a brain teeming with bugs and squirming things like Darwin's was something that I had never hoped for, and now that he is dead I never hope to be able to do it again.

Mark Twain, address delivered to the Nineteenth Century Club, New York, 20 November 1900, http://www.farid-hajji.net/books/en/Twain_Mark/mts-chap052 .html. †Charles William Eliot (1834–1926), American educator, president of Harvard.

My books have had effects, and very good ones, too, here and there, and some others not so good. There is no doubt about that. But I remember one monumental instance of it years and years ago. Professor [Charles Eliot] Norton*, of Harvard, was over here, and when he came back to Boston I went out with [William Dean] Howells* to call on him. Norton was allied in some way by marriage with Darwin.

Mr. Norton was very gentle in what he had to say, and almost delicate, and he said: "Mr. Clemens, I have been spending some time with Mr. Darwin in England, and I should like to tell you something connected with that visit. You were the object of it, and I myself would have been very proud of it, but you may not be proud of it. At any rate, I am going to tell you what it was, and to leave to you to regard it as you please. Mr. Darwin took me up to his bedroom and pointed out certain things there— pitcher-plants, and so on, that he was measuring and watching from day to day—and he said: 'The chambermaid is permitted to do what she pleases in this room, but she must never touch those plants and never touch those books on that table by that candle. With those books I read myself to sleep every night.' 'Those were your own books.' I said: 'There is no question to my mind as to whether I should regard that as a compliment or not. I do regard it as a very great compliment and a very high honor that that great mind, laboring for the whole human race, should rest itself on my books. I am proud that he should read himself to sleep with them.'

Now, I could not keep that to myself—I was so proud of it. As soon as I got home to Hartford I called up my oldest friend—and dearest enemy on occasion—the Rev. Joseph Twichell,† my pastor, and I told him about that, and, of course, he was full of interest and venom. Those people who get no compliments like that feel like that. He went off. He did not issue any applause of any kind, and I did not hear of that subject for some time. But when Mr. Darwin passed away from this life, and some time after Darwin's *Life and Letters* came out, the Rev. Mr. Twichell procured an early copy of that work and found something in it which he considered applied to me. He came over to my house—it was snowing, raining, sleeting, but that did not make any difference to Twichell. He produced the book, and turned over and over, until he came to a certain place, when he said: "Here, look at this letter from Mr. Darwin to Sir Joseph Hooker*." What Mr. Darwin said—I give you the idea and not the very words—was this: I do not know whether I ought to have devoted my whole life to these drudgeries in natural history and the other sciences or not, for while I may have gained in one way I have lost in another. Once I had a fine perception and appreciation of high literature, but in me that quality is atrophied. "That was the reason," said Mr. Twichell, "he was reading your books."

Mark Twain, "Books, Authors, and Hats," address delivered to the Pilgrim's Club, Savoy Hotel, London, 25 June 1907, http://classiclit.about.com/library/bl-etexts/mtwain/bl-mtwain-sp-books.htm. †Joseph Twichell (1838–1918), American pastor, Twain's closest friend.

"Do you know," [Twain] once said, "that I gave Charles Darwin the strength to write some of his most famous and epoch-making volumes? How? I am told that, when the great scientist was utterly fagged out with study, investigation, and with the manifold experiments he was carrying on, he would read my 'Innocents' or 'Tom Sawyer' or, maybe a Harper Magazine story, for half an hour or an hour. Then he would go to work again and later was ready for bed. Only when this here Mark Twain had lulled his nerves into proper condition, Darwin [would] sleep, I am told, but I can't vouch for the truth of this story."

Henry W. Fisher, *Abroad with Mark Twain and Eugene Field* (New York: Nicholas L. Brown, 1922), 117–18.

Secondary Literature: Ron Powers, *Mark Twain: A Life* (New York: Free Press, 2005).

EDWARD BURNETT TYLOR (1832–1917)
ENGLISH ANTHROPOLOGIST

The state of mind in which the explaining powers of a favorite theory are fondly contemplated is, to some extent, antagonistic to the state of mind in which facts are seen, with the eye of impartial criticism, in all their obstinate and uncompromising reality. To be able to preserve the balance between the two opposing tendencies is to give evidence of the most perfect scientific training. Among contemporary writers, Mr. Darwin affords, perhaps, the most striking example of this union of speculative boldness and critical sobriety; and we do not know how we can more aptly express our sense of the thoroughness of Mr. Tylor's scientific culture than by saying that he constantly reminds us of the illustrious author of the *Origin of Species*.

John Fiske, review of *Primitive Culture: Researches into the Development of Mythology, Philosophy, Religion, Art, and Custom,* by Tylor, *North American Review* 114 (1872): 227–31, on 227.

Mr. Darwin saw two Malay women in Keeling Island who held a wooden spoon dressed in clothes like a doll; this spoon had been carried to the grave of a dead man, and becoming inspired at full moon, in fact lunatic, it danced about convulsively like a table or a hat at a modern spirit-séance. [...]

In South America, Mr. Darwin describes the Indians offering their adorations by loud shouts when they came in sight of the sacred tree standing solitary on a high part of the Pampas, a landmark visible from afar. To this tree were hanging by threads numberless offerings such as cigars, bread, meat, pieces of cloth, &c., down to the mere thread pulled from his poncho by the poor wayfarer who had nothing better to give. Men would pour libations of spirits and maté into a certain hole, and smoke upwards to gratify Walleechu, and all around lay the bleached bones of the horses slaughtered as sacrifices. All Indians made their offerings here, that their horses might not tire, and that they themselves might prosper. Mr. Darwin reasonably judges on this evidence that it was to the deity Walleechu that the worship was paid, the sacred tree being only his altar; but he mentions that the Gauchos think the Indians consider the tree as the god itself, a good example of the misunderstanding possible in such cases.

Edward B. Tylor, *Primitive Culture,* 3rd American ed., 2 vols. (New York: Henry Holt, 1889), 2:152, 223.

Secondary Literature: J. W. Burrow, *Evolution and Society: A Study in Victorian Social Theory* (Cambridge: Cambridge University Press, 1966).

JOHN TYNDALL (1820–1893)
Irish Physicist and Philosopher

Mr. Darwin shirks no difficulty; and, saturated as the subject was with his own thought, he must have known better than his critics the weakness as well as the strength of his theory. This of course would be of little avail were his object a temporary dialectic victory instead of the establishment of a truth which he means to be everlasting. But he takes no pains to disguise the weakness he has discerned; nay, he takes every pain to bring it into the strongest light. His vast resources enable him to cope with objections started by himself and others, so as to leave the final impression upon the reader's mind that, if they be not completely answered, they certainly are not fatal. Their negative force being thus destroyed, you are free to be influenced by the vast positive mass of evidence he is able to bring before you. This largeness of knowledge and readiness of resource render Mr. Darwin the most terrible of antagonists. Accomplished naturalists have leveled heavy and sustained criticisms against him—not always with the view of fairly weighing his theory, but with the express intention of exposing its weak points only. This does not irritate him. He treats every objection with a soberness and thoroughness which even Bishop Butler[†] might be proud to imitate, surrounding each fact with its appropriate detail, placing it in its proper relations, and usually giving it a significance which, as long as it was kept isolated, failed to appear. This is done without a trace of ill-temper. He moves over the subject with the passionless strength of a glacier; and the grinding of the rocks is not always without a counterpart in the logical pulverization of the objector.

John Tyndall, *Address delivered before the British Association Assembled at Belfast* (London: Longmans, Green, 1874), 43–44. [†]Samuel Butler (1774–1839), bishop of Lichfield, grandfather of Samuel Butler the novelist.

Secondary Literature: Arthur E. Woodruff, "John Tyndall," in *Encyclopedia of Philosophy,* ed. Paul Edwards, 8 vols. (New York: Macmillan, 1967), 8:167–68; W. H. Brock, N. D. McMillan, and R. C. Mollan, eds., *John Tyndall: Essays on a Natural Philosopher* (Dublin: Royal Dublin Society, 1981).

MIGUEL DE UNAMUNO (1864–1936)
Spanish Philosopher

[1912]

[W]hy be scandalized by the infallibility of a man, of the Pope? What difference does it make whether it be a book that is infallible—the Bible or a society of men—the Church, or a single man? Does it make any essential change in the rational difficulty? And since the infallibility of a book or of a society of men is not more rational than that of a single man, this supreme offense in the eyes of reason had to be posited.

It is the vital asserting itself, and in order to assert itself it creates, with the help of its enemy, the rational, a complete dogmatic structure, and this the Church defends against rationalism, against Protestantism, and against Modernism. The Church defends life. It stood up against Galileo, and it did right; for his discovery, in its inception and until it became assimilated to the general body of human knowledge, tended to shatter the anthropomorphic belief that the universe was created for man. It opposed Darwin, and it did right, for Darwinism tends to shatter our belief that man is an exceptional animal, created expressly to be eternalized. And lastly, Pius IX*, the first Pontiff to be declared infallible, declared that he was irreconcilable with the so-called modern civilization. And he did right.

Miguel de Unamuno, *The Tragic Sense of Life*, trans. J. Crawford Flitch (New York: Dover, 1954), 72.

[I]t is [. . .] quite logical that [Arthur] Schopenhauer*, who deduced pessimism from the voluntarist doctrine or doctrine of universal personalization, should have deduced from both of these that the foundation of morals is compassion. [. . .] Only his lack of social and historical sense, his inability to feel that humanity is also a person, although a collective one, his egoism, in short, prevented him from feeling God, prevented him from individualizing and personalizing the vital and collective Will—the Will of the Universe.

On the other hand, it is easy to understand his aversion from purely empirical, evolutionist, or transformist doctrines, such as those set forth in the works of Lamarck and Darwin which came to his notice. Judging Darwin's theory solely by an extensive extract in *The Times,* he described it, in a letter to Adam Louis von Doss (March 1, 1860), as "downright empiricism" *(platter Empirismus).* In fact, for a voluntarist like Schopenhauer, a theory so sanely and cautiously empirical and rational as that of Darwin left out of account the inward force, the essential motive of evolution. For what is, in effect, the hidden force, the ultimate agent, which impels organisms to perpetuate themselves and to fight for their persistence and propagation? Selection, adaptation, heredity, these are only external conditions. This inner, essential force has been called will on the supposition that there exists also in other beings that which we feel in ourselves as a feeling of will, the impulse to be everything, to be others as well as ourselves yet without ceasing to be what we are.

Miguel de Unamuno, *The Tragic Sense of Life,* trans. J. Crawford Flitch (New York: Dover, 1954), 147.

Secondary Literature: Peter Kestenbaum, "Miguel de Unamuno y Jugo," in *Encyclopedia of Philosophy,* ed. Paul Edwards, 8 vols. (New York: Macmillan, 1967), 8:182–85; Peter G. Earle, "El evolucionismo en el pensamiento de Unamuno," *Cuadernos de la Cátedra Miguel de Unamuno* 14–15 (1964–65): 19–28.

GEORGES VACHER DE LAPOUGE (1854–1936)
FRENCH ANTHROPOLOGIST

A paradoxical result is that the inferior elements reconstitute themselves little by little and each stage in the direction of purity marks a return towards barbarism. Seemingly contrary to Darwin's law, this phenomenon is, in fact, a rigorous application of it. Superior individuals in themselves are relatively inferior whenever they have less chance of success or of leaving posterity, by reason of the social environment in which they struggle for life. [...] An example is the Antilles where the white sector has almost disappeared.

George Vacher de Lapouge, *Les sélections sociales* (Paris: A. Fontemoing, 1896), 66–67, trans. TFG.

Secondary Literature: Jennifer Michael Hecht, "Vacher de Lapouge and the Rise of Nazi Science," *Journal of the History of Ideas* 61 (2000): 285–304.

HANS VAIHINGER (1853–1923)
GERMAN PHILOSOPHER

The idea of evolution became one of the fundamental elements of my mental life. Herder[†] draws special attention to the evolution of spiritual life out of its first animal origins, and he regards man always as linked up with that Nature from which he was gradually evolved. Thus in 1869, when I first heard Darwin's name and when my schoolfriends told me about the new theory of man's animal ancestry, it was no surprise to me, because through my reading of Herder I was already familiar with the idea. In later years, there has been much discussion as to whether Herder can be called a forerunner of Darwin. At any rate in my case Darwin's theory of descent added nothing new to what I had learnt from Herder.

Hans Vaihinger, *The Philosophy of "As If": A System of the Theoretical, Practical, and Religious Fictions of Mankind,* trans. Charles Kay Ogden (London: Routledge & Kegan Paul, 1924), xxiv. †Johann Gottfried Herder (1744–1803), German philosopher.

Secondary Literature: Rollo Handy, "Hans Vaihinger," in *Encyclopedia of Philosophy,* ed. Paul Edwards, 8 vols. (New York: Macmillan, 1967), 8:221–24.

PAUL VALÉRY (1871–1945)
French Author

And this brings us straight to Darwin. I don't like that fellow—he leaves me cold, I can't get inside him. But no historian can pass him over. Just think: if he is right, the whole of history is changed. I mean all thinking about history. And there is no doubt that he has contributed something.

Paul Valéry, *History and Politics,* trans. Denise Folliot and Jackson Mathews (New York: Pantheon, 1962), 516.

Secondary Literature: Reino Virtanen, *The Scientific Analogies of Paul Valéry* (Lincoln: University of Nebraska Press, 1974).

NIKOLAI VAVILOV (1887–1943)
Russian Geneticist

It was not by chance that Darwin was a regular visitor at the Museum of Natural History in London, the location of the fundamental collection of museum exhibitions displaying the breeds of domesticated animals. It is hard to present a better picture of the striking amplitude of variability with respect to the breed of dogs than that located in the British Natural History Museum, which demonstrates the enormous possibilities opened up by reading.

When working during 1913 and 1914 in the personal library of Darwin, which was preserved in its entirety after his death [...] I had the opportunity to see how thoroughly Darwin studied the works of his forerunners with respect to the history of crops and the breeding of plants. Books by the best English, French, and German plant breeders and experts on cultivated plants such as Shireff, Le Coutère, Metzger and Loisleur-Deslongchamps[†] were annotated with remarks by Darwin himself. [...] In the text and at the ends of books, Darwin noted down facts and ideas that were especially important to him. On the basis of his library it is [...] possible to follow the course of the creativity of this great scientist and the enormously laborious work preceding his general conclusions.

In contrast to DeCandolle [Alphonse de Candolle*], who was mainly interested in establishing the true native lands of cultivated plants on the basis of taxonomic-geographical, historical and linguistic data but also on the relationship between cultivated and wild plants, Darwin was first and foremost interested in the evolution of the species, the appearance of subsequent changes to which the species were subjected when taken into cultivation, the amplitude of the variability of the organisms under the influence of the conditions of cultivation and the role of selection. Unlike De Candolle, every plant interested Darwin, not only because of itself but as an integral unit for explaining evolution. Different species were, in essence, considered by Darwin as illustrations for the basic idea he wanted to argue. In contrast to the detailed codex of De Candolle, the book by Darwin on *Variation of Plants and Animals under Domestication* is of characteristic significance as a persistent idea towards a revelation of the dynamics of the evolutionary process.

Nikolai Ivanovich Vavilov, *Origin and Geography of Cultivated Plants*, trans. Doris Love (Cambridge: Cambridge University Press, 1992), 255, 422. [†]Patrick Shireff (1791–1876), farmer, pioneer cereal breeder who produced new varieties of wheat and oats; John Le Couteur (fl. 1830s–1860s), English agriculturist; Johann Christian Metzger (1789–1852), German landscape architect; and Jean Loisleur-Deslongchamps (1774–1849), French botanist.

Secondary Literature: Igor G. Loskutov, *Vavilov and His Institute: A History of the World Collection of Plant Genetic Resources in Russia* (Rome: International Plant Genetic Resources Institute, 1999).

THORSTEIN VEBLEN (1857–1929)
AMERICAN ECONOMIST

This machine technology, with its accompanying discipline in mechanical adaptations and object-lessons, came on gradually and rose to a dominating place in the cultural environment during the closing years of the eighteenth and the course of the nineteenth century; and as fast as men learned to think in terms of technological process, they went on at an accelerated pace in the further invention of mechanical processes, so that from that time the progress of inventions has been of a cumulative character and has cumulatively heightened the disciplinary force of the machine process. This early technological advance, of course, took place in the British community, where the machine process first gained headway and where the discipline of a prevalent machine industry inculcated thinking in terms of the machine process. So also it was in the British community that modern science fell into the lines marked out by technological thinking and began to formulate its theories in terms of process rather than in terms of prime causes and the like. While something of this kind is noticeable relatively early in some of the inorganic sciences, as, e.g., Geology, the striking and decisive move in this direction was taken toward the middle of the century by Darwin and his contemporaries. Without much preliminary exposition and without feeling himself to be out of touch with his contemporaries, Darwin set to work to explain species in terms of the process out of which they have arisen, rather than out of the prime cause to which the distinction between them may be due. (This is the substance of Darwin's advance over Lamarck, for instance.) Denying nothing as to the substantial services of the Great Artificer in the development of species, he simply and naively left Him out of the scheme, because, as being a personal factor, He could not be stated and handled in terms of process. So Darwin offered a tentative account of the descent of man, without recourse to divine or human directive endeavor and without inquiry as to whence man ultimately came and why, or as to what fortune would ultimately overtake him. His inquiry characteristically confines itself to the process of cumulative change. His results, as well as his specific determination of the factors at work in this process of cumulative change, have been questioned; perhaps they are open to all the criticisms levelled against them as well as to a few more not yet thought of; but the scope and method given to scientific inquiry by Darwin and the generation whose spokesman he is has substantially not been questioned, except by that diminishing contingent of the faithful who by force of special training or by native gift are not amenable to the discipline of the machine process. The characteristically modern science does not inquire about prime causes, design in nature, desirability of effects, ultimate results, or eschatological consequences.

Thorstein Veblen, *The Theory of Business Enterprise* (New York: Charles Scribner's Sons, 1904), 368–70.

In the course of modern times conceptions of an evolutionary creation or of genesis come up with increasing frequency, and from an early period in the machine age these conceptions take on more and more of a mechanistic character, but it is not until Darwin that such a genetic process of evolution is conceived in terms of blind mechanical forces alone, without the help of imputed teleological bias or personalized initiative. It may perhaps be an open question whether the Darwinian conception of evolution is in no degree contaminated with teleological fancies, but however that may be it remains true that a purely mechanistic conception of a genetic process in nature had found no lodgment in scientific theory up to the middle of the nineteenth century.

Thorstein Veblen, *The Instinct of Workmanship and the State of the Industrial Arts* (New York: Macmillan, 1914), 328.

Secondary Literature: Douglas F. Dowd, "Thorsten Veblen," in *Encyclopedia of Philosophy,* ed. Paul Edwards, 8 vols. (New York: Macmillan, 1967), 8:237–38; Geoffrey M. Hodgson, "Darwin, Veblen and the Problem of Causality in Economics," *History and Philosophy of the Life Sciences* 23 (2001): 385–423.

JULES VERNE (1828–1905)
French Novelist

My new book will be called either *The Great Forest,* or *The Aerial Village.* In it I am going to study the habits of the monkeys of Equatorial Africa, as Garner[†] did with his researches in Libreville, but the conclusions I shall put forward will be those of a believer, and entirely opposed to the theories of Darwin.

I am trying to reconstruct a race intermediate between the most advanced of the apes and the lowest men. [...] I deal with the question broadly and fancifully, and anyhow, I am far from reaching the same conclusions as Darwin, whose ideas I do not share at all.

Jules Verne, interview by Marcel Hutin, 1901, quoted in *Jules Verne, Inventor of Science Fiction,* by Peter Costello (London: Hodder & Stoughton, 1978), 195–96. †Richard Lynch Garner (1848–1920), American primatologist, linguist, and evolutionist.

It was chance alone, I suppose, that had conducted the *Nautilus* toward the island of Clermont-Tonnerre. [...] Here I had my long-coveted opportunity to study the madreporal system of building, to which the lands in this part of the ocean owe their life.

You must not make the mistake of confusing madrepores with coral. They have a tissue lined with a calcareous crust, and the variations of this structure have led the great naturalist Mr. Milne-Edwards,† my worthy master, to classify them in five sections. The animalculae that the marine polypus secretes live by the billion at the bottom of their cells. Their calcareous deposits become rocks, reefs, and islands. [...]

These polypi are found particularly in the agitated strata of the sea near the surface. And therefore it is from the upper level that they begin their operations downward and by degrees bury themselves in the debris of the secretions that support them. Such, at least, is the theory of Charles Darwin, who thus explains the formation of the atolls. And this is, to my mind, a theory superior to that ordinarily given of the foundation of madreporic works (summits of mountains and volcanoes) which are submerged some feet below the level of ocean.

Jules Verne, *Twenty Thousand Leagues Under the Sea,* trans. Philip Schuyler Allen (New York: Rand McNally, 1922), 159–60. †Henri Milne-Edwards (1800–1885), French zoologist and entomologist, supporter of Darwin.

The morning of the 24th, in 12°5' south latitude, and 94°33' longitude, we observed Keeling Island, a coral formation, planted with magnificent cocos, and which had been visited by Mr. Darwin and Captain Fitzroy. The Nautilus skirted the shores of this desert island for a little distance. Its nets brought up numerous specimens of polypi and curious shells of mollusca.

Jules Verne, *Twenty Thousand Leagues Under the Sea,* trans. Philip Schuyler Allen (New York: Rand McNally, 1922), 237.

Secondary Literature: Peter Costello, *Jules Verne, Inventor of Science Fiction* (London: Hodder & Stoughton, 1978).

CARL VOGT (1817–1895)
German-Swiss Zoologist

The English now begin energetically to study the anatomy of the simian brain; Marshall[†] dissects a chimpanzee; [George] Rolleston*, an orang; Huxley*, an ateles; preparations, drawings, and photographs are made, and—Huxley's assertions remain unshaken. [Richard] Owen* appeals to older drawings of Tiedemann, Van der Kolk, and Vrolik,[‡] to support his negative opinion; whilst his opponents appeal to the same works to prove their positive opinion. But this is too much for the phlegmatic Dutchmen. "Mr. Owen," they write, "has been carried away by his desire to upset the theory of Darwin (of which Messrs. Schroeder Van der Kolk and Vrolik are not over fond), and we believe him to be in error. In order to prove that the Negro brain rises at once without any transitionary stage above that of the anthropoid apes, Mr. Owen asserts that the posterior lobe of the hemispheres, the posterior cornu of the lateral ventricle, and the pes hippocampi minor, which exist in the Negro brain, are absent in the former." The Dutch anatomists now maintain that they have found and delineated all these parts, and that singularly enough, whilst praising the correctness of the sketches, he by a *contradictio in adjecto* denies the existence of these parts; they mention the researches of Huxley, Marshall, and Rolleston, and are glad to observe that they agree with their own. "We also are glad," they say, "that the zoological gardens, now-a-days, so easily furnish us with the necessary materials for comparison. An error, which formerly would have been perpetuated, is now more easily removed; but we feel grieved when we compare the assertions of Mr. Owen with the results obtained by the above eminent naturalists, which confirm our own."

Carl Vogt, *Lectures on Man: His Place in Creation, and in the History of the Earth,* ed. James Hunt (London: Longman, Green, Longman, & Roberts, 1864), 160–61.
[†]John Marshall (1818–91), English anatomist. [‡]Friedrich Tiedemann (1781–1861),

German anatomist; Jacobus Schroeder Van der Kolk (1797–1862), Dutch neurologist; and Willem Vrolik (1801–63), Dutch anatomist.

Secondary Literature: Adrian Desmond and James Moore, *Darwin's Sacred Cause* (Boston: Houghton Mifflin Harcourt, 2009).

RICHARD WAGNER (1813–1883)
German Composer

Cosima Wagner's Diary

1872

30 June: Has a cold and cannot work today; in Darwin *(The Origin of Species)* he reads the case of a dog which, while being dissected, licks the hand of its master, who is doing the operation; the fact that the latter does not stop upsets R. [Richard Wagner] greatly: "One needs always to visualize such things to recognize among what sort of people one is living, to whom one belongs, and to become humble."

4 July: Darwin is giving him pleasure, and he agrees with him that, in comparison with the old world, there is moral progress in the fact that animals are now accepted as part of it.

24 July: In the evening R. read me a chapter from Darwin (on social instinct).

1873

14 February: In the evening we begin Darwin's *Origin of Species* and R. observes the same thing has happened as between Kant and Laplace: the idea came from [Arthur] Schopenhauer* and Darwin developed it, perhaps even without having known Schopenhauer, just as Laplace certainly did not know Kant.

1877

28 September: In the evening he reads us some passages from Darwin's *The Descent of Man,* which he is now beginning to read.

30 September: R. works and tells me: "I was on the point of giving everything up today, picked up my Darwin, but suddenly threw him down, for
during my reading it all came to me, and I was then in such good spirits
that I literally had to force myself to stop, so as not to keep lunch waiting.
It's a mad state."

23 October: R. again had a wretched night; abdominal troubles—he reads
Darwin *(The Descent of Man)*, feels cold.

Cosima Wagner's Diaries, ed. Martin Gregor-Dellin and Dietrich Mack, 2 vols.
(New York: Harcourt Brace Jovanovich, 1978), 1:505, 507, 514, 594, 985, 989.

So far as I can understand the doctrines of its pundits, the upright, cautious Darwin, who pretended to little more than an hypothesis, would
seem to have given the most decisive impetus to the reckless claims of that
historical school by the results of his researches in the province of biology.
To me it also seems that this has chiefly come about through great misun-
derstandings, and especially through much superficiality of judgment in
the all-too-hasty application of the lights there won to the region of Philosophy. The gravest defects I deem the banishment from the new world-
system of the term spontaneous, or spontaneity itself, with a peculiarly
overbearing zeal and at least a thought too early. For we are now told that,
as more change has ever taken place without sufficient ground, so the
most astonishing phenomena [. . .] result from various causes, but which
we shall find uncommonly easy to get at when Chemistry has once laid
hold on logic. Meanwhile, however, the chain of logical deductions not
stretching quite so far as an explanation of the work of Genius, inferior
nature-forces generally regarded as faults of temperament, such as im-
petuosity of will, one-sided energy and stubbornness, are called in to keep
the thing as much as possible upon the realm of Physics.

Richard Wagner, "Public and Popularity" (1878), in *Richard Wagner's Prose Works,*
ed. William Ashton Ellis, vol. 6, *Religion and Art* (London: Routledge & Kegan
Paul, 1897), 75.

Wagner's [. . .] assertion that the natural evolution of each art necessarily
leads it to the surrender of its independence and to its fusion with the

other arts, contradicts so strongly all experience and all the laws of evolution, that it can at once be characterized as delirious. Natural development always proceeds from the simple to the complex—not inversely; progress consists in differentiation, *i.e.,* in the evolution of originally similar parts into special organs of different structure and independent functions, and not in the retrogression of differentiated beings of rich specialization to a protoplasm without physiognomy. [...] What he takes for evolution is retrogression, and a return to a primeval human, nay, to a pre-human stage.

Max Nordau, *Degeneration* (New York: Appleton, 1895), 175–76.

Secondary Literature: Burnett James, *Wagner and the Romantic Disaster* (New York: Hippocrene, 1983).

ALFRED RUSSEL WALLACE (1823–1913)
ENGLISH NATURALIST

Soon after I returned home, in the summer of 1862, Mr. Darwin invited me to come to Down for a night, where I had the great pleasure of seeing him in his quiet home, and in the midst of his family. A year or two later I spent a week-end with him in company with [Henry] Bates*, Jenner Weir,† and a few other naturalists; but my most frequent interviews with him were when he spent a few weeks with his brother, Dr. Erasmus Darwin, in Queen Anne Street, which he usually did every year when he was well enough, in order to see his friends and collect information for his various works. On these occasions I usually lunched with him and his brother, and sometimes one other visitor, and had a little talk on some of the matters specially interesting him. He also sometimes called on me in St. Mark's Crescent for a quiet talk or to see some of my collections.

Alfred Russel Wallace, *My Life: A Record of Events and Opinions,* 2 vols. (London: Chapman & Hall, 1905), 2:1. †See Charles Valentine Riley entry.

I have felt all my life, and I still feel, the most sincere satisfaction that Mr. Darwin had been at work long before me, and that it was not left for me to attempt to write *The Origin of Species.* I have long since measured my own strength, and know well that it would be quite unequal to the task. Far abler men than myself may confess, that they have not that untiring patience in accumulating, and that wonderful skill in using, large masses of facts of the most varied kind—that wide and accurate physiological knowledge, that acuteness in devising and skill in carrying out experiments, and that admirable style of composition, at once clear, persuasive and judicial, qualities, which in their harmonious combination mark out Mr Darwin as the man, perhaps of all men now living, best fitted for the great work he has undertaken and accomplished.

Alfred Russel Wallace, *Contributions to the Theory of Natural Selection* (London: Macmillan, 1875), iv.

Why did so many of the greatest intellects fail, while Darwin and myself hit upon the solution of this problem? [...] On a careful consideration, we find a curious series of correspondences, both in mind and environment, which led Darwin and myself, alone among our contemporaries, to reach identically the same theory.

First (and most important, as I believe) in early life both Darwin and myself became ardent beetle-hunters. Now there is certainly no group of organisms that so impresses the collector by the almost infinite number of its specific forms, the endless modifications of structure, shape, color, and surface—markings that distinguish them from each other, and their innumerable adaptations to diverse environments. [...]

Again, both Darwin and myself had what he terms "the mere passion for collecting," not that of studying the minutiae of structure, either internal or external. I should describe it rather as an intense interest in the variety of living things—the variety that catches the eye of the observer even among those which are very alike, but which are soon found to differ in several distinct characters.

Now it is this superficial and almost child-like interest in the outward form of living things which, though often despised as unscientific, happened to be *the only one* which would lead us towards a solution of the problem of species. For Nature distinguishes her species by just such characters. [...]

Then, a little later [...] we became travelers, collectors, and observers in some of the richest and most interesting portions of the earth; and we thus had forced upon our attention all the strange phenomena of local and geographical distribution, with the numerous problems to which they give rise. Thenceforward our interest in the great mystery of *how* species came into existence was intensified, and—again to use Darwin's expression— "haunted" us.

Finally, both Darwin and myself, at the critical period when our minds were freshly stored with a considerable body of personal observation and reflection bearing upon the problem to be solved, had our attention directed to the system of *positive checks* as expounded by Malthus in his "Principles of Population." The effect of that was analogous to that of friction upon the specially prepared match, producing the flash of insight which led us immediately to the simple but universal law of the "survival of the fittest" as the long-sought *effective* cause of the continuous modification and adaptations of living things.

James Marchant, *Alfred Russel Wallace: Letters and Reminiscences*, 2 vols. (New York: Cassell, 1916), 1:114–15.

Secondary Literature: Ross A. Slotten, *The Heretic in Darwin's Court: The Life of Alfred Russel Wallace* (New York: Columbia University Press, 2006); Peter Raby, *Alfred Russel Wallace: A Life* (Princeton, NJ: Princeton University Press, 2001).

HARRY MARSHALL WARD (1854–1906)
English Botanist

About 1870 he came under the influence of Darwin's writings and teachings, and in 1874 he entered more formally on a scientific career by attending Professor Huxley's* Biology course at South Kensington. His success there, and in the subsequent course in Laboratory Botany then being organized by Mr. Thiselton Dyer,[†] led to his proceeding to the Owens College, and afterwards to Cambridge.

George Washington Moon, *Men and Women of the Time: A Dictionary of Contemporaries*, 13th ed. (London: George Routledge, 1891), 926. [†]William Turner Thiselton-Dyer (1843–1928), English botanist who succeeded Joseph Dalton Hooker as director of Kew Gardens.

He was never satisfied with mere morphological detail. What he always sought was the physiological purpose of which this was the mechanism. I do not doubt that he owed this, in some measure, to his Cambridge training. [...] A curious organism had been sent me from the Eastern counties, which was used in the rustic manufacture of ginger beer. I showed it one day to Marshall Ward, and suggested his examining it. This he did in the most exhaustive way. [...] In all these researches he showed a quality which, I think, amounts to positive genius, and which I have not myself come across, at least to the same extent, in anyone but Mr. Darwin himself. He seemed to have the gift of compelling nature to reveal its most elusive secrets.

W. T. Thiselton-Dyer, "Harry Marshall Ward," *New Phytologist* 6 (1907): 1–8, on 5.

This pronounced variability of the oak was commented upon by the late Charles Darwin, who points out, in the *Origin of Species,* that more than a dozen species have been made by a certain author out of what other botanists regard as mere varieties of the common oak.

H. Marshall Ward, *The Oak: A Popular Introduction to Forest Botany* (New York: Appleton, 1892), 168.

I am convinced that the agriculturist of the future—and the same applies to the horticulturist, planter and forester—will have to concern himself more systematically with the working and the variability of the plant, and particularly with what Darwin termed Variation under Domestication, than has always been the custom in the past. The subject of the plasticity of cultivated plants, and especially of hybrids, is in one sense an old one; but much work is being done which proves, as such work is apt to do, that very much more may be done by well-planned experiments on the selection of new varieties raised by hybridizing and cultivation. [...]

Moreover, the essentials [of sexual reproduction] are found to be the same in the animal kingdom also, and the bearing of all these discoveries on the phenomena of reproduction, variation, and heredity in living organisms has been and is of the highest importance, for they support, control, explain and correct so many of the splendid results of Knight,

Kolreuter, Sprengel,[†] [Friedrich] Hildebrand[‡] and Hermann Müller[*], and in every direction throw side-lights into the crevices of that magnificent structure, the theory of Natural Selection, erected for all time by our countryman, Charles Darwin.

H. Marshall Ward, *Disease in Plants* (London: Macmillan, 1901), 70–72. [†]Charles Knight (1808–91), English–New Zealander civil servant and botanist, lichen expert; Josef Gottlieb Kolreuter (1733–1808), German botanist; Christian Konrad Sprengel (1750–1816), German naturalist, discover of plant sexuality. [‡]See Hermann Müller entry.

Secondary Literature: Peter G. Ayers, *Harry Marshall Ward and the Fungal Thread of Death* (St. Paul, MN: American Phytopathological Society, 2005).

LESTER WARD (1841–1913)
American Sociologist and Biologist

Now, the law of natural selection, or, as Mr. Spencer[*] more clearly terms it, "survival of the fittest," consists in this: that, of all these minute variations which are constantly taking place in all organisms, only those which tend to increase the correspondence between organism and environment, can be successively transmitted to a sufficient number of generations to cause any considerable alteration in the original stock. Darwin, looking at this process from a practical point of view, saw it in the light of a selection by nature of those modifications which proved of economic advantage to the organism. He, therefore, called it "Natural Selection." Herbert Spencer, regarding it from the stand-point of the physicist, saw it to be only another manifestation of the universal tendency of all the forces of the universe to approach the statical condition. He, therefore, dropped it into its appropriate niche in his cosmical system, and named it "Indirect Equilibration."

The Malthusian principle is far broader than its able expounder supposed. Just as Mr. Darwin has been able to apply it to the whole animal world below man, so it is also capable of being applied to supra-economic questions in society. Modern society, in the most civilized countries, is so highly derivative, and rests to so great an extent upon ideal conceptions, that an exceedingly high state of popular intelligence must be constantly kept up to prevent rude tampering with the delicate machinery by which its operations are conducted. Each new individual that is added to its membership must be schooled from the beginning in the principles upon which the social fabric is constructed. He enters it wholly ignorant of them all. The civilized infant is as blank intellectually as the savage infant, and has no longer to live, and immensely more to learn. Each one must be instructed in all this complicated curriculum *in initio,* before he can conceive of his true relations to society. We may therefore say in this connection, as truly as Malthus could say of his economic formula, that, while in the progress of civilization the capacity to acquire knowledge increases only in an arithmetical or some lower ratio, the amount of knowledge necessary to be acquired increases in a geometrical or some higher ratio.

Lester Frank Ward, *Dynamic Sociology* (New York: Appleton, 1920), 178–79, 704–5.

Secondary Literature: Edward C. Rafferty, *Apostle of Human Progress: Lester Frank Ward and American Political Thought* (Lanham, MD: Rowan & Littlefield, 2003).

MARY AUGUSTA WARD (1851–1920)
ENGLISH NOVELIST

In my life I seem to be able to put my finger on two accidental epoch-marking incidents. One was the coming across a certain book at a crucial period of mental development; the other was being invited to deliver an occasional address. When in England in November 1865, shortly after my marriage, I one day chanced upon a copy of John Stuart Mill's* essay on Auguste Comte, at that time just published. My intellectual faculties had then been lying fallow for nearly four years, and I was in a most recipient condition; and that essay of Mill's revolutionized in a single morning my whole mental attitude. I emerged from the theological stage, in which I had been nurtured, and passed into the scientific. I had up to that time never even heard of Darwin. *Inter arma,* etc. From reading that compact little volume of Mill's at Brighton in November, 1865, I date a changed intellectual and moral being beside him, side by side with a volume of Grant Allen's* *Sketches*.

"I didn't know you cared for this sort of thing!"

Robert did not answer for a moment, and a faint flush stole into his face.

"Imagine, Langham!" he said presently, "I had never read even the *Origin of Species* before I came here. We used to take the thing half for granted, I remember, at Oxford, in a more or less modified sense. But to drive the mind through all the details of the evidence, to force oneself to understand the whole hypothesis and the grounds for it, is a very different matter. It is a revelation."

"Yes," said Langham; and could not forbear adding, "but it is a revelation, my friend, that has not always been held to square with other revelations."

Mrs. Humphrey Ward, *Robert Elsmere* (London: Smith Elder, 1901), 170–71.

Secondary Literature: Philip J. Waller, *Writers, Readers, and Reputations: Literary Life in Britain, 1870–1918* (Oxford: Oxford University Press, 2006).

HOWARD C. WARREN (1867–1934)
American Psychologist

Acquaintance with the notion of organic evolution came a year or so before entering college. The Darwinian theory attracted me from the start. The notion of continuity or orderliness in the universe appealed to me. My attitude throughout childhood had been in that direction. It was far easier to accept Darwinian volition than the conservation of energy. For I was imbued with the idea that *doing* work meant *expenditure of energy*, that is, a loss of force, of power, of something.

Carl Murchison, ed., *A History of Psychology in Autobiography,* vol. 1 (New York: Russel & Russell, 1930), 446–47.

Secondary Literature: Paul Charles Kemeny, *Princeton in the Nation's Service: Religious Ideals and Educational Practice, 1868–1928* (New York: Oxford University Press, 1998).

ROBERT PENN WARREN (1905–1989)
American Poet and Novelist

Home was, I guess, more important than school for me. My father would read poetry and history aloud, etc. Also people did their home work or caught hell. Lots of books around. Father gave me Darwin, for instance, as a birthday present when I was twelve or thirteen. That was tough going. I would have preferred a .22.

Selected Letters of Robert Penn Warren, ed. Randy Hendricks and James A. Perkins, 4 vols. (Baton Rouge: Louisiana State University Press, 2008), 4:496.

Secondary Literature: Floyd C. Watkins, *Then & Now: The Personal Past in the Poetry of Robert Penn Warren* (Lexington: University Press of Kentucky, 1982).

BOOKER T. WASHINGTON (1865–1915)
African American Educator

I am not a student of biology and cannot answer with any definiteness the questions you ask.

It seems to me that the question of evolution is very largely a question for experts and I am willing to leave it to them to settle.

As to the general ideas of the progress of mankind, which have been associated with the name of Darwin in most of our minds, I confess that I share them, though I have never taken much stock in the theories that people have advanced from time to time as to the way in which this progress has come about. As far as my experience goes, I have seen but very little progress either moral or material that was not pretty closely associated with work. It is the evolution brought about by human beings who work that I believe in.

Washington to Elmer Kneale, 29 November 1911, in *The Booker T. Washington Papers*, ed. Louis R. Harlan and Raymond W. Smock, 14 vols. (Urbana: University of Illinois Press, 1972–89), 11:378–79.

Secondary Literature: Robert J. Norrell, *Up from History: The Life of Booker T. Washington* (Cambridge, MA: Harvard University Press, 2009).

ERICH WASMANN, SJ (1859–1931)
German Jesuit Biologist

Our only reason for quoting the *Descent of Man* is to show, that in his endeavor to derive the mental faculties of man from the psychic faculties of the animal, Charles Darwin was preoccupied by the principles of pseudo-psychology, which is unable to distinguish correctly between sense perception and intelligence. Darwin considers it self-evident that animals have intelligence, because he takes for intelligence any combination of sense representations which is brought about by individual experience. Consequently [Wilhelm] Wundt's* verdict on the want of critical method in pseudo-psychology applies equally well to the *Descent of Man* by Charles Darwin.

Erich Wasmann, *Instinct and Intelligence in the Animal Kingdom,* 2nd ed. (St. Louis: Herder, 1903), 16.

The name Darwinism is applied in a general way to *the theory of evolution.* [. . .] This confusion of ideas has done much harm in many ways. If, for instance, a serious student, engaged in scientific research, finds in his special department what he regards as evidence of the development of species, he is at once called a Darwinist, and as such is assailed by another party. In the same way [. . .] the advance of the theory of evolution as a scientific hypothesis is quite wrongly appropriated as an outcome of Darwinism, as Haeckel* especially has done. This explains the great applause which Darwinism has received in the widest circle and down even to the lowest classes.

In 1859 came the moment when a powerful wave, starting from England, assailed us like a deluge. It increased in strength and power until the foam flecked the very pinnacles of the rock. It is true that this wave no longer bears the name of Darwin and of the Darwinian system in the narrower sense, but is the theory of evolution which is waging war upon the theory of permanence, and has hitherto been victorious in the strife, and will probably remain so to the end.

Erich Wasmann, S.J., *The Berlin Discussion of the Problem of Evolution* (London: Kegan, Paul, Trench, Trübner, 1909), 41, 83.

Secondary Literature: A. J. Lustig, "Erich Wasmann, Ernst Haeckel, and the Limits of Science," *Theory in Biosciences* 121 (2002): 252–59.

HEWETT COTTRELL WATSON (1804–1881)
English Botanist

Even earlier than Huxley*, H. C. Watson wrote warmly accepting natural selection. In a letter [to Darwin], which is dated November 21st, 1859, he said:

"Your leading idea will surely become recognized as an established truth in science—*i.e.* 'Natural Selection.' It has the characteristics of all great natural truths, clarifying what was obscure, simplifying what was intricate, adding greatly to previous knowledge. You are the greatest revolutionist in natural history of this century, if not of all centuries."

Edward B. Poulton, *Charles Darwin and the Theory of Natural Selection* (London: Cassell, 1896), 144.

Much new light has been brought upon this debatable matter [the nature of species] by the theory of Charles Darwin, so rapidly become known to all naturalists, and still so much discussed among them, although with an increasing tendency to acceptance. It bears very importantly on phytogeography in various ways; and thus some consideration of that theory has become unavoidable with every student in this department of botany. According to the Darwinian theory, there is no absolute difference between species and varieties. They are supposed to be the same in kind, differing only in age and degree. [. . .] So far, many other naturalists had held views closely similar to those lately announced by Mr. Darwin. Among animals and plants they have seen variations arise, and augment, and accumulate, until some of the descendants have thus gradually become widely unlike their ancestors, certainly known or fairly supposed to be such. And they have thought and argued, if such changes can be seen produced in [the] course of a human life or of human history—then, why not all the changes which geology shows to have occurred between the former and the present, the extinct and the living, animals of the globe? [. . .] Before Darwin's theory was formed and announced, all may justly be said to have failed in their attempts to show satisfactorily *how* the changes from past to present species could actually have been brought about. According to present appearance, Charles Darwin has made a really good advance beyond his predecessors; although he too, after much ingenious effort, has left the most important part of the problem for some success to elucidate. [...]

In a controversial pamphlet published so long ago as 1836, [I] intimated [my] own leaning in favor of the transition-of-species theory, as it was then designated. [...] At the same time, a full conviction of the truth of the transition-of-species theory is not declared [...] being evidently withheld from lack of the one key explanation which Mr. Darwin had brought to its support, whether successfully or unsuccessfully so brought, under the designation of *Natural Selection*. [...]

What is the real advance? [...] The Darwinian theory asserts that a [process like selective breeding] is always operative in nature but it is one not truly identical in action though supposed to be analogous in its effects. [...] From one point of view, and to a limited extent, this is such an obvious truth as to be simply a truism, which nobody would dispute, and nobody had taken the trouble to describe in detail or to endow with a special name. [...] By giving a special name to the agency, describing its operation very clearly, illustrating it in copious detail, showing its universality, and connecting its phenomena together, Mr. Darwin has rendered vast service to the progress of natural science, and fairly made the previously rude recognition of this agency into a new and explanatory theory of organic life. [...] The reaction against a first skepticism has been great and rapid in favor of the Darwinian doctrines. The danger now is, that Mr. Darwin will be supposed to have discovered and established much more than he has truly done.

Hewett Cottrell Watson, *A Compendium of Cybele Britannica, or, British Plants in their Geographical Relations* (London: Longmans, Green, Reader & Dyer, 1870), 43–52.

Secondary Literature: Frank N. Egerton III, "Hewett Cottrell Watson," in *Dictionary of Scientific Biography*, ed. Charles Coulston Gillispie, 16 vols. (New York: Scribner's, 1970–80), 14:189–91.

JOHN B. WATSON (1878–1958)

AMERICAN PSYCHOLOGIST

Damn Darwin. The Neo-Darwinians and Neo-Lamarckians, etc. are in a worse hole than psychologists! I am terribly at sea as to finding a proper place and scope for psychology. What are our similar presuppositions and what are we good for? [...] I have come out of this—one chapter will have Behavior a biological problem—the scientific determination of modes of behavior and the *modus operandi* of behavior—a part of the problem of natural selection—the second the psychological implications in ideas of behavior. My interests are all in the first where an objective standard of determination is possible and where interpretation takes the line of the *importance* of the *observed facts*—for the theory of selection—facts—and interpretation possible without mentioning consciousness or deviating from a wider biological point of view. What is there left? Am I a physiologist? Or am I just a mongrel? I don't know how to get on.

Watson to Robert Yerkes, 29 October 1909, quoted in *Mechanical Man: John Broadus Watson and the Beginnings of Behaviorism,* by Kerry W. Buckley (New York: Guildford, 1989), 71.

Secondary Literature: Ernest Keen, *A History of Ideas in American Psychology* (New York: Praeger, 2001); Kerry W. Buckley, *Mechanical Man: John Broadus Watson and the Beginnings of Behaviorism* (New York: Guildford, 1989).

ALFRED WEBER (1868–1958)

GERMAN ECONOMIST

[Materialism's] alliance with political and religious radicalism gained for it the sympathies of the public, and it receives support from a number of recent discoveries and scientific theories. It appeals to the transformistic theory of Lamarck and Charles Darwin against the miracle of creation; to the anatomical study of anthropoid apes, against the view that there is an insurpassable gulf between animals and man, matter and mind; to the advance of chemical synthesis, against the phantom of *vital principle;* to the theory of the equivalence and transformation of forces and electrological discoveries, against the hypothesis of separate force for the explanation of thought; to the geological theory of gradual evolutions and imperceptible changes, against the theory of cataclysms, behind which, according to materialism, lurks the belief in the arbitrary intervention of a supernatural power; finally, to the many conclusive facts which prove, beyond the shadow of a doubt, that a relation exists between the brain and thought, against the spiritualistic distinction between soul and body.

Of all these innovations, the Darwinian theory is the one which materialism appropriated most readily, and to which it is most indebted. The theory answers the following cardinal question, which had remained unsolved until the days of Darwin: How can the purposiveness which is revealed in the structure and arrangement of our organs be produced without the intervention of an intelligent creative cause, and through the purely mechanical action of unconscious forces? Or, rather: How can we explain finality [...] without final causes? Darwinism provides materialism with a satisfactory answer to the main objection of theistic spiritualism, and thereby becomes its indispensable ally. So close in this alliance that Darwinism and materialism are regarded as synonymous terms.

Alfred Weber, *History of Philosophy,* trans. Frank Thilly (New York: Charles Scribner's, 1899), 563–64.

Secondary Literature: Frederick M. Bernard, "Alfred Weber," in *Encyclopedia of Philosophy,* ed. Paul Edwards, 8 vols. (New York: Macmillan, 1967), 8:281–82; Derek Gregory, "Alfred Weber and Location Theory," in *Geography, Ideology and Social Concern,* ed. D. R. Stoddart (Oxford: Blackwell, 1981).

MAX WEBER (1864–1920)
German Sociologist

Why do the German day-laborers move away? [...] The process is a mass-psychological one: the German agricultural laborers can no longer adapt to the *social* conditions of life in their homeland. [...] Why is it the *Polish* peasants who are gaining ground? [...] The small Polish peasant gains because he is prepared to eat even grass, as it were—in other words not *despite* but rather *because* of his low physical and intellectual standard of living.

Thus what we see taking place seems to be a *process of selection*. Over a long period both nationalities have been placed in the same conditions of existence. The consequence of this has *not* been, as vulgar materialists imagine, that they have acquired the same physical and psychological qualities, but rather that one group yields to the other, that the victorious nationality is the one possessing the greater ability to adapt itself to the given economic and social conditions of life.

I think I hardly need to observe that the disputes in natural science over the significance of the principles of selection, or over the general application *in natural science* of the concept of "selective breeding," and all the discussions relating to it in this area (with which I am not familiar), have no relevance to these remarks. However, the *concept* of "selection" is as much a commonplace today as, say, the heliocentric hypothesis, and the idea of "breeding" human beings is already to be found in Plato's *Republic*. [...] More difficult to answer is the question of how much lasting value should be attached to the latest attempts of anthropologists to extend the principle of selection, as understood by Darwin and [August] Weismann*, to the field of economic investigation. They are ingenious, but arouse considerable reservations as to method and factual results, and are no doubt mistaken in a number of exaggerated claims. Nevertheless, the writings of Otto Ammon† *(Natural Selection in Man, The Social Order and its Natural Basis)*, for example, deserve more attention than they have been given, irrespective of all the reservations that have to be made. An error made by most attempts of natural scientists to throw light on the problem of our science consists in their misdirected ambition to "disprove" socialism. In their enthusiasm to attain this goal, they involuntarily turn what was intended to be a "natural-scientific theory" of the social order into an apologia for it.

Max Weber, "The National State and Economic Policy" (1895), in Weber, *Political Writings*, ed. Peter Lassman and Ronald Speirs (Cambridge: Cambridge University Press, 1994), 1–28, on 8–10. †Otto Ammon (1842–1916), German nationalist anthropologist.

Thus the capitalism of to-day, which has come to dominate economic life, educates and selects the economic subjects which it needs through a process of economic survival of the fittest. But here one can easily see the limits of the concept of selection as a means of historical explanation. In order that a manner of life so well adapted to the peculiarities of capitalism could be selected at all, i.e. should come to dominate others, it had to originate somewhere, and not in isolated individuals alone, but as a way of life common to whole groups of men. This origin is what really needs explanation.

Max Weber, *The Protestant Ethic and the Spirit of Capitalism,* trans. Talcott Parsons (1930, reprint, London: Unwin Hyman, 1990), 55.

Secondary Literature: Stephen Kalberg, *Max Weber's Comparative-Historical Sociology* (Chicago: University of Chicago Press, 1994); Peter Ghosh, "Max Weber," in *Encyclopedia of Historians and Historical Writing,* ed. Kelly Boyd, 2 vols. (London: Fitzroy Dearborn, 1999), 2:1286–88.

AUGUST WEISMANN (1834–1914)
German Biologist

Many and diverse were the discoveries made by Charles Darwin in the course of a long and strenuous life, but none of them has had so far-reaching an influence on the science and thought of his time as the theory of selection. I do not believe that the theory of evolution would have made its way so easily and so quickly after Darwin took up the cudgels in favor of it, if he had not been able to support it by a principle which was capable of solving, in a simple manner, the greatest riddle that living nature presents to us—I mean the purposiveness of every living form relative to the conditions of its life and its marvelously exact adaptation to these.

Everyone knows that Darwin was not alone in discovering the principle of selection, and that the same idea occurred simultaneously and independently to Alfred Russel Wallace*. At the memorable meeting of the Linnean Society on 1st July, 1858, two papers were read (communicated by Lyell* and Hooker*) both setting forth the same idea of selection. One was written by Charles Darwin in Kent, the other by Alfred Wallace in Ternate, in the Malay Archipelago. It was a splendid proof of the magnanimity of these two investigators, that they thus, in all friendliness and without envy, united in laying their ideas before a scientific tribunal: their names will always shine side by side as two of the brightest stars in the scientific sky. [. . .]

The idea of selection set forth by the two naturalists was at the time absolutely new, but it was also so simple that Huxley* could say of it later,

"How extremely stupid not to have thought of that." As Darwin was led to the general doctrine of descent, not through the labors of his predecessors in the early years of the century, but by his own observations, so it was in regard to the principle of selection. He was struck by the innumerable cases of adaptation, as, for instance, that of the woodpeckers and tree-frogs to climbing, or the hooks and feather-like appendages of seeds, which aid in the distribution of plants, and he said to himself that an explanation of adaptations was the first thing to be sought for in attempting to formulate a theory of evolution.

August Weismann, "The Selection Theory," in *Essays in Commemoration of the Centenary of the Birth of Charles Darwin and of the Fiftieth Anniversary of the Publication of the Origin of Species,* ed. A. C. Seward (Cambridge: Cambridge University Press, 1909), 18–65.

Secondary Literature: Gloria Robinson, "August Friedrich Leopold Weismann," in *Dictionary of Scientific Biography,* ed. Charles Coulston Gillispie, 16 vols. (New York: Scribner's, 1970–80), 14:232–39.

H. G. WELLS (1866–1946)
English Novelist

I was told that while Huxley* lectured [at the Royal College of Science] Charles Darwin had been wont at times to come through those very curtains from the gallery behind and sit and listen until his friend finally had done. In my time, Darwin had been dead for only a year or so (he died in 1882).

These two were very great men. They thought boldly, carefully and simply, they spoke and wrote fearlessly and plainly, they lived modestly and decently; they were mighty intellectual liberators. It is a pity that so many of the younger scientific workers of to-day, ignorant of the conditions of mental life in the early nineteenth century and standing for the most part on ground won, cleared and prepared for themselves by these giants, find a perverse pleasure in belittling them. [...] [A]ny little Mr. Whippersnapper who chooses to use the vastly greater resources of to-day can find statements made by them that were insufficient or slightly erroneous, and theoretical suggestions that have been abandoned and disproved, and he can catch a bit of personal publicity from the pulpit or the reactionary press by saying that Darwin has been discredited or Huxley superseded. Great joy for Mr. (and Mrs.) Whippersnapper it is, naturally enough, to realize that he knows clearly things that Darwin never heard of, or is able to tatter some hypothesis of Huxley's. Little men will stand on the shoulders of giants to the end of time and small birds foul the nests in which they were hatched. Darwin and Huxley knew about one per cent of the facts about variation and mutation that are accessible to Mr. Whippersnapper. That does not alter the fundamental magnificence of Darwin's and Huxley's achievement. They put the fact of organic evolution upon an impregnable base of proof and demonstration so that even the Roman Catholic controversialists at last ceased to vociferate, after the fashion of Bishop [Samuel] Wilberforce* of the Anglican church on a memorable occasion, "Yah! Sons of apes! You *look* it," and discovered instead that the Church had always known all about Evolution and the place of man in Nature, just as it had always known all about the place of the solar system in space. Only it had said nothing about these things, because it was wiser so. Darwin and Huxley, in their place and measure, belong to the same aristocracy as Plato and Aristotle and Galileo, and they will ultimately dominate the priestly and orthodox mind as surely, because there is a response, however reluctant, masked and stifled, in every human soul to righteousness and a firmly stated truth.

H. G. Wells, *Experiment in Autobiography* (New York: Macmillan 1934), 161–63, referring to the years 1884–85.

Now what is called the scientific method in the physical sciences rests upon the ignoring of individualities; and like many mathematical conventions, its great practical convenience is no proof whatever of its final truth. Let me admit the enormous value, the wonder of its results in mechanics, in all the physical sciences, in chemistry, even in physiology,—but what is its value beyond that? Is the scientific method of value in biology? The great advances made by Darwin and his school in biology were not made, it must be remembered, by the scientific method, as it is generally conceived, at all. His was historical research. He conducted research into pre-documentary history. He collected information along the lines indicated by certain interrogations; and the bulk of his work was the digesting and critical analysis of that. For documents and monuments he had fossils and anatomical structures and germinating eggs too innocent to lie. But, on the other hand, he had to correspond with breeders and travelers of various sorts; classes entirely analogous, from the point of view of evidence, to the writers of history and memoirs. I question profoundly whether the word "science," in current usage anyhow, ever means such patient disentanglement as Darwin pursued. It means the attainment of something positive and emphatic in the way of a conclusion, based on amply repeated experiments capable of infinite repetition, "proved," as they say, "up to the hilt."

H. G. Wells, *First and Last Things: A Confession of Faith and a Rule of Life* (New York: Putnam, 1908), 50–51.

It is Mr. Belloc's[†] brilliant careless way to begin most of his arguments somewhere about the middle and put the end first. His opening peroration, so to speak, is a proclamation that this "Natural Selection"—whatever it is—is "an old and done-for theory of Darwin and Wallace*." "It is a laughing-stock for half a generation among competent men." Mr. George Bernard Shaw* does not believe in it! G. B. S. among the Fathers! That wonderful non-existent "latest European work" which plays so large a part in Mr. Belloc's dialectic is summoned briefly, its adverse testimony is noted, and it is dismissed to the safe again. And then there is a brief statement of how these two vile fellows, Darwin and Wallace, set out upon this reprehensible theorizing. What a ruthless expose it is of the true motives of scientific people!

H. G. Wells, *Mr. Belloc Objects to "The Outline of History"* (London: Watts, 1926), 29–30. [†]Hilaire Belloc (1870–1953), Anglo-French writer.

Secondary Literature: Roslynn D. Haynes, *H. G. Wells, Discoverer of the Future: The Influence of Science on His Thought* (London: Macmillan, 1980).

EDITH WHARTON (1862–1937)
AMERICAN NOVELIST

He [Egerton Winthrop] it was who gave me Wallace's* *Darwin and Darwinism,* and *The Origin of Species* and made known to me Huxley*, Herbert Spencer*, [George John] Romanes*, Haeckel*, [Edward] Westermarck,[†] and the various popular exponents of the great evolutionary movement. But it is idle to prolong the list, and hopeless to convey to a younger generation the first overwhelming sense of cosmic vastness which such "magic casements" let into our little geocentric universe.

Edith Wharton, *A Backward Glance* (New York: Appleton; London: Century, 1934), 94. †Edward Westermarck (1862–1939), Finnish sociologist, anthropologist, and philosopher of morality.

Secondary Literature: Paul Ohler, *Edith Wharton's "Evolutionary Conception": Darwinian Allegory in Her Major Novels* (New York: Routledge, 2006).

WILLIAM WHEWELL (1794–1866)
English Philosopher and Polymath

It still appears to me that in tracing the history of the world backwards, so far as the Paleontological sciences enable us to do so, all the lines of connection stop short of a beginning explicable by natural causes; and the absence of any conceivable natural beginning leaves room for, and requires, a supernatural origin. Nor do Mr Darwin's speculations alter this result. For when he has accumulated a vast array of hypotheses, still there is an inexplicable gap at the beginning of his series. To which is to be added, that most of his hypotheses are quite unproved by fact. We can no more adduce an example of a new species, generated in the way which his hypotheses suppose, than Cuvier could. He is still obliged to allow that the existing species of domestic animals are the same as they were at the time of man's earliest history. And though the advocates of uniformitarian doctrines in geology go on repeating their assertions, and trying to explain all difficulties by the assumption of additional myriads of ages, I find that the best and most temperate geologists still hold the belief that great catastrophes must have taken place; and I do not think that the state of the controversy on that subject is really affected permanently. I still think that what I have written is a just representation of the question between the two doctrines.

Whewell to Rev. D. Brown, 26 October 1863, in *William Whewell: An Account of His Writings,* by I. Todhunter, 2 vols. (London: Macmillan, 1876), 2:433–34.

Secondary Literature: Richard R. Yeo, *Defining Science: William Whewell, Natural Knowledge, and Public Debate in Early Victorian Britain* (Cambridge: Cambridge University Press, 1993).

ANDREW DICKSON WHITE (1832–1918)
American Educator

[1868–1874]

Perhaps the most comical of these attacks [on "atheists"] was one made by a clergyman of some repute before the Presbyterian Synod at Auburn in western New York. This gentleman, having attended one or two of the lectures by Agassiz* before our scientific students, immediately rushed off to this meeting of his brethren, and insisted that the great naturalist was "preaching atheism and Darwinism" at the university [Cornell]. He seemed about to make a decided impression, when there arose a very dear old friend of mine, the Rev. Dr. Sherman Canfield,† pastor of the First Presbyterian Church in Syracuse, who, fortunately, was a scholar abreast of current questions. Dr. Canfield quietly remarked that he was amazed to learn that Agassiz had, in so short a time, become an atheist, and not less astonished to hear that he had been converted to Darwinism; that up to that moment he had considered Agassiz a deeply religious man, and also the foremost—possibly, indeed, the last—great opponent of the Darwinian hypothesis. He therefore suggested that the resolution denouncing Cornell University brought in by his reverend brother be laid on the table to await further investigation. It was thus disposed of, and, in that region at least, it was never heard of more. Pleasing is it to me to chronicle the fact that, at Dr. Canfield's death, he left to the university a very important part of his library.

Andrew Dickson White, *Autobiography*, 2 vols. (New York: Century, 1907), 1:423.
†Sherman Bond Canfield (1810–71), Presbyterian minister.

Secondary Literature: Glenn C. Altschuler, *Andrew D. White: Educator, Historian, Diplomat* (Ithaca, NY: Cornell University Press, 1979).

ALFRED NORTH WHITEHEAD (1861–1947)
English Mathematician and Philosopher

The Malthusian doctrine, in its popular rendering, affirmed that as a law of nature the masses of mankind could never emerge into a high state of well-being. Still worse, biological science drew the conclusion that the destruction of individuals was the very means by which advance was made to higher types of species. This was the famous doctrine of Natural Selection, promulgated in 1859, by Charles Darwin. This exclusive reliance upon Natural Selection was not characteristic of Darwin's own theory. For him, it was one agency among many others. But, in the form in which his doctrine reigned in thought from that day to this, Natural Selection was the sole factor to be seriously considered. As applied to human society this theory is a challenger to the whole humanitarian movement.

The contrast between the dominant theories of Lamarck and Darwin made all the difference. Instead of dwelling on the brotherhood of man, we are now directed to procure the extermination of the unfit. Again the modern doctrines of heredity, gained partially from the experience of breeders of stock, partly from practical horticulturists, partly from the statistical researches of Francis Galton*, Karl Pearson*, and their school, partly from the laws of heredity discovered by Mendel, the Austrian Abbot, who published his unnoticed researches contemporaneously with the publication of Darwin's *Origin of Species*—these doctrines have all weakened the stoic-Christian ideal of democratic brotherhood.

Alfred North Whitehead, *Adventures of Ideas* (New York: Macmillan, 1954), 44–45.

Secondary Literature: Dorothy M. Emmet, "Alfred North Whitehead," in *Encyclopedia of Philosophy,* ed. Paul Edwards, 8 vols. (New York: Macmillan, 1967), 8:290–96; Michael Hampe, *Alfred North Whitehead* (Munich: Beck, 1998).

WALT WHITMAN (1819–1892)
American Poet

In due time, the evolution theory will have to abate its vehemence, cannot be allow'd to dominate everything else, and will have to take its place as a segment of the circle, the cluster—as but one of many theories, many thoughts, of profoundest value—and readjusting and differentiating much, yet leaving the divine secrets just as inexplicable and unreachable as before—maybe more so.

Walt Whitman, *Notes Left Over* (ca. 1888), http://etext.virginia.edu/toc/modeng/public/WhiPro3.html (accessed 20 May 2009).

With W. [Whitman] for full half hour, though I was on my way to Philadelphia and intended to stay only a few minutes. He was on his bed. "I was just about to get up—go to my chair"—doing so now (the cane always on the bed beside him). "Well, what is new? what do you bring me?" I had a copy of Darwin. Gazed at the portrait long. "What a grand head! Do you suppose the crag-like eyebrow was there? Oh! The great Darwin! None greater in our time! Big—big—big! I for one am grateful to have lived as one of his contemporaries."

Horace Traubel, *With Walt Whitman in Camden,* vol. 8 (Oregon House, CA: W. L. Bentley, 1961), 21, entry for 20 May 1891.

Secondary Literature: Roger Asselineau, *The Evolution of Walt Whitman* (Iowa City: University of Iowa Press, 1999).

JOSIAH DWIGHT WHITNEY (1819–1896)
AMERICAN GEOLOGIST

Oddly enough, the appearance of the *Origin of Species* in 1859 hurt the [California geological] survey. To the general position of Darwin, Whitney was an early convert, but his thoroughly scientific habit of mind as little inclined him to follow Haeckel* on the one hand as Agassiz* on the other. Like his old master Lyell*, he believed that the first task was to help the new doctrine to a hearing on its merits. For this reason, in his address to the legislature early in 1862, he went somewhat out of his way to set forth the ideas of the *Origin*. Nothing could have been more moderate in tone. Whitney dwelt upon Darwin's high repute in the scientific world, assumed that the matter was one in which an intelligent legislature would naturally take an interest, but committed himself only to the opinion that "the discussion of this interesting subject [. . .] will be of essential service to the progress of science."

Whitney promptly discovered that this by no means radical conviction was very far from being shared by the California clergy. The life of the survey coincided almost precisely with the controversy over Evolution, a controversy whose bitterness we of these easy-going days find it hard to realize. Inevitably, therefore, the church-going portion of the community became still further exacerbated against the survey, when after the discovery of the Calaveras skull in 1866, Whitney became the foremost American advocate of Tertiary man.

Edwin Tenney Brewster, *Life and Letters of Josiah Dwight Whitney* (Boston: Houghton Mifflin, 1909), 299–300.

Secondary Literature: Gerald D. Nash, "Josiah Dwight Whitney," in *Dictionary of Scientific Biography*, ed. Charles Coulston Gillispie, 16 vols. (New York: Scribner's, 1970–80), 14:315–16.

WILLIAM DWIGHT WHITNEY (1827–1894)
AMERICAN LINGUIST

This little work [W. H. Bleek,[†] *On the Origin of Language*] is written with much apparent profundity, but it seems to be one of a class, not quite unknown in German literature, in which a minimum of valuable truth is wrapped up in a maximum of [high] sounding phraseology. [. . .] We do not feel tempted to yield our opinions either to his guidance or to that of his cousin and editor, Professor Haeckel* of Jena, who also has a good deal to say within the same covers. The latter gentleman, particularly, appears to be one of those headlong Darwinians who take the whole process of development by natural selection as already proved and unquestionable, and go on with the fullest and most provoking confidence to draw out its details. [. . .] We have great faith in the substantial truth of the central Darwinian idea, and would no more regard the analogies and correspondences of form among different kinds and races as meaningless sports of nature, than the fossils in the rocks, which used to be interpreted as such—and are still by many, from whose knowledge and spirit those of the scientific and half-scientific denouncers of Darwin are not perhaps so far removed as they imagine. But we cannot think the theory yet converted into a scientific fact; and those are perhaps the worst foes to its success who are over-hasty to take it and use it as a proved fact. [. . .]

The eminent linguistic scholar [August] Schleicher* was also sorely infected with Darwinism, and sought to bring the science of language into relation with it in a couple of noted essays, which are far the weakest and most valueless of all his productions, though here referred to with high approval by his colleague Haeckel; and it is a part of Dr. Bleek's aim, as well, to connect the development of speech with his particular mode of development of our race—although we hardly see how he would bring it about, since his theories seem to require only that man should have been, at some indefinite epoch of the past, a creature without language.

William Dwight Whitney, "Dr. Bleek and the Simious Theory of Language," review of *On the Origin of Language*, by W. Bleek (New York, 1869), in Whitney, *Oriental and Linguistic Studies: First Series* (New York: Scribner's, 1893), 292–97, on 292–94. †Wilhelm Bleek (1827–75), German linguist.

Secondary Literature: Stephen G. Alter, *William Dwight Whitney and the Language of Science* (Baltimore: Johns Hopkins University Press, 2005).

JULIUS VON WIESNER (1838–1916)
GERMAN BOTANIST

It seemed as if the last traces of teleology would have been effaced from biology by Darwin's theory of selection. No one has expressed this more clearly than Schleiden,† who, always a tireless opponent of every ideological conception, speaking concerning Darwin's doctrine, cried out in triumph at the close of his activity as a student: "Teleology belongs no more to science, but has its place now only in mere talk." His opposition to teleology started from a one-sided, pedantic philosophy, but his disputatious arguments gained great weight with the majority of the botanists of his time, and his influence in this direction has remained, sporadically to be sure, up till our time. Most of the botanists of his time were so overawed by him that scarcely one of them dared to speak of the purposes of organs or of purposeful arrangements in organisms and so on. And this, as a result, worked a desolation in morphology, and made more difficult its union with physiology.

In this also, however, Schleiden by his hypercriticism overshot the mark. For it was just this great scientific movement, which Darwin set up through the rehabilitation of the doctrine of descent that of necessity placed teleology in its right place. And this teleology, enriched by an immense number of facts, contributed materially to the advance of the biological sciences. It has also brought it about that eminent and scientifically educated philosophers, such as [Wilhelm] Wundt*, enforced again the recognition of teleology together with causality.

In this the reaction of natural science upon philosophy is only slightly indicated. It extended, however, much further, for the rehabilitation of the theory of knowledge is the result of the advancement of natural science; and the cooperation of eminent scientists, such as Boltzmann,[‡] [Ernst] Mach*, [Wilhelm] Ostwald*, Reinke,[§] and other also scientifically trained philosophers, shows, in the building up of the theory of perception, how science entered this field to its advancement. That which in the teleological conception concerns transcendentalism we leave to the specialists in the theory of knowledge.

We stand on the ground of experience, and permit of metaphysics, as we have said above, only as a source of helpful ideas, which, however, may be permitted only when they do not negate experience, and only so long as they prove themselves useful in opening up to us new directions for inductive research. If through this kind of scientific operation the clear area within which we move appears to be limited within narrow confines, yet our advance within them is the more certain.

Julius Wiesner, "The Development of Plant Physiology under the Influence of the Other Sciences," in *Congress of Arts and Science: Universal Exposition, St. Louis, 1904,* ed. Howard J. Rogers (Boston: Houghton Mifflin, 1906), 119. [†]Matthias Schleiden (1804–81), German botanist, cofounder (with Theodor Schwann) of cell theory. [‡]Ludwig Boltzmann (1844–1906), Austrian physicist. [§]Johannes Reinke (1849–1931), German botanist and philosopher.

Secondary Literature: Richard Biebl, "Julius von Wiesner," in *Dictionary of Scientific Biography,* ed. Charles Coulston Gillispie, 16 vols. (New York: Scribner's, 1970–80), 14:349–50.

ALBERT WIGAND
See EMIL DU BOIS-REYMOND and ALBERT WIGAND

SAMUEL WILBERFORCE (1805–1873)
ENGLISH BISHOP

[1860]

Mr. Darwin writes as a Christian, and we doubt not that he is one. We do not for a moment believe him to be one of those who retain in some corner of their hearts a secret unbelief which they dare not vent; and we therefore pray him to consider well the grounds on which we brand his speculations with the charge of such a tendency. First, then, he not obscurely declares that he applies his scheme of the action of the principle of natural selection to man himself, as well as to the animals around him. Now, we must say at once, and openly, that such a notion is absolutely incompatible not only with single expressions in the word of God on that subject of natural science with which it is not immediately concerned, but, which in our judgment is of far more importance, with the whole representation of that moral and spiritual condition of man which is its proper subject matter. Man's derived supremacy over the earth; man's power of articulate speech; man's gift of reason; man's free will and responsibility; man's fall and man's redemption; the incarnation of the Eternal Son; the indwelling of the Eternal Spirit—all are equally and utterly irreconcilable with the degrading notion of the brute origin of him who was created in the image of God, and redeemed by the Eternal Son assuming to himself His nature. Equally inconsistent, too, not with any passing expressions, but with the whole scheme of God's dealings with man as recorded in His word, is Mr. Darwin's daring notion of man's further development into some unknown extent of powers and shape, and size, through natural selection acting through that long vista of ages which He casts mistily over the earth upon the most favored individuals of His species.

Samuel Wilberforce, "On Darwin's Origin of Species," in Wilberforce, *Essays Contributed to the Quarterly Review*, 2 vols. (London: John Murray, 1874), 1:92–99.

May 20 (Sunday).—Up in good time and prepared sermon on "All are yours." Preached at St. James's, great crowd; collected 1/6/. Then back to my rooms and finished [Darwin review]. Walked across the Park with [William] Gladstone*, he rather subdued; he said, "If the next twenty years alter as much the position of those who govern England, &c."

From June 27 to July 3 the British Association was at Oxford. [...] [The bishop spoke] in the Zoology and Botany Section, where a discussion took place on the soundness or unsoundness of the Darwinian theory. The Bishop, who, as the last-quoted Diary entry shows, had just reviewed Mr. Darwin's work *On the Origin of Species by means of Natural Selection*, made a long and eloquent speech condemning Mr. Darwin's theory as unphilosophical and as founded on fancy, and he denied that any one instance had been produced by Mr. Darwin which showed that the alleged change from one species to another had ever taken place. In the course of this speech, which made a great impression, the Bishop said, that whatever certain people might believe, he would not look at the monkeys in the Zoological as connected with his ancestors, a remark that drew from a certain learned professor the retort, "I would rather be descended from an ape than a bishop."

Samuel Wilberforce, quoted in *Life of the Right Reverend Samuel Wilberforce,* by Reginald G. Wilberforce, 3 vols. (London: John Murray, 1881), 1:450–51.

Secondary Literature: Standish Meacham, *Lord Bishop: The Life of Samuel Wilberforce, 1805–1873* (Cambridge, MA: Harvard University Press, 1970); Owen Chadwick, *The Victorian Church,* 2 vols. (New York: Oxford University Press, 1966–70).

OSCAR WILDE (1854–1900)
English Author

The English mind is always in a rage. The intellect of the race is wasted in the sordid and stupid quarrels of second-rate politicians or third-rate theologians. It was reserved for a man of science to show us the supreme example of that "sweet reasonableness" of which [Matthew] Arnold* spoke so wisely, and alas! to so little effect. The author of the *Origin of Species* had, at any rate, the philosophic temper. [. . .] Ethics, like natural selection, makes existence possible. Aesthetics, like sexual selection, make life lovely and wonderful, fill it with new forms, and give it progress, and variety and change.

Oscar Wilde, "The Critic as Artist [1891]," in *The Artist as Critic: Critical Writings of Oscar Wilde,* ed. Richard Ellmann (Chicago: University of Chicago Press, 1982), 406.

Secondary Literature: Philip J. Waller, *Writers, Readers, and Reputations: Literary Life in Britain, 1870–1918* (Oxford: Oxford University Press, 2006).

WOODROW WILSON (1856–1924)
American President

No government, of course, is a mechanism; no mechanical theory will fit any Government in the world, because Governments are made up of human beings, and all the calculations of mechanical theory are thrown out of adjustment by the intervention of human will. Society is an organism, and every government must develop according to its organic forces and instincts. I do not wish to make analysis tedious; I will merely ask you, after you go home, to think over this proposition: that what we have been witnessing for the last hundred years is the transformation of a Newtonian constitution into a Darwinian constitution. The place where the strongest will is present will be the seat of sovereignty.

Woodrow Wilson, address at the annual banquet of the Economic Club of New York, 23 May 1912, in *The Papers of Woodrow Wilson*, ed. Arthur S. Link, 69 vols. (Princeton, NJ: Princeton University Press, 1966–94), 24:413–34, on 416.

I once timidly enquired of Woodrow Wilson in our Baltimore boarding house, "Is society an organism?" And he took me to one side, holding my coat lapel, and whispered: "Yes, but keep it in the dark!"

Frederick Jackson Turner to Andrew C. McLaughlin, 8 June 1927, in *The Historical World of Frederick Jackson Turner*, by Wilbur R. Jacobs (New Haven, CT: Yale University Press, 1968), 142.

Secondary Literature: W. Barksdale Maynard, *Woodrow Wilson: Princeton to the Presidency* (New Haven, CT: Yale University Press, 2008).

ALEXANDER WINCHELL (1824–1891)
American Geologist and Paleontologist

Dr. M'Cosh [James McCosh]* declares "there is nothing irreligious in the idea of development, properly understood"; and Bishop E. S. Foster frankly confesses: "It would not appall our faith if it should be discovered that all the forms of life below man could be traced to a spontaneous generation from the unliving monads, and that from unity they were developed into diversity, given that the spontaneous movement, from its inception to its ultimatum, emanated from and was guided by the Divine factor." Similar views are entertained by many orthodox theologians of the present day.

Nor is it to be supposed that the advocates of these theories are generally willing to regard themselves shut out from the fold of theistic believers. It is better to be content with ignorance of a man's religious faith than to assign him a creed which he has not avowed. Whatever be the views of such writers as [Carl] Vogt* and [Ludwig] Büchner* and Haeckel*, Mr. Darwin sincerely believes that his theory ought not to "shock the religious feelings of any one"; and he speaks of life "having been originally breathed by the Creator into a few forms or only one."

Alexander Winchell, *The Doctrine of Evolution* (New York: Harper, 1874), 118.

The author now approaches the critical point of his discussion. Having admitted that the scientist often feels himself impelled to pass beyond the field of physical phenomena, and from phenomena to induce an abstract generalization under which an entire category of phenomena may be ranged—as in the case of the force of gravitation—it is not strange that Lucretius should have reached the generalization that his atoms were endowed with life; or that Darwin should have permitted himself to be understood as abstracting creative power, exercised in a limited number of initial cases, as the antecedent and cause of the series of organized beings. Darwin, our author thinks, should speak with clearness at this juncture, and assume the responsibility of carrying derivative development back, not only to one primitive stock, but to unorganized matter itself. At the same time, he admits that the doctrine of spontaneous generation is not yet proved, though he seems to regard that achievement as not very remote.

Alexander Winchell, *Reconciliation of Science and Religion* (New York: Harper, 1877), 236.

Secondary Literature: Philip Harrold, "Alexander Winchell's Science with a Soul: Piety, Profession, and the Perils of Nineteenth-Century Popular Science," *Methodist History* 36 (1998): 97–112.

NICHOLAS PATRICK WISEMAN (1802–1865)
English Prelate

[1863]
I spoke to [Richard] Owen*, who was a guest like myself [at the Astronomer's "Club"], to say I would come and see him at the Museum. I want to talk about Lyell*, Huxley*, Darwin, etc.

Wiseman to Canon Walker, Easter Sunday 1863, in Wilfrid Philip Ward, *The Life and Times of Cardinal Wiseman* (London: Longmans, Green, 1897), 495.

Secondary Literature: Brian Fothergill, *Nicholas Wiseman* (London: Faber & Faber, 1963); Owen Chadwick, *The Victorian Church,* 2 vols. (New York: Oxford University Press, 1966–70).

LUDWIG WITTGENSTEIN (1889–1951)
Austrian-English Philosopher

A tendency which has come into vogue with the modern sciences is to explain certain things by evolution. Darwin seemed to think that an emotion got its importance from one thing only, utility. A baby bares its teeth when angry because its ancestors did so to bite. Your hair stands on end when you are frightened because hair standing on end served some purpose for animals.

Ludwig Wittgenstein, from 1932–33 lecture notes, in *Wittgenstein's Lectures,* ed. Alice Ambrose (Oxford: Blackwell, 1979), 2–4.

Secondary Literature: Norman Malcolm, "Ludwig Joseph Johann Wittgenstein," in *Encyclopedia of Philosophy,* ed. Paul Edwards, 8 vols. (New York: Macmillan, 1967), 8:327–40; Oskari Kuusela, *The Struggle against Dogmatism: Wittgenstein and the Concept of Philosophy* (Cambridge, MA: Harvard University Press, 2008).

THOMAS VERNON WOLLASTON (1822–1878)
English Entomologist

[H]owever important an element, in the eradication of species, submergence may be, we must not entirely omit to notice other methods also, through the medium of which genera may become well-defined. We should recollect that the removal of a *very few* links from an endemic cluster is sufficient to cause its disjunction from the type to which it is next akin, and that where the creatures which unite in composing it are of slow diffusive powers, or sedentary habits, the elimination of such links is (through the smallness of the areas which have been overspread) a comparatively easy operation. The accidental introduction of organic beings amongst others to the interests of which they are hostile, may be a powerful means, as Mr. Darwin has suggested, of keeping the latter in check, and of finally destroying them. The gradual upheaval of a tract which has been well-stored with specific centers of radiation, created expressly for itself, may (through the climatal changes which have been brought about) succeed in extirpating races innumerable—those only surviving which are able to adapt themselves to the altered conditions; and which would *now* be consequently looked upon as abrupt topographical assemblages. The overwhelming effect of a volcanic eruption, in a region where the aborigines of the soil have not wandered far from their primeval haunts, may, as Sir Charles Lyell* has well remarked, put an end to others, and so effect the separation of their allies from the central stock. And, lastly, the intervention of man, with all the various concomitants which civilization, art, and agriculture bring in his train, is the most irresistible of every agency in the extensive (though often accidental) demolition of a greater or less proportion of the animate tribes.

Thomas Vernon Wollaston, *On the Variation of Species with Special reference to the Insecta, followed by an Inquiry into the Nature of Genera* (London: John van Voorst, 1856), 178.

It would appear from information given me by Mr. [Hewett Cottrell] Watson*, Dr. Asa Gray*, and Mr. Wollaston, that generally when varieties intermediate between two other forms occur, they are much rarer numerically than the forms which they connect. [. . .]

The species which inhabit ocean islands are few in number compared to those of equal continental areas: Alp. [Alphonse] De Candolle* admits this for plants, and Wollaston for insects. [. . .] He who admits the doctrine of the creation of each separate species, will have to admit, that a sufficient number of the best adapted plants and animals have not been created on oceanic islands; for many have unintentionally stocked them from various sources far more fully and perfectly than nature.

Darwin, *Origin of Species* (London: John Murray, 1859), 176, 389–90.

Secondary Literature: David B. Stamos, *Darwin and the Nature of Species* (Albany: State University of New York Press, 2007).

VIRGINIA WOOLF (1882–1941)
ENGLISH AUTHOR

"Peter Walsh," said Clarissa.

That meant nothing.

Clarissa had asked her. It was tiring; it was noisy; but Clarissa had asked her. So she had come. It was a pity that they lived in London— Richard and Clarissa. If only for Clarissa's health it would have been better to live in the country. But Clarissa had always been fond of society.

"He has been in Burma," said Clarissa.

Ah. She could not resist recalling what Charles Darwin had said about her little book on the orchids of Burma.

(Clarissa must speak to Lady Bruton.)

No doubt it was forgotten now, her book on the orchids of Burma, but it went into three editions before 1870, she told Peter. She remembered him now. He had been at Bourton (and he had left her, Peter Walsh remembered, without a word in the drawing-room that night when Clarissa had asked him to come boating).

Virginia Woolf, *Mrs. Dalloway,* http://ebooks.adelaide.edu.au/w/woolf/virginia/w91md/ (accessed 16 November 2008).

I'm beginning Sense & Sensibility—& reading about Apes. That reminds me—to do a C[ommon]. R[eader]. on Darwin V. of the Beagle one section: Downe the other.

The Diary of Virginia Woolf, ed. Anne Olivier Bell, vol. 5, *1936–1941* (New York: Harcourt Brace Jovanovich, 1984), 274, entry for 24 March 1940.

Secondary Literature: John Keith Johnstone, *The Bloomsbury Group: A Study of E. M. Forster, Lytton Strachey, Virginia Woolf, and Their Circle* (London: Secker & Warburg, 1954).

CHAUNCEY WRIGHT (1830–1875)
AMERICAN PHILOSOPHER

It must have been in 1859 or 1860 that Chauncey first felt the influence which was to be more powerful than any other giving direction and color to his intellectual life. This was the publication of Darwin's *Origin of Species*. We read it and re-read it aloud together, and talked over it and the reviews that appeared of it interminably. The ground had been prepared for the seed by Chauncey's interest in theoretic geology, and the argument for the sufficience of causes now in operation to explain past changes in the condition of the earth; by the discussion which had gone on for years in Cambridge between Agassiz* and [Asa] Gray* concerning the true nature of the terms "genus" and "species"; and by the fruitfulness, already shown, of the historical method in dealing with social phenomena, I think I am not mistaken in putting the publication of [Henry Sumner] Maine's* "Ancient Law"—my interest which was shared by Chauncey—very near that of the *Origin of Species*.

Up to this time, however, the abstract theory of evolution had not found favor in Chauncey's mind. In illustration of this, I recall, years previously, a talk with him about the "Vestiges of Creation," into which, I think, he had barely dipped, and how lightly he regarded the thesis itself, as well as the arguments. I remember, too, how decided were his leanings for Cuvier as against [Étienne] Geoffroy St. Hilaire [Saint-Hilaire], and how destitute of attraction for him had been the nebular hypothesis. To his mind, no theory of evolution would have commended itself on *a priori* grounds; but the cumulative argument, based on observations and experiment, of the *Origin of Species*, in harmony as it was with his own habits of thought, carried with it complete conviction.

E. W. Gurney, in *Letters of Chauncey Wright*, ed. James Bradley Thayer (Cambridge: John Wilson, 1878), 367–68.

Few scientific theories have met with such a cordial reception by the world of scientific investigators, or created in so short a time so complete a revolution in general philosophy, as the doctrine of the derivation of organic species by Natural Selection; perhaps in this respect no other can compare with it when we consider the incompleteness of the proofs on which it still relies, or the previous prejudice against the main thesis implied in it, the theory of the development or transmutation of species. The Newtonian theory of gravity, or Harvey's theory of the circulation of the blood, in spite of the complete and overwhelming proofs by which these were soon substantiated, were much longer in overcoming to the same degree the deeply-rooted prejudices and preconceptions opposed to them. In less than a decade the doctrine of Natural Selection had conquered the opposition of the great majority of the students of natural history, as well as of the students of general philosophy; and it seems likely that we shall witness the unparalleled spectacle of an all but universal reception by the scientific world of a revolutionary doctrine in the lifetime of its author; though by the rigorous tests of scientific induction it will yet hardly be entitled to more than the rank of a very probable hypothesis. How is this singular phenomenon to be explained? Doubtless in great part by the extraordinary skill which Mr. Darwin has brought to the proof and promulgation of his views.

Chauncey Wright, "Limits of Natural Selection [1870]," *North American Review* III, no. 2 (October 1870): 282–311.

If you can imagine me enthusiastic—absolutely and unqualifiedly so, without a *but* or criticism—then think of my last evening's and this morning's talks with Mr. Darwin as realizing that beatific condition. Mr. Horace Darwin (whom I like very much, and mean to visit at his college in Cambridge before I sail for home) was at home; and I had several hours of pleasant discussion with him, while his father was taking the rests he always needs after talking a while. Who would not need rest after exercising such powers of wise, suggestive, and apt observation and criticism, with judgments so painstaking and conscientiously accurate—unless, indeed, he should be sustained by an Olympian diet? [. . .]

It would be quite impossible to give by way of report any idea of these talks before and after dinner, at breakfast, and at leave-taking; and yet I dislike the egotism of "testifying," like other religious enthusiasts, without any verification, or hint of similar experience; though what I have said must be to you a confirmation of what you already know. One point I may mention, however, of our final talk. I am some time to write an essay on matters covering the ground of certain common interests and studies, and in review of his *Descent of Man,* and other related books, for which the learned title is *Psychozoology*—as a substitute for "Animal Psychology," "Instinct," and the like titles—in order to give the requisite subordination [...] of consciousness in men and animals, to their development and general relations to nature. So, if you ever see that learned word in print, you will know better than other readers when and where it was born. [...]

I also found Mrs. Darwin and her daughter very agreeable; and I repent now, as I have regretted all along, that indolence has kept me so many weeks from making acquaintance with so charming a household.

Chauncey Wright to Sara Sedgwick, 5 September 1872, in *Letters of Chauncey Wright,* ed. James Bradley Thayer (Cambridge: John Wilson, 1878), 248–49.

A few months ago [...] we took occasion to point out and emphasize a division, very fundamental and important in our view, in books on this subject, namely, between those which treat of it as a theorem of natural history from a Baconian or scientific point of view, either mainly or exclusively (confining themselves to scientific considerations of proof), and those which treat of evolution as a philosophical thesis deductively and as a part of a system of metaphysics. Such a division separates the names of Darwin and Spencer* (which are popularly so often pronounced together) as widely as any two names could be separated on real grounds of distinction.

[I]n Germany, where the theory first got the name of Darwinism, it is much more of an "ism," or connects itself much more intimately with general philosophical views, than in England or America, except where in these countries it has got confounded with Mr. Spencer's speculations. It is the significance of this fact—the character of Darwinism in Germany— that we wished especially [...] to call attention, as an interesting phenomenon in the history of modern speculation, determining the true place and essential influence of Bacon and the Baconian philosophy. German systematic historians of philosophy were never able to make out where to place Bacon's so-called philosophy, or indeed to discover that he had a philosophy, or, what appeared to their minds as the same thing, a "system." And indeed he had no system.

The sun of Baconism has not yet even shone fully on the German mind. [...]

That such a system as [Lorenz] Oken's *Naturphilosophie,* with its vague and meaningless abstractions, was an influence at the beginning of the present century, is not, however, so surprising as perhaps it would be if Mr. Spencer's system (bearing a much greater resemblance to it than any theories of Darwin), had not got such a footing with English-speaking readers as it appears to have. There is, however, at present in Germany an ascetic school of experimental and inductive science, which deprives itself of the aid and guidance of theoretical and deductive considerations, in order the more effectually to protect itself from undue influence. These *Gelehrten* are not true Baconians; but their method might appropriately be named "experimentalism." [...] It is a little incongruous that one so preeminently cautious and painstaking [i.e., Darwin], so little speculative or metaphysical in the range of his researches, should be hailed as chief by so large a constituency of what really amounts to a philosophical school; albeit they are the brightest minds of Germany, and pre-eminently men of science.

Chauncey Wright, "German Darwinism," *Nation* 9 (September 1875), reprinted in Wright, *Philosophical Discussions* (New York: H. Holt, 1877), 398–405, on 398, 403, 405.

My Dear Dr. Darwin,—I am sorry to have to send you intelligence which I know will cause you regret. Ten days ago our friend Chauncey Wright died suddenly, apparently without suffering, from congestion of the brain. He was found in the morning seated at his desk in an easy attitude, unconscious, but still breathing. He died in a few minutes. He had apparently not been in bed during the night, but had been writing. He was busy with an article on your last book, which lay open beside him. A week before I had brought down to him from Ashfield a box full of specimens of *Drosera* which my little girl Sally, knowing his interest in the plant, had secured for him. The last letter any of us had from him was a pleasant note to Sally telling her what he had been observing of the habits of the Sundew. He had stayed with us here for more than a week in August, and had seemed uncommonly well. The last time I saw him was about ten days before his death at Cambridge, when he came to read me the proof of his article in the *Nation* on "German Darwinism," and to tell me of some changes which he proposed to make in this essay which he had read to me in manuscript some time before. I was particularly struck with his animation and his cheerfulness, and with his readiness to be interested in other subjects than that which for the time was chiefly occupying his thought. I believe that he had contemplated the probability of such a death as it was his good fortune to die by. He had no shrinking from death, and no desire to die. He was far too much of a philosopher to form wishes about life or death. And, so far as he is concerned, there is no reason to regret his death. The prospect of his life was not unclouded. But to his friends his death is an irreparable, lifelong loss. [. . .] His death, following so soon on that of [Jeffries] Wyman*, is a great blow, not merely to Cambridge, but to the interests of sound thought and scientific inquiry throughout the country. But he was not widely known, and there are but few persons who will know what a great loss we have suffered. [. . .] P.S. I ought, perhaps, to add that besides Wright's interest as a scientific man in your work, he had a strong moral and personal interest in it. Your work was the illustration and exhibition of the spirit which he sought in scientific enquiry. I think that in his late years, he had no greater gratification than the recognition you gave to his work, and the occasional receipt of a letter or note from you. He was radically modest, but he was pleased with your expression of interest in or approval of what he wrote. His visit to you was the most prized experience of his stay in Europe, and from that time his feeling toward you was one in which a certain shy affection gave a still deeper character to the complete respect he had long cherished.

Charles Eliot Norton to Darwin, 22 September 1875, in *Letters of John Ruskin to Charles Eliot Norton*, ed. Charles Eliot Norton, 2 vols. (Boston: Houghton Mifflin, 1905), 2:57–59.

Secondary Literature: Edward H. Madden, *Chauncey Wright and the Foundations of Pragmatism* (Seattle: University of Washington Press, 1963); idem, "Chauncey Wright," in *Encyclopedia of Philosophy*, ed. Paul Edwards, 8 vols. (New York: Macmillan, 1967), 8:348–49.

GEORGE FREDERICK WRIGHT (1838–1921)
American Geologist and Evangelical Minister

The works in philosophy occupying attention at that time were those of Sir William Hamilton,[†] John Stuart Mill[*], and President Noah Porter.[‡] Careful study of these works, together with the scientific discussions aroused by Darwin's *Origin of Species* and Lyell's[*] *Antiquity of Man,* led me, toward the close of my pastorate in Bakersfield, to prepare an article on the "Ground of Confidence in Inductive Reasoning," which was published in the *New Englander* for October, 1871. This, as I afterwards learned from him, was approved for publication by President Porter. It received high commendation from one of the Scotch philosophical periodicals, and was the means of attracting to me the attention of Professor Asa Gray[*], with whom an acquaintance was formed which ripened into a life-long friendship, indeed, I may say partnership, in which he assisted me in the preparation of the first book which I published, in which I was asked to discuss the relations of theology to current speculations concerning the origin of species. He, in turn, sought my aid in the preparation of various of his publications having the same end in view, especially in the last chapter of his *Darwiniana.*

Then Professor Park[§] wished me to prepare a series of articles stating the arguments for and against Darwinism, and showing the bearing of that theory upon the doctrine of design in nature, and upon theological opinions in general. Fortunately my readiness to undertake this work was greatly facilitated by the friendship, to which I have already referred, of Professor Asa Gray, who in addition to his regular work in botany had been foremost in contending that the doctrine of design in nature was not at all endangered by Darwinism, and who, as already remarked, after reading my article in the *New Englander,* on the "Ground of Confidence in Inductive Reasoning," had requested my acquaintance. This I cheerfully granted, and he became from that time like a father to me in the work in which I was engaged. It was enough for me that these articles on Darwinism in the Bibliotheca Sacra met his approval, and were indebted to him for much of their form of statement. It was gratifying, also, to have a letter from Darwin, written in his own hand, in which he said that the statement of his theory "was powerfully written and most clear," and requested me to send him the following article in which objections were to be presented. These articles maintained what has been more and more evident as attention has been given to the subject, that the observed variations in both plants and animals are much greater than Darwin had supposed, and that so many correlated variations had to take place at once to make any one variation an advantage, that nothing less than design either wrought into the original plan, or added by way of increment, could account for the facts. From the theological side it was maintained that Calvinism and Darwinism had so many points in common that theologians could not consistently cast stones at the men of science favoring a scheme in which "predestination and foreordination" were salient features. In fact, from a philosophical point of view, Darwinism has all the unlovely characteristics of hyper-Calvinism without any of the redeeming remedial features inherent in the Calvinistic system. Pure Darwinism leaves no place for the gospel. These essays were subsequently republished in a volume, together with the essay on the "Ground of Confidence in Inductive Reasoning," and an essay on "The Antiquity of Man," dealing especially with the evidence of glacial man in America, also an essay on "The Relation of the Bible to Science." From a copy owned by Henry Ward Beecher*, which has fallen into my hands, I have found from his annotations that he had read the book carefully, and been duly influenced by it.

George Frederick Wright, *Story of My Life and Work* (Oberlin, OH: Bibliotheca Sacra, 1916), 116, 137–38. †William Rowland Hamilton (1805–65), Irish physicist whose reformulation of Newtonian mechanics is now called Hamiltonian mechanics. ‡Noah Porter (1811–92), American philosopher, president of Yale. §Edwards Amasa Park (1808–1900), American Congregational theologian.

We are no longer shut up to the conception of the infinitesimal rate of variation in species with which Darwin carried on his speculations. In reference to the production of different races of mankind, one needs but make a simple calculation to see how easily these all may have been brought about in two or three thousand years through the simple operation of Darwin's law of natural selection.

George Frederick Wright, *Origin and Antiquity of Man* (Oberlin, OH: Bibliotheca Sacra, 1912), 483.

Secondary Literature: David Livingstone, *Darwin's Forgotten Defenders: The Encounter between Evangelical Theology and Evolutionary Thought* (Grand Rapids, MI: Eerdmans, 1987).

WILHELM WUNDT (1832–1920)
German Physiologist and Psychologist

And this leads us to the principle which Darwin enunciated as of prime importance for the development of instinct and for the course of evolution in general—the principle of adaptation to environment. There can be no doubt that this adaptation and voluntary action constitute the two universal determinants of the development of animal impulses. The first supplements the second; volition must have an object towards which it is directed. The converse, of course, need not necessarily be the case. In the vegetable kingdom specific alterations are gradually effected by the sole operation of the environment, influencing the functions of growth or favoring certain peculiarities which are thus more readily and certainly perpetuated. And this passive adaptation will naturally be found among animals as well, since they share with plants all the physiological functions which are capable of modification by it. But Darwin's explanation of the development of instinct as being mainly the result of passive adaptation seems to contradict the facts. Instinctive action is impulsive, that is, voluntary action; and however far back we may go, we shall not find anything to derive it from except similar, if simpler, acts of will. The development of any sort of animal instinct, that is to say, is altogether impossible unless there exists from the first that interaction of external stimulus with affective and voluntary response which constitutes the real nature of instinct at all stages of organic evolution. We may possibly succeed in deriving a complicated form of instinct from a more simple one; but we can never explain instinct in terms of something which is as yet neither instinct nor impulse.

Wilhelm Wundt, *Lectures on Human and Animal Psychology,* trans. J. E. Creighton and E. B. Titchener (London: George Allen, 1912), 408–9.

Secondary Literature: Albert Wellek, "Wilhelm Wundt," in *Encyclopedia of Philosophy,* ed. Paul Edwards, 8 vols. (New York: Macmillan, 1967), 8:349–51; Serge Nicolas, *La psychologie de W. Wundt (1832–1920)* (Paris: Harmattan, 2003).

JEFFRIES WYMAN (1814–1874)
AMERICAN NATURALIST

In these days it is sure to be asked how an anatomist, physiologist, and morphologist like Professor Wyman regarded the most remarkable scientific movement of his time, the revival and apparent prevalence of doctrines of evolution. As might be expected, he was neither an advocate nor an opponent. He was not one of those persons who quickly make up their minds, and announce their opinions, with a confidence inversely proportionate to their knowledge. He could consider long, and hold his judgment in suspense. How well he could do this appears from an early, and so far as I know, his only published presentation of the topic, in a short review of [Richard] Owen*'s "Monograph of the Aye-Aye" (in *American Journal of Science*, September 1863)—the paper in which Professor Owen's acceptance of evolution but not of natural selection, was promulgated. Dr. Wyman compares Owen's view with that of Darwin (to whom he had already communicated interesting and novel illustrations of the play of natural selection); and he adds some acute remarks upon the rather earlier speculation by Mr. Agassiz*, in which the latter suggests that the species of animals might have been created as eggs rather than as adults.

Scientific Papers of Asa Gray, ed. Charles Sprague Sargent, 2 vols. (Boston: Houghton Mifflin, 1889), 2:398.

Secondary Literature: Toby A. Appel, "Jeffries Wyman, Philosophical Anatomy, and the Scientific Reception of Darwin in America," *Journal for the History of Biology* 21 (1988): 69–94.

Y

EDMUND HODGSON YATES (1831–1894)
ENGLISH JOURNALIST AND NOVELIST

"It is better so [to be reclusive]," says Mr. Darwin, "than to be interviewed and harassed with questions which cannot be answered without some appearance of vanity. Moreover it strikes me as not proper that a man should communicate anything to the author of a biographical notice. He should behave as if already dead."

Edmund Hodgson Yates, *Celebrities at Home,* 2nd ser. (London: The World, 1878).

Secondary Literature: Peter David Edwards, *Dickens's "Young Men": George Augustus Sala, Edmund Yates and the World of Victorian Journalism* (Aldershot: Ashgate, 1997).

WILLIAM BUTLER YEATS (1865–1939)
IRISH POET

[ca. 1881] I planned some day to write a book about the changes through a twelvemonth among the creatures of some hole in the rock, and had some theory of my own, which I cannot remember, as to the color of sea-anemones; and after much hesitation, trouble, and bewilderment, was hot for argument in refutation of Adam and Noah and the Seven Days. I had read Darwin and Wallace*, Huxley* and Haeckel*, and would spend hours on a holiday plaguing a pious geologist who [. . .] came with a hammer to look for fossils in the Howth Cliffs. "You know," I would say, "that such-and-such human remains cannot be less, because of the strata they were found in, than fifty thousand years old." "O!" he would answer, "they are an isolated instance." And once when I pressed hard my case against [Bishop

James] Ussher's chronology, he begged me not to speak of the subject again. "If I believed what you do," he said, "I could not live a moral life."

I began occasionally telling people that one should believe whatever had been believed in all countries and periods, and only reject any part of it after much evidence, instead of starting all over afresh and only believing what one could prove. But I was always ready to deny or turn into a joke what was for all that my secret fanaticism. When I had read Darwin and Huxley and believed as they did, I had wanted, because an established authority was upon my side, to argue with everybody.

W. B. Yeats, *Autobiographies*, ed. William H. O'Donnell and Douglas N. Archibald (New York: Scribner, 1999), 77, 89.

Secondary Literature: Terence Brown, *The Life of W. B. Yeats: A Critical Biography* (Oxford: Blackwell, 1999).

EDWARD LIVINGSTON YOUMANS (1821–1887)
American Science Writer and Publisher

Things are going here furiously. I have never known anything like it. Ten thousand *Descent of Man* have been printed, and I guess they are nearly all gone. Five or six thousand of [Huxley's*] *Lay Sermons* have been printed, while [St. George Jackson] Mivart* [*On the Genesis of Species*] is reprinted and has fallen stillborn. The progress of liberal thought is remarkable. Everybody is asking for explanations. The clergy are in a flutter. [James] McCosh* told them not to worry, as whatever might be discovered he would find design in it and put God behind it.

Twenty-five clergymen of Brooklyn sent for me to meet them of a Saturday night and tell them what they should do to be saved. I told them they would find the way of life in the Biology and in the *Descent of Man*. They said, "Very good," and asked me to come again at the next meeting of the clerical club, to which I went and was again handsomely resoluted. My warrant for attempting to enlighten these gentlemen is that they know nothing whatever about the subject, while I was in wonderfully sympathetic nearness to them.

Youmans to Herbert Spencer, 21 April 1871, in *Edward Livingston Youmans: Interpreter of Science for the People,* by John Fiske (New York: Appleton, 1894), 266.

Darwin was in town this week and called on Mr. W. H. Appleton, who presented me. I was invited to lunch Wednesday at his brother's, Mr. Erasmus Darwin. There were Mr. D., Mrs. D., Miss D., and Master D. It was altogether juicy and jolly. This was fresh material—no dregs of '68—and it went with a rush. I took Mrs. D. in to lunch. They were all curiosity about America. Mr. D. had just resolved to send two of his boys across the Atlantic, and they leave the last of August. I told him about my lecturing the Brooklyn clergymen on evolution. "What!" said he, "clergymen of different denominations all together? How they would fight if you should get them together here!" They were greatly amused with a spiritualistic paper they had received from Chicago, which stated that if it were known that God were dead [Henry Ward] Beecher* would be unanimously elected by the American people to fill his place. [. . .] Well, my point is this: that Mr. Darwin insisted on having my [science-book-series] project brought up at the British Association and indorsed [*sic*] there.

Youmans to his sister Catherine, 15 July 1871, in *Edward Livingston Youmans: Interpreter of Science for the People,* by John Fiske (New York: Appleton, 1894), 276.

Secondary Literature: Barry Werth, *Banquet at Delmonico's: Great Minds, the Gilded Age, and the Triumph of Evolution in America* (New York: Random House, 2009).

Z

JOHN ZAHM (1851–1921)
American Priest and Science Professor

To judge from the declarations of some of the most ardent champions of Evolution, it must be admitted that orthodoxy had reason to be at least suspicious of the theory that was heralded forth with such pomp and circumstance. For it was announced with the loudest flourish of trumpets, not only that Evolution is a firmly established doctrine, about whose truth there can no longer be any doubt, but it was also boldly declared, by some of its most noted exponents, to be subversive of all religion and of all belief in a Deity. Materialists, atheists, and anarchists the world over loudly proclaimed that there is no God, because, they would have it, science had demonstrated that there is no longer any *raison d'etre* for such a Being. Evolution, they claimed, takes the place of creation, and eternal, self-existent matter and force exclude an omnipotent personal Creator. "God," we are told, "is the world, infinite, eternal, and unchangeable in its being and in its laws, but ever-varying in its correlations." A glance at the works of Haeckel*, [Carl] Vogt*, [Ludwig] Büchner*, and others of this school, is sufficient to prove how radical and rabid are the views of these "advanced thinkers." [. . .]

But all evolutionists have not entertained, and do not entertain, the same opinions as those just mentioned. America's great botanist, Prof. Asa Gray*, was not so minded. One of the earliest and most valiant defenders of Darwinism, as well as a professed Christian believer, he maintained that there is nothing in Evolution, or Darwinism, which is incompatible with Theism. In an interesting chapter on Evolution and Theology, in his *Darwiniana*, he gives it as his opinion, arrived at after long consideration, that "Mr. Darwin has no atheistical intent," and that, as respects the test question of design in nature, his view may be made clear to the theological mind by likening it to that of the "believer in general, but not in particular,

Providence." So far, indeed, was Darwin from having any "atheistical intent," that when interrogated regarding certain of his religious views he replied: "In my most extreme fluctuations I have never been an atheist in the sense of denying the existence of God." And the late Dr. [James] McCosh* declared that he had "never been able to see that religion, and in particular that Scripture, in which our religion is embodied, is concerned with the absolute immutability of species."

John Augustine Zahm, *Evolution and Dogma* (Chicago: D. H. McBride, 1896), 209–12.

Secondary Literature: Mariano Artigas, Thomas F. Glick, and Rafael Martínez, *Negotiating Darwin: The Vatican Confronts Evolution, 1877–1902* (Baltimore: Johns Hopkins University Press, 2006).

CHAIM ZHITLOVSKY (1865–1943)
Jewish Socialist, Philosopher, and Critic

The past explains how we became what we are. [. . .] But it also strengthens our pride and our respect for ourselves. And there is truly that in our past of which to be proud. [. . .] Here I only want to observe that we cannot blame Moses for not having read Darwin, or Joshua for not being an expert on Copernicus. In order to evaluate any cultural product of the past one must be able to identify with the soul of that period, breathe its very atmosphere. [. . .] Where did it lead, to progress in the important phases of human life or to reaction? And the reforms, the progress, the reaction, must be measured not with the yardstick of today but of those ancient circumstances.

Chaim Zhitlovsky, *Id un mensh: Tsvey farlezunge* [Jew and man: Two lectures] (New York: Komite far der aroysgabe fun H. Zhitlovski's shriften, 1910), second lecture, quoted in Max Rosenfeld, "Zhitlovsky: Philosopher of Jewish Secularism," *Jewish Currents*, http://www.csjo.org/pages/essays/essayzhitlovsky.htm (accessed 28 January 2009).

Secondary Literature: Matthew Hoffman, "From *Pintele Yid* to *Racenjude:* Chaim Zhitlovsky and Racial Conceptions of Jewishness," *Jewish History* 19 (2005): 65–78.

EMILE ZOLA (1840–1902)
FRENCH NOVELIST

Étienne was now studying Darwin. He had read fragments, summarized and popularized in a five-*sou* volume; and out of this ill-understood reading he had gained for himself a revolutionary idea of the struggle for existence, the lean eating the fat, the strong people devouring the pallid middle class. But Souvarine furiously attacked the stupidity of the Socialists who accept Darwin, that apostle of scientific inequality, whose famous selection was only good for aristocratic philosophers. His mate persisted, however, wishing to reason out the matter, and expressing his doubts by an hypothesis: supposing the old society were no longer to exist, swept away to the crumbs; well, was it not to be feared that the new world would grow up again, slowly spoilt by the same injustices, some sick and others flourishing, some more skilful and intelligent, fattening on everything, and others imbecile and lazy, becoming slaves again? But before this vision of eternal wretchedness, the engine-man shouted out fiercely that if justice was not possible with man, then man must disappear. For every rotten society there must be a massacre, until the last creature was exterminated. And there was silence again.

Was Darwin right, then, and the world only a battlefield, where the strong ate the weak for the sake of beauty and continuance of the race?

Emile Zola, *Germinal* (1885), trans. Havelock Ellis (1894; reprint, New York: Boni & Liverright, 1924), 454, 529.

There are two modes of heredity: either the baby resembles its father or mother, or there results a mixture of the two. [...] That which is inborn, but not inherited, on the other hand, is combinatory: a chemical combination whereby two bodies introduced into each other's presence can constitute a new body, completely different from the ones that gave rise to it. This was the result of a considerable register of observations, not only in anthropology, but also in zoology, arboriculture and horticulture. It was difficult to sort out these multiple facts, to synthesize them all and formulate a theory that explains everything; for such a theory would stand on the shifting sands of hypothesis, which each new discovery transforms.

And if hypothesis does not lead to a solution, then the human spirit is capacious enough to leave the question open. Such was the case with the gemmules in Darwin's theory of pangenesis. It was true of Haeckel's* perigenesis, and the same applies to [Francis] Galton* for whom the sum total of the gemmules comprised what he called the stirp, from Latin *stirpes*. Dr. Pascal, however, had an intimation of the theory that [August] Weismann* would advance later on: the idea of extremely small and complex bodies, the germinative plasma, one part of which is always kept in reserve in each new being, whence it is transmitted from generation to generation, invariable and immutable.

Émile Zola, *Le Docteur Pascal* (Paris: Bibliothèque Charpentier, 1893), 37–38, trans. TFG.

Henri Céard, the novelist, who had once been a medical student, introduced Zola to the works of Claude Bernard and Darwin, as well as to the doctrines of experimental determinism.

These unfortunates [the nobility] turn their back on the light, on progress; they have never read Darwin, Spencer* or Claude Bernard, Zola—and [Georges] Clemenceau* too for that matter—was convinced that no titled person had ever read Darwin and Claude Bernard, or even [Charles] Letourneau† or Haeckel*. All nobles are brought up by the Jesuits—and, as everyone knows, the latter never teach any history, science or sociology. They are still attached to those reactionary authors, those pillars of noneducation, Homer, Virgil and Racine.

Memoirs of Léon Daudet, ed. Arthur Kingsland Griggs (New York: Dial, 1925), 19, 146. †Charles Letourneau (1831–1902), radical French Darwinian anthropologist.

Secondary Literature: Louise Lyle, "Le Struggle for Life: Contesting Balzac through Darwin in Zola, Borget, and Barres," *Nineteenth-Century French Studies* 36 (2007–8): 305–19.

INDEX

Page numbers for individual entries appear in **boldface**.

Academy, 191

Acland, Henry Wentworth, 366, 367

Adams, Henry, **1–2**

adaptation, 209, 226, 235, 322, 329, 331, 367, 471, 495; critics of Darwin's view, 341; as finalism, 218; functional, 401

affinity groups, xxiii–xxiv

Agassiz, Alexander, 170

Agassiz, Louis, xxvii, xxix, **2–3**, 13, 76, 77, 116, 119, 152, 183, 205, 211, 279, 325, 477; accused of Darwinism, 475; argues with Chevis, 240; discusses Darwin with Verrill, xxvi; *Essay on Classification,* 35, 257; and Longfellow, 193; scientific achievement, 241–42; on species, 391, 488, 496; students of debate Darwin, 390; supported by Parsons, 338; theology, 337

agnosticism, 281

Alcott, Bronson, xxiii, **3**, 77; introduced to *Origin,* xxvii

Aleichem, Sholem, **3–4**

Allen, Grant, **4–5**, 434, 462; and Kingdon's circle, xxiv

Allingham, William, **5–6**, 33; characterizes Darwin, xxix; visits Down House, 5

altruism, 393

Ameghino, Florentino, 394

American Academy of Arts and Sciences, xxviii, 314

American Entomologist, 362

American Journal of Science, xxviii, 338, 365, 496

American Naturalist, 11

Ammon, Otto, 469

Anaxagoras, 293

Anaximander, 293

animal intelligence, 464

Annals of Natural History, 296

anthropologists and Darwin: Benedict, 20; Boas, 24–25; Broca, 32–33; Mead, 281; L. Morgan, 298; Sergi, 388–90; Topinard, 435–36; Tylor, 441–42; Vacher de Lapouge, 446; Winchell, 483–84

Aquinas, Thomas, 35

Aristotle, 45, 164, 234, 293, 472

Arnold, Matthew, **6–7**, 69, 482

Ashburton, Lord, 54

Asquith, Herbert Henry, **7**

associationalism, threatened by *Origin,* 343

Athenaeum (journal), 67, 348

Athenaeum Club, 54

Atlantic Monthly, 193, 325

Audubon, John Jacob, 368

Augustine, 35, 123–24

Aveling, Edward, **8**

Babington, Churchill, 317

Bacon, Sir Francis, 339, 490; negative impact, 34

Baer, Karl Ernst von, 200

Bagehot, Walter, **9–10**

Bain, Alexander, xxii, 4, 6, **10**, 136, 287

Baird, Spencer Fullerton, 194

Baldwin, James Mark, **11–12**

Balfour, Arthur James, **12–13**; characterizes Darwin, xxix

Bancroft, George, **13–14**

Bangs, John K., **14**

Barbier, Edmond, visits Darwin, 379–80

Barnum, P. T., **15**

Bartlett, William Francis, 314

Bates, Henry, **16**, 230–31

Bateson, William, **16–17**, 36

Baudelaire, Charles, 137

Beagle. See H.M.S. *Beagle*

Becker, Lydia E., **17–18**

Beecher, Henry Ward, xxviii, **18–19**, 202, 494, 499

Beethoven, Ludwig von, 236, 360

beetles, and variation, 457

Bell, Clive, xxiii, **19**

Bell, Julian, 109

Bell, Thomas, visits Darwin, 190

Belloc, Hilaire, 473

Bellow, Saul, **20**

Benedict, Ruth, **20**

Bentham, George, 58
Bentham, Jeremy, 220
Bergson, Henri, 21–22, 375; critique of Darwin, 245
Berkeley, George, 186
Berlin, Isaiah, 22
Bernard, Claude, 503
Bernhardi, Friedrich, 290
Besant, Annie, 22–23
Bewick, William, 373
Bishop, Elizabeth, 23–24
Bismarck, Otto von, 24, 344
Blaine, James G., 136
Blatchford, Robert, 239
Bleek, Wilhelm, 478
"Bloomsbury Group," xxiii
Boas, Franz, 24–25
Bodin, Jean, 45
Bodington, Alice, 25–26
Bölsche, Wilhelm, 26–27
Boltzmann, Ludwig, 480
Bonar, James, 27
Boott, Francis, xxvii
Boston Index, 60
Boston Natural History Society, xxviii
Bourget, Paul, 129
Bowen, Francis, 28–29, 314; attacks Darwin, xxviii
Brace, Charles Loring, xxvii, 29–31; characterizes Darwin, xxix
breeders/breeding, xxi, 111, 416, 422, 469, 476
Brentano, Franz, 31–32
Bridges, Robert, 32
British Association for the Advancement of Science, 157, 266, 318, 342, 349, 366, 413, 437, 481, 499
Broca, Paul, 32–33
Brodie, Benjamin, 198, 318
Bronn, Heinrich Georg, 383
Browning, Robert, 33–34
Brownson, Orestes, 34–35
Bruhl, Carl, 130
Brunetière, Ferdinand, 35–36
Bruno, Giordano, 339
Bryan, William Jennings, 36–37, 282
Bryant, William Cullen, 37–38
Bryce, James, 38–39
Büchner, Ludwig, 39–40, 110, 249–50, 484, 500; Kraft und Stoff, 234
Buckland, William, 125, 357
Buckle, Henry Thomas, xxiii, 40, 45, 99, 433

Buckley, Arabella, 40–41
Buffon, Count, 200, 329
Bulwer-Lytton, Edward, 42–43, 52
Bunbury, Charles, xxii, 224
Burbank, Luther, 43–44, 147
Burgon, John William, 129–30, 155
Burke, Edmund, 220, 348
Burroughs, John, 44–45
Bury, J. B., 45–46
Bussy, Dorothy, 413
Butler, Joseph, 226
Butler, Samuel, 46–49, 238, 340; "Darwin among the Machines" (1863), 47; The Way of All Flesh, 126, 238
Butler, Samuel (bishop), 443
Butler, Thomas, 49
Bywater, Ingram, 155, 156

Cabot, Elizabeth, xxviii, 50
Cabot, James Elliot, reads Origin, xxviii
California Geological Survey, 477
Calvinism and Darwinism, 494
Cambridge University, 65
Cameron, Julia, photograph of Darwin, 183
Campbell, George (Duke of Argyll), 50–52, 116, 291, 305; at Darwin's funeral, 397; discusses Darwin with Liddon, 248
Candolle, Alphonse de, 52–53, 448, 486–87; characterizes Darwin, xxix, 52
Canfield, Sherman, 475
Cánovas del Castillo, Antonio, 356
Carlyle, Alexander, 55
Carlyle, Jane, xxiii, 54
Carlyle, John A., 54
Carlyle, Thomas, 53, 54–56, 137; as connector of intellectuals, xxiv; converses on Darwin and Haeckel, 56; on Darwin's intellect, 374; inability to cope with Darwinism, 4; on Spencer, 375; as writer, 327
carnivorous plants, 429, 436, 440
Carpenter, William B., 56–58, 67; "converts" to Darwinism, 56; and P. Gosse, 153; inability to cope with Darwinism, 4
Carroll, Lewis, 58–59; Through the Looking Glass, 183
Castelar, Emilio, 59
Cather, Willa, 59
Catholic Church and science, 123, 201, 282, 315–16, 444, 472
Caverni, Rafaello, 201
Céard, Henri, 503

cell-organization, 330

Chadwick, John W., **60**

Chambers, Robert, 158. See also *Vestiges of the Natural History of Creation*

chance in evolution, 31, 220, 299; meaning of, 63

Charcot, Jean-Martin, 335

Chekhov, Anton, **60–61**

Chernishevsky, Nikolai, 99

Chesterton, G. K., **61–62**; Hardy's poem on, 175

chestnut, extinction of, 43

Chevis, Langdon, 240

Churchill, Winston, **63**

Clark, Austin H., 238

classification, natural, 180, 388

class struggle, 218, 273

Claus, Carl, **63–64**; *Text Book of Zoology*, 328

Clemenceau, Georges, **64–65**, 503

Clemens, Samuel. *See* Twain, Mark

Clifford, William Kingdon, 6, **65–66**, 314; as connector of intellectuals, xxiv

Clodd, Edward, 65, **66–67**, 149

Cobbe, Frances Power, **67–71**, 399

Cobbett, William, 220

Cohn, Ferdinand, **71–75**

Coleoptera, 362

Coleridge, Samuel Taylor, 220

Collier, John, 324

Collingwood, Cuthbert, 317

"Colloquy of the Round Table," 275

comparative anatomy, 64, 299

comparative embryology, 285

Comte, Auguste, 45, 148, 461

Conklin, Edwin Grant, **75–76**

Constitution (U.S.), 483

Contemporary Review, 412

contrivance, 13, 226, 290, 307, 367

Conway, Moncure D., xx, **76–80**, 408

Coolidge, Calvin, **80–81**

Cope, Edward Drinker, 11, **81–82**, 369

Copernicus, 123, 501

Cornell University, 475

Cornhill Magazine, 106

Courthope, William John, 66

Creighton, Mandell, **82–83**

Crosse, Andrew, 350

Cruz e Sousa, João, **83**

Cuvier, Georges, xvii, 200, 228, 288, 488; as adaptationist, 57; types, 427

Dalton, John, 427

Dana, James Dwight, **84–85**

Dana, Richard Henry, **85–86**

Danilevsky, Nikolai, **86–87**

Darwin, Charles: on aesthetics, 434; and L. Agassiz, 252; appears in a dream, 210; and Athenaeum Club, 55; attends T. Huxley's lectures, 471–72; and *Beagle* voyage, 169; on *bee orchis*, 298; and beetles, 20; called "missing link," 5; caricatures of, 170–71; caution of, 181, 214, 249, 320; and children, 30; as church elder, 429; on climbing plants, 17, 73; compared to Newton, 17, 26, 329; compared to Shakespeare, 44; complains about Herschel, xxii; conversation with Farrar, 115; on coral reefs, 451; creates philosophy of natural history, 347; on crossing plants, 177; divergence of character, 103; on earthworms, 53; as ethnologist, 442; experiments, 144, 431, 432, 433, 439, 457; explains method, xxx, 96, 410; field trip to Wales, 352; followers of, 7, 144, 322; and French language, 379–80; funeral, 79–80, 116, 397–98; Galton's questionnaire, 364; at Geological Society, 357, 372; geology, 364–65; German-type mind of, 247; and P. Gosse, 152; health of, 109, 189, 314, 361, 370; on T. Huxley, 269; impact on geology, 141–42; and inductive method, 242, 259–60, 387, 393, 414, 472, 489; and insectivorous plants, 155; on interviews, 497; "interviews" visitors, 190; on Lewes, 119; library of, 433, 447–48; Linnean Society contribution, 316, 390, 470; and Lyell, xxx; meets A. Gray, 156; meets Tegetmeier, 421; and Mill, 343; and music, 98; as naturalist, 407; on orchids, 41, 209, 487; on origin of music, 174; pangenesis, 167, 280, 503; personality/persona, xix, xxix–xxx, 30, 39, 55, 74, 119, 166, 183, 238, 268, 360, 361, 372, 408; philosophical forerunners, 186–87; physical appearance of, 72, 138, 166, 183, 184, 210, 361, 378–79, 410–11, 430–31; plain speech of, 325, 379; on plant physiology, 431; and poetry, 16, 20, 144, 433–34; on primal hordes, 130, 267; questioned by Ruskin, 372; rashness of his critics, 349; reading the mail, 30–31, 78–79; reads Malthus, xxxi; receives or requests information or specimens, 18, 58, 70, 114, 296, 297, 313; and religion, 185, 186; on Royer, 371; in John Sanderson's lab, 139; on sea-captains, 209; as scoundrel, 282; and slavery, 432; on Spencer, 402; triumphalism of, 387; on Trollope, 438; on Twain, 172, 184, 439; use of creation language, 271;

Darwin, Charles (cont'd.)
  works, read as novels, 327; on worms, 249;
  as writer of Tennyson's poems, 418; writes to
  J. Geike, 142; writes to G. Wright, 493; writing style, 22, 52–53, 273, 327, 369, 413, 457.
  See also natural selection
Darwin, Charles, visitors: Cobbe, 68–70; A. Gray,
  156; Hague, 168–71; Macaulay, 259; C. Norton, 326; Ramsay, 357; Stephen, 408; Wallace,
  456; Youmans, 499
  —at Down House: Allingham, 5, 55; Balfour,
  12; Barbier, 379–80; Bates, 456; T. Bell,
  190; Brace, 30; Bryce, 39; Buckley, 40;
  S. Butler, 48; Candolle, 52; T. Carlyle, 55;
  Cohn, 71–75; Conway, 78–79; Grant Duff,
  154–55; C. Eliot, 439; Falconer, 190; Fiske,
  119–20; Forbes, 190; Galton, 123–24;
  Geddes, 139–40; Gladstone, 149, 375;
  A. Gray, 157; Haeckel, 166–67; Hague,
  171–72; Higginson, 171–72, 183–84;
  J. Hooker, 5, 119, 157, 189–90; H. James,
  204; Lockyear, 155; Lowe, 154; Lubbock
  (son), 155; Marsh, 269; F. M. Müller, 304;
  Nevill, 314; North, 323–24; C. Norton,
  440; Paget, 336; Richter, 360; Riley, 361;
  A. Ritchie, 363; Romanes, 367; Rothery,
  155; Ruskin, 372; Sarcey, 378–80; W. Spottiswoode, 155; Stephen, 408; Timiriazev,
  429–33; Tyndall, 157; Wallace, 456; Waterhouse, 190; Weir, 361, 456; C. Wright,
  489–90, 492
Darwin, Charles, works: Autobiography, 16, 20;
  Climbing Plants, 207; Expression of the Emotions in Man and Animals, 58, 485; Fertilisation of Orchids, 226, 297, 363; Formation of
  Vegetable Mould through the Action of Worms,
  207; Life and Letters, 16, 32, 64, 440; Life of
  Erasmus Darwin, 379; Power of Movement in
  Plants, 431; Variation of Plants and Animals
  under Domestication, 68, 205, 299, 308, 421,
  448; Voyage of the Beagle, 178, 369, 381, 418
  —Descent of Man, 64, 68, 227, 267, 291, 373;
  American edition, 498; confirmed by
  ethnology, 358; distressing to Pusey, 353;
  in L. Morgan's library, 298; Nehru on,
  312; press response, 170, 205; reviewed by
  C. Wright, 490
  —Origin of Species, 64, 68; advertised, 14;
  compared to Principia, 199, 208; earliest
  reception, xxi–xxiv, xxvii–xxix; effect on
  zoology, 300; "greatest book of the century," 221; as humorous book, 341; merits
  of, 43; reception in Edinburgh, 364; and
  religion, 233; reviews, xxi, 191, 365, 394;
  A. Sedgwick on, 386–87; Shaler reads,
  390; sixth edition, 17; translations, 66, 143,
  371, 383; and Wallace, 456; as "wonder
  book," 44
Darwin, Elizabeth (Bessie) (daughter), 119
Darwin, Emma (wife), 73, 96, 120, 172, 313, 357,
  360, 430, 490; lunch with Youmans, 499;
  reads L. Carroll aloud, 183
Darwin, Erasmus (brother), xxiii, 68, 189; lends
  Martineau copy of Origin, 272
Darwin, Erasmus (grandfather), 73, 133, 172
Darwin, Francis (son), 52, 119, 172, 412, 430, 433
Darwin, George (son), 12, 120, 172
Darwin, Henrietta (Litchfield) (daughter), 119,
  120; on Haeckel's poor English, 167
Darwin, Horace (son), 120, 489
Darwin, William (son), xxviii, 78, 167, 253
Darwinism: and Calvinism, 494; and charity,
  392; and compatibility with theism, 500;
  confused with Lamarckism, 364; Darwin
  Memorial, 248; distinct from "evolutionism,"
  348, 423; eclipse of, xxv; economics, 27; as
  empiricism, 445; in France, xxvi; in Germany, 234, 490–91; Haeckel's admiration
  for, 165; as "ingenious dream," 176; as laissez-
  faire, 220, 232; limitations of, 166; as marking
  an epoch, 105, 199, 377; meaning of, 123; mis-
  applied to race struggle, 162, 217; as natural
  theology, 196, 224; necessary consequences
  of, 307; as "new totemism," 284; as ontology,
  xxvi; as orthodox biology, 173; and religion,
  164, 202, 206, 208, 240, 241, 354; in Russia,
  227; synonym of materialism, 468; as theory
  of knowledge, xxvi, as topic of conversation,
  38, 48, 170, 196, 240, 256, 327, 384, 405, 413;
  and utilitarianism, 408
Daubeny, Charles: critiques Origin, 366; meets
  Darwin, 372
Daudet, Léon, 336, 503
Davies, G. R., 329
Dawson, John William, 87–88; reviews Origin,
  349
De la Beche, Henry, 125, 357
Delpino, Federico, 308
Derby, Lord, 116, 397
Descartes, René, 238
design, 11, 28, 29, 101, 131, 206, 226, 271, 450, 493,
  500; argument from, 329

Devonshire, Duke of, 116, 397
DeVries, Hugo, **89–90**, 168
Dewey, John, **90–91**
Dickens, Charles, 53, **91–93**, 268
Dickie, George, 279
Dickinson, Emily, **93–94**
Disraeli, Benjamin, **94–95**
division of labor, 103, 270
Doddington, George Bubb, 34
Dodel, Arnold, *Moses or Darwin?* (1889), 331
Dohrn, Anton, **95–96**; asks Darwin about method, xxx, 96; characterizes Darwin, xxix
Donnelly, Ignatius, **96–97**
Dostoevsky, Fydor, **97–98**
Down House: Darwin's study, 53, 73, 433; description of, 171, 430; greenhouse, 53, 149
Doyle, Arthur Conan, **98–99**
Draper, John William, 159, 193; his "atheistic rigmarole," 317–18
Drosera rotundifolia, 149, 432, 491
Dubnow, Simon, **99**
Dubois, Eugène, **100**, 215
Du Bois-Reymond, Emil, **101**, 281
DuGard, Thomas, 49
Duhem, Pierre, **101–2**
Dühring, Eugen, 111
Du Maurier, George, **102**
Duns Scotus, 186
Durkheim, Émile, **102–3**

Ebattson, Roger, xix, xxvi
economists and Darwinism: Bonar, 27; Jevons, 208–9; Keynes, 220–21; A. Marshall, 270–71; Schumpeter, 384–85; Veblen, 449–50; A. Weber, 467–68
Edinburgh Review, 250
Edison, Thomas A., 144, 147
Eimer, Theodore, 11, **104**
Einstein, Albert, **104–5**
Eliot, Charles W., 439
Eliot, George, **105–6**, 112, 204; reads *Origin,* xxii, xxiv
Eliot, T. S., **106–7**
Ellis, Alexander, 183
Ellis, Havelock, **107–8**
Elphinstone, Montstuart, 259
Emerson, Ralph Waldo, xxiii, 77, 78, **108**
emotions, expression of, 31
Empson, William, **109**
Engels, Friedrich, **110–12**, 274
Entomological Society, 362

Eohippus, 137
Esher, Reginald, **112–13**
ethics, utilitarian, 68
Eton, 255
eugenics, 75, 203, 281
Evans, T. S., 198
evolution: of art, 455; conflated with Darwinism, 464; gradual, 468; as machine process, not cause, 450; and purpose, 376; and revolution, 403

Fabre, Jean Henri, **114–15**
Falconer, Hugh, xxx, 157; visits Darwin, 190
Faraday, Michael, xxii, **115**, 116, 225, 289
Farrar, Frederic W., **115–17**; at Darwin's funeral, 397–98; on Huxley-Wilberforce exchange, 199
Fawcett, Henry, 287, 318
Ferenczi, Sandor, **117–18**
Ferri, Enrico, 252
Fichte, Johann Gottlieb, 187
finches, 381
Fiske, John, **118–20**, 442; characterizes Darwin, xxix; teleology of, 294
fitness, meaning of, 426
Fitzgerald, Edward, **120–21**
Fitzgerald, F. Scott, **121**
FitzRoy, Robert, 156, 170, 318, 381, 451
Flaubert, Gustave, xxvi, **121–22**; reads Haeckel, 122
Flourens, Pierre, 233
Flower, Benjamin, **122**
Fogazzaro, Antonio, **122–24**
Forbes, Edward, **124–25**, 357; visits Darwin, 190
Ford, Ford Maddox, **126**
Forster, E. M., xix, **126–27**
Fosdick, Harry Emerson, **127–28**
Foster, E. S., 483
France, Anatole, xxvi, **128–29**
Franke, Hermann, 360
Fraser's Magazine, 266
Freeman, Edward A., **129–30**
freethought, 76
Freud, Sigmund, xxvi, 116, **130–31**
Frost, Robert, **131–32**
fundamentalism, 328

Galileo, 123, 444, 472
Galton, Francis, 79, **133–34**, 203, 229, 364, 476, 503
Gandhi, Mahatma, **134–35**
Garfield, James A., **135–36**

Garland, Hamlin, **136**

Garner, Richard Lynch, 459

Garofalo, Raffaelo, 252

Garrison, Wendell Philips, **137**

Garvey, Marcus, **137–38**

Gary, Thomas, 77

Gaskell, Elizabeth, **138–39**, 325

Gavarret, Jules, 371

Geddes, Patrick, **139–40**

Gegenbaur, Carl, **140–41**

Geikie, Archibald, **141–42**

Geikie, James, **142–43**

Geoffroy Saint-Hilaire, Étienne, 39, 57, 233, 488

Geoffroy Saint-Hilaire, Isidore, **143**

geographical distribution, 160, 168, 240

geologists and Darwinism: J. Dana, 84; Dawson, 87–88; A. Geike, 141–42; J. Geike, 142; J. Jones, 211; Judd, 214–15; Leconte, 240–43; Lyell, 255–58; Murchison, 310–11; J. Powell, 351–52; Ramsay, 357; W. Rogers, 364–65; A. Sedgwick, 386–87; Shaler, 390

geology, scriptural, 288

La Gerbe, 164

Giard, Alfred, **143–44**

Gibbons, James Cardinal, **144–45**

Gibbons, Tom, xxv

Gide, André, xxvi, **145**

Gilbert, Joseph H., 431

Gilbert, W. S., **146–47**

Gilman, Charlotte Perkins, **147–48**

Gissing, George, **148–49**

Gissing, Thomas, 148

Gladstone, William, 78, 80, **149–51**, 248, 375, 481; and T. Huxley, 281

Glasgow, Ellen, **151**

Glyptodon, 125

Godwin, William, 220

Goethe, Johann Wolfgang von, 74, 130, 180, 238, 279, 379, 426

Gordon, George, xxii

Goschen, George, 248

Gosse, Edmund, 65, **152–53**

Gosse, Philip Henry, and natural selection, 152–53

Gould, John, xxx

Gould, Stephen Jay, **153–54**

Grant Duff, Mountstuart E., **154–56**, 304

Gray, Asa, xxi, xxviii–xxix, 31, 116, 152, **156–57**, 170, 176, 257, 371, 434, 486, 496; *Darwiniana*, 493, 500; lends copy of *Origin*, xxvii; and

Muir, 303; reviews *Origin*, xxviii, 325, 365; on species, 488; and G. Wright, 493

Gray, John Edward, xxx

Green, John Richard, **157–59**; T. Huxley on, 198

Grote, John, 386

Grove, William Robert, 224

Gulick, John T., **160–61**

Gumplowicz, Ludwig, **161–63**, 217

Ha'am, Ahad, **164–65**

Haeckel, Ernst, 66, **165–68**, 222, 234, 314, 329, 331, 353, 356, 382, 473, 477, 478, 484, 497, 503; equates evolution with Darwinism, 464; and Gegenbaur, 140–41; as materialist, 500; mien, 26; perigenesis, 503; Schleicher's letter to, 383–84

Hague, James Duncan, **168–72**

Haldane, J. B. S., **173**

Hall, G. Stanley, **173–74**

Hamilton, William, 493

Hamlet, 176, 236

Harding, Warren G., 137

Hardy, Thomas, xix, **174–75**; and Kingdon's circle, xxiv

Harvard College, xxvii, xxviii, 118, 325

Harvey, William, circulation of blood, 489

Harvey, William H., **175–76**

Hawthorne, Nathaniel, 325

Hays, Willet M., **176–77**

Hearn, Lafcadio, **178**

Hegel, George Wilhelm Friedrich, 27, 59, 232, 273, 294, 404; *Philosophie der Geschichte*, 82

Heisenberg, Werner, **179**

Helmholtz, Hermann von, **180**, 199, 404

Henslow, John Stevens, **180–82**, 318; on *Origin of Species*, 181

Herder, Johann Gottfried, 446

Herndon, William H., reads *Origin* to Lincoln, 251

Herschel, John, xxii, xxvii, 67, 80, 116, **182**, 289

Herzl, Theodor, **182–83**

Higginson, Thomas Wentworth, **183–85**

Hildebrand, Friedrich, 308, 460

H.M.S. *Beagle*, 11, 49, 69, 156, 260, 318, 407; Darwin's collections while on, 362; Darwin's recollection of, 169

Hobbes, Thomas, 186, 233, 273

Hodge, Charles, **185**

Hodgkin, Thomas, **185–86**

Hoffding, Harald, **186–87**

Holmes, Oliver Wendell, Sr., 80, **187–88**

Holmes, Oliver Wendell, Jr., **188–89**, 238

Holmgren, Frithiof, 71

Holyoake, George, xxiii, 271

Hooker, Joseph Dalton, xxi, xxvii, 76, 79, 116, 152, **189–91**, 255, 316, 342, 436, 441, 470; characterizes Darwin, xxix; as Darwin's aide, 345; reaction to *Origin*, 363; visits Down House, 5, 119

Hooker, William, 156

Hoole, J. W., discusses Darwin with Pater's circle, 338

Hopkins, Gerard Manley, **191–92**

Hopkins, William, 266

Hort, Fenton, **192–93**

horticulture, 276

Howells, William Dean, **193**, 440

Hudson, William Henry, **193–94**

Hughes, Thomas, xv, 265

Hugo, George, 335

Hugo, Victor, 379, 380

Humboldt, Alexander von, 271

Hume, David, 186, 220

Hunter, John, 156, 330

Huskisson, William, 220

Huxley, Aldous, **195–96**

Huxley, Julian, 195, **196**

Huxley, Leonard, 195

Huxley, Thomas Henry, xv, xxi, xxx, **196–200**, 314–15; as battler for Darwin, 240, 412; biology course, 458; and Conway's circle, 408; criticized, 148–49; on Darwin, 154, 199; at Darwin's funeral, 397; dogmatism of, 7; exchange with Wilberforce, xxv, 158–59, 197–98, 266, 318, 343, 349, 366, 412; and Gladstone, 281; interesting to read, 369; and Kingdon's circle, xxiv; *Lay Sermons*, 498; lectures at Royal College of Science, 471; letters, 6; and natural selection, 465, 470–71; and natural theology, 224; and Paget, 336; on Peabody Museum (Yale), 184; reviews *Origin*, xxi, 364; and H. Rogers, 365; and T. Roosevelt, 368; as scoundrel, 282; teleology of, 294; and *Vestiges*, 364; as writer, 22

The Ibis, 437

Ichthyosaurus, 219

Index of Prohibited Books, **201**, 382

Ingersoll, Robert G., 126, **202**; and Gladstone, 151

Jacobs, Joseph, **203–4**

James, Constantine, refutation of Darwin, 346

James, Henry, **204–5**, 410; characterizes Darwin, xxix; and C. Norton, xxviii, 3

James, William, 11, **205–6**; and chance, 329

Jeffries, Richard, **206–8**

Jena, 167, 356

Jenkin, Fleeming, 158

Jenyns, Leonard, 182

Jevons, W. Stanley, **208–9**

Jewett, Sarah Orne, **209–10**

Jones, Ernest, **210**

Jones, John, **211**

Jordan, David Starr, **211–12**

Joule, James, **212**

Journal of the Linnean Society, 316

Jowett, Benjamin, **212–13**

Joyce, James, **213–14**

Judd, John Wesley, 7, **214–15**

Jung, Carl, **215–16**

Kammerer, Paul, **217**

Kant, Emmanuel, 59, 119, 187, 294, 329, 331, 411, 454; *Grundlegung der Sitten*, 70; *Kritik der Urtheilskraft*, 186

Kautsky, Karl, **217–18**

Keeling Island, 451

Kendall, May, **218–19**

Kepler, Johannes, 382, 383

Ker, William Paton, 434

Kessler, Karl Federovich, 230

Keynes, John Maynard, xxiii, **220–21**

Kidd, Benjamin, **221–22**; Principles of Western Civilization, 148

King, Martin Luther, Jr., **222–23**

Kingsley, Charles, xv, 66, **223–26**, 265; discusses *Origin* with Bunbury, xxii, 224; last words, 305; receives copy of *Origin*, xxi, 223

Kingsley, Henry, xv, **226–27**, 265

Kipling, Rudyard, **227**

Knight, Charles, 459

Knight, William Angus, 434

Knight-Darwin Law, 308

Knowles, James, 230

Kolreuter, Josef, 460

Kovalevsky, Alexander, 196, **227–28**, 432

Kovalevsky, Vladimir, 432

Kravchinsky, Sergei, **228–29**

Kropotkin, Alexander, 230

Kropotkin, Peter, **229–31**
Kugelmann, Ludwig, 274

Lacan, Jacques, **232**
Lafargue, Paul, **233–34**, 274
Laforgue, Jules, 195
Lamarck, Jean-Baptiste, xvii, xxvi, 25, 39, 64,
    128, 145, 200, 211, 233, 240, 289, 350, 356, 396,
    449, 467; compared to Galileo, 26; and cre-
    ation, 263; method, 393; not a good read,
    369; *Philosophie Zoologique,* 390
Lamarckians, 11
Lamarckism, 16, 257. *See also* Neo-Lamarckism
Lane, Edward Wickstead, 10
Lange, Friedrich Albert, **234–35**, 274
language, evolution of, 115, 305, 383
Lanier, Sidney, **235–37**
Lankester, E. Ray, **237–38**
Laski, Harold, **238–39**
Lawrence, D. H., xix, **239–40**
The Leader, 148
Le Conte, Joseph, **240–43**
LeCouteur, John, 448
Leibniz, Gottfried, 238, 404
Lenin, Vladimir, **243**
Lepsius, Karl Richard, 304
LeTourneau, Charles, 503
Lewes, George Henry, xx, 113, 178, 226, **244–45**;
    History of Philosophy, 148; left behind by
    Darwinism, 4; *Problems of Life and Mind,* 119;
    C. Wright on, 407; writes to Darwin, 106
Lewis, C. S., **245–46**
Lewis, George Cornewall, **247**; borrows *Origin*
    from Lowe, 247
Liddon, Henry Parry, **248–49**
Lieber, Francis, **249–50**
Liebig, Justus, 110
Lincoln, Abraham, 80, **250–51**
Linnean Society, xxi, 437, 470; journal of, 316
Linneaus, Carl von, 116, 362
Literary Gazette, 125
Littré, Émile, 83
Lockyer, Joseph Norman, 155
Loisleur-Deslongchamps, Jean, 448
Lombroso, Cesare, 26, **251–52**
London Quarterly, 13
London Times, xxi, 412, 445; T. Huxley's review,
    412
London Zoological Society, 108
Longfellow, Henry Wadsworth, 77, **252–53**; and
    L. Agassiz, 193; and Tennyson, xxviii

Lowe, Robert (Viscount Sherbrooke), 154,
    262–63
Lowell, James Russell, xxvii, 116, 155, **253–54**,
    325, 397
Lubbock, Sir John (father), 357
Lubbock, John (son), 55, 79, 116, 149, 154, 206,
    255, 304, 318, 397, 408
Lucretius, 128, 378, 484; and Arnold, 7
Lumb, Edward, 260
Lushington, Vernn, 324
Lyell, Charles, xxii, xxiii, xxvii, xxx, 1, 4, 31, 53,
    59, 64, 67, 76, 152, 158, **255–58**, 288, 316, 345, 357,
    364, 433, 470, 477, 485; *Antiquity of Man,* 2, 35,
    493; and Dawson, 349; esteemed in Ger-
    many, 96; on Falconer, 157; and geographic
    isolation, 486; his caution, 234; on Inquisi-
    tion, 429; and natural theology, 224; *Prin-
    ciples of Geology,* xxx, 2, 256, 281; recommends
    Hague to Darwin, 168; and Ticknor, xxviii,
    365; urges Darwin to publish, 170

Macaulay, Thomas B., 52, 256, **259**
MacDonald, Ramsay, **259–60**; on Darwin's
    method, xxx
Macdonell, Anne Lumb, **260–61**
Mach, Ernst, **261–62**, 480
MacLeay, William Sharp, **262–65**; quinary sys-
    tem, 333
Macmillan, Alexander, **265–67**; hosts discus-
    sion of *Origin of Species,* xv, xxiii, 265; and
    Tennyson, xxi
Macmillan's Magazine, cv, 181
Le Magasin Pittoresque, 143
Maine, Henry, **267–68**, 488
Malthus, Thomas Robert, xxxi, 6, 27, 32, 113,
    135, 186, 220, 273, 308, 321, 395, 460–61, 475;
    checks, 23, 458; *Essay on Population,* 358;
    "population fantasy," 274
Mandelstam, Osip, **268**
Mantegazza, Paolo, 136
Mao Tse-tung, **268–69**
Marsh, Othniel C., 137, 172, **269**
Marshall, Alfred, **270–71**
Marshall, John, 452
Martin, Richard B., 424
Martineau, Harriet, xxiii, **271–73**
Marx, Karl, 8, 112, 218, **273–74**, 376; *Capital,* 243
Masson, David, xv, 265, **275**
Masters, Maxwell T., **275–76**
materialism, 234, 372, 467, 500; synonymous
    with Darwinism, 468

Maudsley, Henry, 276–77
Maurice, Frederick Denison, xv, 224, 265, 277–78
Maxwell, James Clerk, 292
McCarthy, Justin Huntly, 278–79
McCaul, James, 288
McCosh, James, 279–80, 483, 498, 500
McDougall, William, 280–81
Mead, Margaret, 281–82
Mencken, H. L., 282
Mendeleev, Dmitri, 404
Mendelism, 11, 195
Menzbir, Mikhail, 283
Mercer, John Edward, 283–84
Merivale, Herman, 284–85
Meta, Rudolph, 222
Metchnikoff, Elie, 285–86
Metchnikoff, Léon, 286
Metzger, Johann Christian, 448
Miers, John, 405
Mill, John Stuart, 64, 69, 99, 287–88, 320, 327, 343, 493; essay on Comte, 461
Miller, Gerrit S., 238
Miller, Hugh, 158
Miller, William Hallows, 155
Milman, Henry Hart, 256, 288–89
Milne-Edwards, Henri, 451
mind, evolution of, 21, 24, 25, 29, 277, 401
missing link, 62
Mitchell, Peter Chalmers, 108, 289–90
Mivart, St. George Jackson, 290–92; On the Genesis of Species, 192, 298, 498
Moleschott, Jacob, 110, 228, 234
monogenesis, 346
Monro, Cecil James, 292
Montesquieu, 341, 348
Montessori, Maria, 292–93
Moore, Aubrey, 293–94
Moore, C., 157
Moore, G. E., 294–95
Moorhouse, M. B., and Pater, 338, 339
morality: Christian, 378; evolution of, 35, 68, 70, 421, 426
More, Alexander Goodman, 296–97
Morgan, C. Lloyd, 11
Morgan, Lewis Henry, 298–99
Morgan, Thomas Hunt, 299–300
Morris, Francis Orpen, 317
Morris, Robert Tuttle, 402
Morse, Edward S., xxv, 300–301
Moscow University, 229
Mosso, Angelo, 301–2

Mozart, Wolfgang Amadeus, 360
Mozley, James B., 302–3
Muir, John, 303
Müller, Friedrich Max, 304–5; left behind by Darwinism, 4
Müller, Fritz, 305–7, 308; For Darwin, 285
Müller, Hermann, 308–9, 460
Müller, Johannes, 200
Mumford, Lewis, 309–10
Murchison, Roderick Impey, xxii, 225, 263, 310–11; and L. Agassiz, 252
Murray, George, visits Darwin, 139
Murray, John, xxi, 257, 275, 398, 399
Museum of Natural History, 447
Musil, Robert, 311
mutual aid, 230, 310

Nägeli, Karl Wilhelm von, 166–67
Naples, Marine Biology Station, 96
The Nation, 491
National Review, 56, 58
The Naturalist, 317
natural selection: acts on individuals, 154; application to economics, 469; applied to man, 481; Belloc opposes, 473; and chance, 12, 29; contrary to humanitarian principles, 476; conversion to, 211; critics of Darwin's view, 175, 341; elasticity of term, 51; evidentiary basis of, 28, 63; as historical explanation, 470; impact on philosophy, 479–80; as indirect equilibrium, 460; and Lamarck's besoin, 263; as law, 107; logic of, xxi; as mechanical principle, 237; as non-rational, 13; and religion, 9; as revelation, 317; Romanes on, 367; A. Sedgwick on, 387; Shaw opposes, 473; Spencer on, 400; as "ultra-Lamarckian," 310; unproved, 328, 478; as vera causa, 175, 256, 287, 317
Natural Theology, 279
Nature, 161
nebular hypothesis, 488
Nehru, Jawaharlal, 312
Neo-Lamarckism, 25, 229–30, 404
Nevill, Lady Dorothy, 313–14
Newcomb, Simon, 314–15; attends Darwinism debate, xxviii
New Englander, 493
Newman, Edward, 362
Newman, John Henry Cardinal, 121, 315–16
Newton, Alfred, 316–20, 437; hears Linnean Society papers, xxi; at Oxford meeting, 349

Newton, Isaac, 80, 116, 123, 203, 225, 238, 255, 288, 331, 375, 382, 420, 489; Spencer compared to, 402

New York Times, 403, 418

Nietzsche, Friedrich, 282, **320–21,** 375; stigmatizes Darwin and Spencer, 164; superman, 215

Nineteenth Century, 230

Nordau, Max, **321–23,** 456

Norman, George, 323

North, Marianne, **323–24**

North American Review, 205

Norton, Charles Eliot, xxvii, 78, 170, 205, **325–26,** 408; as link between Americans and English, xxix; in London, 204

Norton, Grace, 254

novelists and Darwinism: Aleichem, 3–4; Bellow, 20; Bulwer-Lytton, 42; S. Butler, 46–48; Cather, 59; Chekhov, 60–61; Chesterton, 61; Dickens, 91–93; Donnelly, 96–97; Dostoevsky, 97–98; Doyle, 98–99; Du Maurier, 102; G. Eliot, 105–9; F. Fitzgerald, 121; Flaubert, 121–22; Forster, 126–27; France, 128; Garland, 136; Garrison, 137; Gaskell, 138–39; Gide, 145; G. Gissing, 148–49; Glasgow, 151; H. James, 204–5; Jewett, 209; Joyce, 213–14; C. Kingsley, 223–25; H. Kingsley, 226–27; Kipling, 227; Lawrence, 239; Musil, 311; Pérez Galdós, 344; Reade, 358; Sienkiewicz, 392; Singer, 395–96; Steinbeck, 406–7; Stevenson, 409–10; Tolstoy, 434–35; Trollope, 438; Twain, 439–41; Verne, 450–51; M. Ward, 461–62; R. Warren, 463; Wells, 471–73; Wharton, 473–74; Wilde, 482; Woolf, 487; Zola, 502

Occam, William of, 186

Ogilvy, Mabel, 313

Oken, Lorenz, 491

Oliphant, Margaret, **327–28**

O'Neill, Eugene, **328**

Ormerod, Eleanor, **328–29**

orthogenesis, 11, 104

Osborn, Henry F., 11, **329–30, 427**

Osler, William, **330–31**

Osterroth, Nikolaus, **331–32**

Ostwald, Wilhelm, **332–33,** 480

Owen, Richard, xxx, 35, 56, 124, 152, 156, 157, 219, 224, 225, 266, **333–34,** 357; and Astronomer's Club, 485; and Athenaeum Club, 55; and Aye-Aye, 496; at British Association meeting, 343, 366, 412; on "creation," 350; "diluted Owenism," 317; "malignant attitude" of, 67;

On the Homologies of the Vertebrate Skeleton, 51; *On the Nature of Limbs,* 50; on race, 452; reviews *Origin,* 334; seen as "old school," 278

Oxford University, 38, 462

Paget, James, **327–28**

paleontology, 474

Paley, William, xvii, 218, 220, 226, 294

Palgrave, Francis T., xxii, 1, **336–37;** loans *Origin* to Tennyson, 424

Park, Amasa, 493

Parker, Theodore, xxiii, xxviii, **337**

Parsons, Theophilus, **328**

Pasteur, Louis, 336

Pater, Walter, **338–40**

Pearson, Charles H., **340–41,** 476

Pearson, Karl, **341**

Pedro I (Brazilian emperor), **341–42**

Peirce, Benjamin, xxviii, 314, **342–43**

Peirce, Charles Sanders, **343–44**

Penn, Granville, 288

Pérez Galdós, Benito, **344**

Philips, John, 257

philosophers and Darwinism: Bain, 10; Bergson, 21–22; Berlin, 22; Bowen, 28–29; Brentano, 31; Büchner, 39–40; Dewey, 90–91; Duhem, 101–2; Emerson, 108; Fiske, 118–19; Lange, 234–35; Lewes, 244–45; Mach, 261–62; Martineau, 271–73; Mill, 287–88; G. Moore, 294–95; Nietzsche, 320–21; C. Peirce, 343; Renan, 360; D. Ritchie, 363–64; Royce, 370; Royer, 371–72; B. Russell, 375–76; Santayana, 377–78; Schopenhauer, 384; Stirling, 410–13; Tyndall, 443; Unamuno, 444–45; Vaihinger, 446–47; Whewell, 474; Whitehead, 475–76; Wittgenstein, 485; C. Wright, 488–92

phytogeography, 465

pigeons, 9, 235, 405, 422

Piltdown Man, 375

Pisarev, Dmitry, 61, 99, **344–46**

Pithecanthropus, 62, 215, 375

Pitt Rivers, Augustus, **346–47**

Pius IX (pope), **347,** 444

Pius X (pope), 124

Plato, 63, 164, 469, 472; cosmology, 377–78

poets/poems and Darwinism: Bishop, 23, 32; Bridges, 32; Browning, 33–34; Courthope, 66; Cruz e Sousa, 83; Dickinson, 93; T. S. Eliot, 105–6; Empson, 109; Forbes, 124–25; Frost, 131–32; W. Gilbert, 146–47; E. Gosse, 152–53; Hardy, 174–75; G. Hopkins, 191–92; Kendall,

218–19; Longfellow, 252; Lowell, 253; Mandelstam, 268; Masson, 275; Merivale, 284; Pound, 350; M. Savage, 382–83; Swinburne, 418; Tennyson, 424; Yeats, 504–5

Pollock, Frederick, **347–48**; and Kingdon's circle, xxiv

Polybius, 45

polygenesis, 345

Porter, Noah, 493

Poulton, Edward Bagnall, 52, 188, **348–50**, 465

Pound, Ezra, **350**

Powell, Baden, **350–51**

Powell, John Wesley, **351–52**

Powis, Lord (Edward Herbert), 49

pragmatism, 206, 370, 375

pre-Adamites, 288–89

predestination, 494

Prestwich, Joseph, xxx

Price, John, **352**

primogeniture, as opposed to selection, 363

Primrose, Archibald (Lord Roseberry), 248

Princeton University, 11

progress, belief in, 375

Proust, Marcel, 19

psychology, 467

psychozoology, 490

Punch, 66, 102, 170

Pupin, Michael, **352–53**

purposiveness, 470

Pusey, Edward Bouverie, 248, 316, **353–54**

Putnam, James Jackson, **354**

Quarterly Review, 67, 359, 386

Quatrefages, Armand de, **355**

Quetelet, Adolphe, *homme moyen*, 292–93

race, 161–62, 446, 452

Raleigh, Walter (professor), 413

Rambler, 394

Ramón y Cajal, Santiago, **356**

Ramsay, Andrew Crombie, 142, **357**

Ranke, Leopold von, *Weltgeschichtliche Bewegung*, 83

Ratzenhofer, Gustav, 163

Ray, John, 116

Reade, Winwood, **358–59**

Reinke, Johannes, 480

Renan, Ernest, 127, 239, **360**, 420

Revue des Deux Mondes, 36

Ricardo, David, 111, 220

Richter, Hans, **360**

Riley, Charles Valentine, xix, **361–63**

Ritchie, Anne Thackeray, **363**

Ritchie, David George, **363–64**

Robertson, J. M., 239

Robinson, Henry Crabb, 55

Rogers, Henry Darwin, xxii, **364–65**; writes to T. Huxley, 365

Rogers, William Barton, xxii, xxiii, xxvii, 197, **364–65**; predicts victory of Darwin, xxviii

Rolleston, George, 354, **366–67**, 452; at Oxford meeting, 349

Romanes, George John, 11, **367–68**, 473; and Kingdon's circle, xxiv

Roosevelt, Theodore, **368–69**

Rosetti, Dante Gabriel, 204

Rothery, Henry Cadogan, 155

Rothsmsted, 431

Rousseau, Jean-Jacques, 220

Roux, Wilhelm, 167

Royal Institution: T. Huxley at, 412, 413; Rolleston's lecture, 367

Royce, Josiah, **370**

Royer, Clémence, 143, 274, **371–72**

Runkle, John David, 303

Ruskin, John, 204, **372–74**, 430

Russell, Arthur, **374**

Russell, Bertrand, **375–76**

Russell, Mary, xxiii, 54

Russell, Odo, 112

Sanborn, Franklin B., xxvii

Sand, George, 122

Sanford, Edward, 160

Santayana, George, **377–78**

Sarcey, Francisque, **378–80**

Sarmiento, Domingo Faustino, **381**

Saturday Club (Boston), 188

Savage, Minot Judson, **382–83**

Savage, Richard, 37

Savigny, Friedrich von, 348

Schleicher, August, **383–84**, 478

Schleiden, Matthias, 479

Schmiedeberg, Oswald, 385

School of Mines, 412

Schopenhauer, Arthur, **384**, 444; Darwin's debt to, 454

Schumpeter, Joseph A., **384–85**

Schwalbe, Gustav, 385

Schweitzer, Albert, **385–86**

science, defined by technological thinking, 449

Scopes Trial, 404

Sebright, John, 9

Sedgwick, Adam, xxvii, 120, 225, **386–87**; and
L. Agassiz, 252; opposes *Origin*, xxii, 67;
reviews *Vestiges*, 386; writes to Darwin, xxii

Sedgwick, Adam (grandnephew), 328

Sedgwick, Sara, xxviii, 78

Seeck, Otto, 45

Semper, Karl Gottfried, **388**

Sergi, Giuseppe, 252, **388–90**

sexual selection, 37, 117, 174, 220, 264, 309, 373,
398–99; and aesthetics, 482

Shakespeare, William, 151, 236, 375, 413

Shaler, Nathaniel Southgate, **390–91**

Shaw, George Bernard, 282, **391–92**

Shireff, Patrick, 448

Sidgwick, William, at Oxford meeting, 349

Sienkiewicz, Henry, **392–93**

Simpson, George Gaylord, **393–94**

Simpson, Richard, 318, **394**

Singer, Isaac Bashevis, **395–96**

Sitwell, Osbert, **396–97**

Skertchly, Sydney, 142

Smalley, George Washburn, **397–98**

Smith, Adam, 154, 186

socialism, 363

Socrates, 74

Somerville, Mary, **398–99**

Sorby, Henry Clifton, 155

Southey, Robert, 49

Spectator, 148

Spencer, Herbert, 314–15, **400–402**; on aesthetics, 434; critique by Holmes Jr., 189; dinner
for Fiske, 119; and laissez-faire, 220; and
natural selection, 460; nature of his theories,
208, 241; on origin of music, 174; as scoundrel, 282; use of "higher" and "lower," 295;
C. Wright on, 407. *See also* survival of the
fittest

Spinoza, Baruch, 74, 187, 203, 339, 378

Spottiswoode, J., 248

Spottiswoode, William, 55, 115, 116, 155, 397

Sprague, Franklin Monroe, **403**

Sprengel, Christian, 460

Stalin, Joseph, **403–5**

Stanley, Arthur Penrhyn, 398

Stebbing, Thomas, **405–6**

Steinbeck, John, **406–7**

Stendhal (Marie-Henri Beyle), **19**

Stephen, Leslie, 254, **407–9**; characterizes Darwin, xxix–xxx; and Kingdon's circle, xxiv

Stevenson, Robert Lewis, **409–10**

Stimson, William, **390–91**

Stirling, James Hutchison, **410–13**

Strachey, Lytton, xxiii, **413–14**

Strauss, David Friedrich, **414–15**; *Alte und Neue
Glaube*, 212–13, 364

Strong, George Templeton, **416–17**; buys *Origin*, xxviii

Sullivan, Arthur, **417**

Sunday, Billy, **417–18**

survival of the fittest: applied to nations, 376;
critique of, 160, 237, 276, 321; as effective
cause, 458; Germans obsessed with, 425;
inadequacy of, 351; as justification for imperialism, 312; as tautology, 309

Swedenborg, Emmanuel, 3, 240

Swinburne, Algernon, **418**

symbiosis, 310

sympathy, 41

Synge, J. M., **419**

Taine, Hippolyte, **420**

Taylor, James Monroe, **421**

Taylor, Richard, 156

Tegetmeier, William Bernhardt, **421–22**

Teilhard de Chardin, Pierre, **423–24**

teleology, 13, 57, 168, 186, 223, 225, 273, 293, 332,
450, 479

Temple, Frederick, 318, 338

Tennyson, Alfred Lord, xx, xxi, 1, 5, 121, 204,
208, 263, 336, **424–25**; as connector of intellectuals, xxiv

Tertiary man, 478

Thayer, William Roscoe, **425–26**

Thistleton-Dyer, William, **458–59**

Thompson, D'Arcy Wentworth, **426–28**

Thomson, William (Lord Kelvin), 125, 212

Thoreau, Henry David, xxiii, **428**; introduced
to *Origin*, xxvii

Ticknor, George, xxiii, 257, **428–29**; and Lyell,
xxviii, 365

Tiedemann, Friedrich, 452

Timiriazev, Kliment, xx, **429–33**; characterizes
Darwin, xxix

Titchener, Edward, **433–34**

Tolstoy, Leo, **434–35**

Topinard, Paul, **435–36**

Torrey, John, xxvii, 325

transcendentalists, xxiii

Treat, Mary, **436**

Treitschke, Heinrich von, 403
Tristram, Henry Baker, xxi, 66, 318, **437–38**
Trollope, Anthony, **438**
Trotsky, Leon, **438–39**
Tuckwell, W., at Oxford meeting, 349
Twain, Mark, 172, **439–41**; Darwin's fondness for, 184
Twichell, Joseph, 440–41
Twistleton, Edward, 50, 256
Tylor, Edward Burnett, 399, **441–42**
Tyndall, John, 1, 6, 79, 135, 155, 178, 278, 303, 304, 314–15, 328, 336, **443**; address, 191; and Conway's, circle, 408; as writer, 22
type, unity of, 57, 427

Unamuno, Miguel de, **444–45**
uniformitarianism, 474
Unitarianism, 60
L'Universe, 80
University of Strasbourg, 385
Ussher, James, 498
Utilitarianism, 408–9. *See also* ethics

Vacher de Lapouge, Georges, **446**
Vaihinger, Hans, **446–47**
Valéry, Paul, 145, **447**
Van der Kolk, Jacobus Schroeder, 452
variation, 111, 189, 238, 389, 422, 427, 447, 459, 465, 494
Vavilov, Nikolai, **447–48**
Veblen, Thorsten, **449–50**
Verne, Jules, xxvi, **450–52**
Verrill, Addison Emery, xxvii–xxviii
Vestiges of the Natural History of Creation, xxii, 76, 77, 105, 120, 152, 158, 211, 250, 251, 263, 275, 350, 364, 386, 390, 424, 488
Villier, Lady Constance, 55
Virchow, Rudolph, 336
vitalism, xxv, 168
Vivian, Philip, 239
Vogt, Carl, 110, 234–35, **452–53**, 484, 500
Vrolik, Willem, 452

Wagner, Moritz, 160, 168, 303
Wagner, Richard, 360, **454–56**
Walker, Francis, 362
Wallace, Alfred Russel, 11, 58, 59, 81, 116, 122, 147, 152, 309, 397, 421, **456–58**, 497; and evolution of mind, 25; and human evolution, 389; Linnean Society communication,

316, 390, 437, 470; trust in Darwin, 345; as writer, 22
Wallace, Alfred Russel, works: *Contributions to the Theory of Natural Selection*, 298; *Darwin and Darwinism*, 473; *The Malay Archipelago*, 298
Walsh, Benjamin Dann (classmate of Darwin), 362
Ward, Henry Marshall, **458–60**
Ward, Lester, 147, 163, **460–61**
Ward, Mary Augusta, **461–62**
Warren, Howard C., **462**
Warren, Robert Penn, **463**
Washington, Booker T., **463**
Wasmann, Erich, SJ, 114, **464–65**
Waterhouse, George, 362; visits Darwin, 190
Watson, D. M. S., 245
Watson, Hewett Cottrell, 288, 296, **465–66**, 486
Watson, John B., **467**
Weber, Alfred, **467–68**
Weber, Max, **468–70**
Wedgwood, Hensleigh, 68, 78
Weekes, William Henry, 350
Weir, J. Jenner, 361
Weismann, August, 11, 104, 167, 398, 469, **470–71**, 503; controversy with Spencer, 229; immortality of unicellular animals, 331; on origin of music, 174
Wells, H. G., **471–73**
West, Hilborne, 368
Westermark, Edward, 473
Westminster Review, 250, 400, 412
Westwood, J. O., 349
Wharton, Edith, **473–74**
Whewell, William, xxii, 67, 349, 386, **474**
White, Adam, 362
White, Andrew Dickson, **475**
Whitehead, Alfred North, **475–76**
Whitman, Walt, 301, **476–77**
Whitney, Josiah Dwight, **477–78**
Whitney, William Dwight, **478–79**
Wiesner, Julius von, **479–80**
Wigand, Albert, **101**
Wilberforce, Samuel, 67, **480–82**; exchange with T. Huxley, xxv, 158–59, 197–98, 266, 318, 343, 349, 366, 412, 472; reviews *Origin*, 481
Wilde, Oscar, **484**
Wilkinson, J. Garth, *Oannes*, 33
Wilson, Woodrow, **483**
Winchell, Alexander, **483–84**
Wiseman, Nicholas, **485**
Wittgenstein, Ludwig, **485**

Wollaston, Thomas Vernon, 152, **485–87**
woman, evolution of, 147
Wood, Robert Henry, discusses Darwin with Pater's circle, 338
Woodward, Samuel P., 211
Woolf, Virginia, xxiii, xxiv, 149, **487–88**
Wren, Jenny, 42
Wright, Chauncey, 343, 407–8, **488–92**
Wright, George Frederick, **493–95**
Wundt, Wilhelm, 216, 464, 479, **495**
Wyman, Jeffries, 193, **496**; Darwinian meeting in his lab, xxvii, 325; death, 492

Yarrell, William, 421
Yates, Edmund H., **497**
Yates, John, 260
Yeats, William Butler, **497–98**
Yiddish, 3–4
Youmans, Edward L., **498–99**

Zahm, John, **500–501**
Zaizeff, B., 228
Zhitlovsky, Chaim, **501**
Zola, Emile, xxvi, **502–3**
zoology, 237, 300

CREDITS

The author and the publisher are grateful to use copyrighted material by the following authors:

Sholem Aleichem: "Progress in Kasrilevke," from *My First Love and Other Stories,* reprinted by permission of Curt Leviant.

Herbert Henry Asquith: Excerpt from *Letters to Venetia Stanley,* edited by Michael and Eleanor Brock, reprinted by permission of Oxford University Press.

Saul Bellow: Excerpt from *More Die of Heartbreak,* © 1987 the Estate of Saul Bellow, reprinted by permission of the Wylie Agency, LLC.

Elizabeth Bishop: Quoted by Zachariah Pickard in "Natural History and Epiphany: Elizabeth Bishop's Darwin Letter," *Twentieth-Century Literature* 50 (2004): 121–38, reprinted by permission of Hofstra University Press.

William Empson: "Invitation to Juno," from *The Atlantic Book of British and American Poetry,* reprinted by permission of Little Brown; Letter from William Empson to John Hayward, from *Selected Letters of William Empson,* reprinted by permission of Oxford University Press.

Sigmund Freud: Excerpt from *An Autobiographical Study,* translated by James Strachey, © 1952 by W. W. Norton & Company, Inc., renewed © 1980 by Alix Strachey, © 1935 by Sigmund Freud, renewed © 1963 by James Strachey, reprinted by permission of W. W. Norton & Company, Inc.

Robert Frost: "Accidentally On Purpose," from *The Poetry of Robert Frost,* edited by Edward Connery Lathem, © 1969 by Henry Holt and Company, © 1960, 1962 by Robert Frost, © 1988 by Alfred Edwards, reprinted by arrangement with Henry Holt and Company, LLC.

Stephen Jay Gould: "Spin Doctoring Darwin," from *Natural History,* reprinted by permission of Art Science Research Laboratory, Inc.

Gerard Manley Hopkins: Excerpt from *Gerard Manley Hopkins: Selected Letters,* edited by Catherine Phillips, reprinted by permission of Oxford University Press.

Carl Jung: Excerpt from *The Symbolic Life* reprinted by permission of Princeton University Press.

D. H. Lawrence: Letter from D. H. Lawrence to Reverend Robert Reid, from *The Letters of D. H. Lawrence,* reprinted by permission of Pollinger Limited and the Estate of Frieda Lawrence Ravagli.

Lewis Mumford: Excerpt from *Technics and Civilization,* © 1934 by Houghton Mifflin Harcourt Publishing Company and renewed 1961 by Lewis Mumford, reprinted by permission of the publisher.

John Henry Newman: Excerpt from *The Letters and Diaries of John Henry Newman,* Vol. XXV, reprinted by permission of Oxford University Press.

Hans Richter: Quote by Christopher Fifield from *True Artist and True Friend: A Biography of Hans Richter* reprinted by permission of Oxford University Press.

Bertrand Russell: Excerpt from *Autobiography of Bertrand Russell* reprinted by permission of Routledge.

Joseph A. Schumpeter: Except from *History of Economic Analysis* reprinted by permission of Oxford University Press.